AF815985

Methods in Molecular Biology

Series Editor
John M. Walker
School of Life and Medical Sciences
University of Hertfordshire
Hatfield, Hertfordshire, UK

For further volumes:
http://www.springer.com/series/7651

For over 35 years, biological scientists have come to rely on the research protocols and methodologies in the critically acclaimed *Methods in Molecular Biology* series. The series was the first to introduce the step-by-step protocols approach that has become the standard in all biomedical protocol publishing. Each protocol is provided in readily-reproducible step-by step fashion, opening with an introductory overview, a list of the materials and reagents needed to complete the experiment, and followed by a detailed procedure that is supported with a helpful notes section offering tips and tricks of the trade as well as troubleshooting advice. These hallmark features were introduced by series editor Dr. John Walker and constitute the key ingredient in each and every volume of the *Methods in Molecular Biology* series. Tested and trusted, comprehensive and reliable, all protocols from the series are indexed in PubMed.

Protein-Protein Docking

Methods and Protocols

Edited by

Agnieszka A. Kaczor

Faculty of Pharmacy, Medical University of Lublin, Lublin, Poland

Humana Press

Editor
Agnieszka A. Kaczor
Faculty of Pharmacy
Medical University of Lublin
Lublin, Poland

ISSN 1064-3745 ISSN 1940-6029 (electronic)
Methods in Molecular Biology
ISBN 978-1-0716-3984-9 ISBN 978-1-0716-3985-6 (eBook)
https://doi.org/10.1007/978-1-0716-3985-6

© The Editor(s) (if applicable) and The Author(s), under exclusive license to Springer Science+Business Media, LLC, part of Springer Nature 2024, Corrected Publication 2025
This work is subject to copyright. All rights are solely and exclusively licensed by the Publisher, whether the whole or part of the material is concerned, specifically the rights of translation, reprinting, reuse of illustrations, recitation, broadcasting, reproduction on microfilms or in any other physical way, and transmission or information storage and retrieval, electronic adaptation, computer software, or by similar or dissimilar methodology now known or hereafter developed.
The use of general descriptive names, registered names, trademarks, service marks, etc. in this publication does not imply, even in the absence of a specific statement, that such names are exempt from the relevant protective laws and regulations and therefore free for general use.
The publisher, the authors and the editors are safe to assume that the advice and information in this book are believed to be true and accurate at the date of publication. Neither the publisher nor the authors or the editors give a warranty, expressed or implied, with respect to the material contained herein or for any errors or omissions that may have been made. The publisher remains neutral with regard to jurisdictional claims in published maps and institutional affiliations.

This Humana imprint is published by the registered company Springer Science+Business Media, LLC, part of Springer Nature.
The registered company address is: 1 New York Plaza, New York, NY 10004, U.S.A.

If disposing of this product, please recycle the paper.

Preface

Protein-protein complexes are involved in a number of cellular phenomena, including signaling, immune recognition, transmembrane transport, intercellular interactions, post-translational modification, molecular switching, ubiquitination, and maintenance of cellular organization. Modelling structures of protein assemblies is one of key problems in structural biology. Among computational approaches used for this purpose, protein-protein docking has several advantages: it is easy to use and requires limited time and computer resources.

This volume covers wide range of topics on protein-protein docking: from the fundamentals of the method and its recent developments to docking tools and examples of protein-protein docking applications. These broad topics should be useful for readers of different background: structural biologists, bioinformaticians, molecular modelers, and medicinal chemists.

The first part of the book is devoted to the fundamentals of the protein-protein docking. It starts with the description of docking foundations, from rigid to flexible docking (Chapter 1), followed by the discussion of sampling and scoring in protein-protein docking (Chapter 2), and concluded by the topic of free docking and template-based docking, i.e., physics versus knowledge-based docking (Chapter 3).

The second part of the volume focuses on novel developments in the field. One important topic is how the protein backbone flexibility is treated (Chapter 4). With the progress in the availability of computer resources, it is now possible to model more challenging protein-protein complexes and to address the flexible protein regions behavior such as loops. Nowadays docking techniques need to properly treat the conformational changes between unbound and bound structures. This can also be supported by modern molecular dynamics approaches which are often used in combination with protein-protein docking techniques (Chapter 6). Most challenging protein-protein docking problems involve large-scale modeling of protein interaction networks which requires protein modeling followed by high-throughput docking. This issue is mentioned in Chapter 5 and can be facilitated by using artificial intelligence approaches, AI (Chapter 7).

The third part of the book is constituted by practical applications of protein-protein docking tools and databases (Chapter 8), with special emphasis on software for predicting binding free energy of protein–protein complexes and their mutants (Chapter 9). This part is concluded by Chapter 10 on application of deep learning methods for assessment of protein-protein docking models. Selection of the most appropriate model is the central problem in protein-protein docking and can be successfully approached by AI techniques.

The last, fourth, part of the volume is a selection of applications of protein-protein docking approaches to different systems. In Chapter 11, challenges and strategies of molecular docking of intrinsically disordered proteins is discussed. It is followed by description of application of protein-protein docking to membrane systems (Chapter 12) where the presence of the cell membrane constitutes an additional condition which needs to be considered in modelling and should be taken into account, e.g., in G protein-coupled receptors (GPCRs modeling), in particular in modeling GPCR dimers and oligomers

(Chapter 14), but also in correct identification of GPCR-intracellular protein complexes models (Chapter 13). In Chapter 16, the role of features for antigens-antibodies interaction site prediction is discussed. The remaining three chapters focus on methodological problems with wider applicability. In Chapter 15, refinement of docked protein-protein complexes using repulsive scaling replica exchange simulations is presented. Chapter 17 summarizes different computational approaches to model protein-protein complexes. Finally, Chapter 18 addresses in silico design, synthesis, and evaluation of PROTAC against hematopoietic prostaglandin D synthase.

The field of protein-protein docking is nowadays very active and needs to address a number of challenges. With the development of AI power and increasing computer resources, the determination of static interactomes, both in terms of interacting proteins and the structure of formed complexes, should be soon possible. Next, thanks to additional application of experimental approaches, the focus of modeling will go beyond the static models toward the more realistic, dynamic picture of proteins placed in the context of cellular environment. AI, as mentioned in Chapters 7 and 10, ruling over the major branches of science, is expected to take over protein-protein docking field very shortly.

Lublin, Poland ***Agnieszka A. Kaczor***

Contents

Part IV Application of Protein–Protein Docking

Contributors

RUBEN ABAGYAN • *Skaggs School of Pharmacy and Pharmaceutical Sciences, University of California San Diego, La Jolla, CA, USA*

AYESHA ASIM • *Department of Synthesis and Chemical Technology of Pharmaceutical Substances with Computer Modeling Laboratory, Faculty of Pharmacy, Medical University of Lublin, Lublin, Poland*

DAMIAN BARTUZI • *Science for Life Laboratory, Department of Cell and Molecular Biology, Uppsala University, Uppsala, Sweden; Department of Synthesis and Chemical Technology of Pharmaceutical Substances with Computer Modelling Lab, Medical University of Lublin, Lublin, Poland*

DHRUVIL CHAVDA • *Applied Phycology and Biotechnology Division, CSIR Central Salt and Marine Chemicals Research Institute, Bhavnagar, Gujarat, India; Academy of Scientific and Innovative Research (AcSIR), Ghaziabad, Uttar Pradesh, India*

DOMINIKA CIEŚLAK • *Laboratory of Plant Protein Phosphorylation, Institute of Biochemistry and Biophysics, Polish Academy of Sciences, Warsaw, Poland*

YOSUKE DEMIZU • *Division of Organic Chemistry, National Institute of Health Sciences, Kawasaki-shi, Kanagawa, Japan; Graduate School of Medical Life Science, Yokohama City University, Yokohama-shi, Kanagawa, Japan; Graduate School of Medicine, Dentistry and Pharmaceutical Sciences, Okayama University, Kita-ku, Okayama, Japan*

CARLO FERRARI • *Department of Information Engineering, University of Padua, Padua, Italy*

REEMA GABRANI • *Centre for Emerging Diseases, Department of Biotechnology, Jaypee Institute of Information Technology, Noida, Uttar Pradesh, India*

RITU GHILDIYAL • *Centre for Emerging Diseases, Department of biotechnology, Jaypee Institute of Information Technology, Noida, Uttar Pradesh, India*

YUNHAN HUANG • *Department of Computer Science, Purdue University, West Lafayette, IN, USA*

ISHRAT JABEEN • *School of Interdisciplinary Engineering and Sciences (SINES), National University of Sciences and Technology (NUST), Islamabad, Pakistan*

PRIYANJAL JAIN • *Centre for Emerging Diseases, Department of Biotechnology, Jaypee Institute of Information Technology, Noida, Uttar Pradesh, India*

MAŁGORZATA JAROŃCZYK • *National Medicines Institute, Warsaw, Poland*

MICHAŁ K. JASTRZĘBSKI • *Medical University of Lublin, Lublin, Poland*

IVO KABELKA • *Science for Life Laboratory, Department of Cell and Molecular Biology, Uppsala University, Uppsala, Sweden*

SHALMALI KHARCHE • *CSIR-National Chemical Laboratory, Pune, India*

DMYTRO KHYLYUK • *Chair and Department of Organic Chemistry, Medical University of Lublin, Lublin, Poland*

YUSRA SAJID KIANI • *School of Interdisciplinary Engineering and Sciences (SINES), National University of Sciences and Technology (NUST), Islamabad, Pakistan*

DAISUKE KIHARA • *Department of Computer Science, Purdue University, West Lafayette, IN, USA; Department of Biological Sciences, Purdue University, West Lafayette, IN, USA*

MAGDALENA A. KRUPA • *Institute of Computer Science, Polish Academy of Sciences, Warsaw, Poland*

PAWEŁ KRUPA • *Institute of Physics, Polish Academy of Sciences, Warsaw, Poland*
KAMIL J. KUDER • *Department of Technology and Biotechnology of Drugs, Faculty of Pharmacy, Jagiellonian University Medical College, Kraków, Poland*
MOUTUSI MANNA • *Applied Phycology and Biotechnology Division, CSIR Central Salt and Marine Chemicals Research Institute, Bhavnagar, Gujarat, India; Academy of Scientific and Innovative Research (AcSIR), Ghaziabad, Uttar Pradesh, India*
DARIUSZ MATOSIUK • *Department of Synthesis and Chemical Technology of Pharmaceutical Substances with Computer Modeling Laboratory, Faculty of Pharmacy, Medical University of Lublin, Lublin, Poland*
ILONA MICHALIK • *Department of technology and Biotechnology of Drugs, Faculty of Pharmacy, Jagiellonian University Medical College, Kraków, Poland*
AMIT NAGLEKAR • *CSIR-National Chemical Laboratory, Pune, India; Academy of Scientific and Innovative Research (AcSIR), Ghaziabad, India*
KEYUR N. PATEL • *Applied Phycology and Biotechnology Division, CSIR Central Salt and Marine Chemicals Research Institute, Bhavnagar, Gujarat, India; Academy of Scientific and Innovative Research (AcSIR), Ghaziaba, Uttar Pradesh, India*
VIJETA PRAKASH • *Centre for Emerging Diseases, Department of Biotechnology, Jaypee Institute of Information Technology, Noida, Uttar Pradesh, India*
MICHELA QUADRINI • *School of Science and Technology, University of Camerino, Camerino, Italy*
YASMIN SAREMI NANJI • *Physics Department, Center of Functional Protein Assemblies, Technical University of Munich, Garching, Germany*
DURBA SENGUPTA • *CSIR-National Chemical Laboratory, Pune, India; Academy of Scientific and Innovative Research (AcSIR), Ghaziabad, India*
SRISHTI SHARMA • *Centre for Emerging Diseases, Department of Biotechnology, Jaypee Institute of Information Technology, Noida, Uttar Pradesh, India*
NORIHITO SHIBATA • *Division of Biochemistry, National Institute of Health Sciences, Kawasaki-shi, Kanagawa, Japan*
TILL SIEBENMORGEN • *Physics Department, Center of Functional Protein Assemblies, Technical University of Munich, Garching, Germany*
ARCHANA SONAWANI • *School of Biotechnology and Bioinformatics, D.Y. Patil Deemed to be University, CBD Belapur, Navi Mumbai, India*
MAXIM TOTROV • *Molsoft, LLC, La Jolla, CA, USA*
XIAO WANG • *Department of Computer Science, Purdue University, West Lafayette, IN, USA; Okayama University, Kita-ku, Okayama, Japan*
HIDETOMO YOKOO • *Division of Organic Chemistry, National Institute of Health Sciences, Kawasaki-shi, Kanagawa, Japan; Medical Chemistry, Graduate School of Medical Science, Kyoto Prefectural University of Medicine, Sakyo-ku, Kyoto, Japan*
MARTIN ZACHARIAS • *Physics Department, Center of Functional Protein Assemblies, Technical University of Munich, Garching, Germany*
YUANYUAN ZHANG • *Department of Computer Science, Purdue University, West Lafayette, IN, USA*
ZICONG ZHANG • *Department of Computer Science, Purdue University, West Lafayette, IN, USA*
AGATA ZIĘBA • *Department of Synthesis and Chemica Technology of Pharmaceutical Substances with Computer Modeling Laboratory, Faculty of Pharmacy, Medical University of Lublin, Lublin, Poland*

Part I

Fundamentals of Protein–Protein Docking

Chapter 1

Docking Foundations: From Rigid to Flexible Docking

Kamil J. Kuder

Abstract

Despite the development of methods for the experimental determination of protein structures, the dissonance between the number of known sequences and their solved structures is still enormous. This is particularly evident in protein–protein complexes. To fill this gap, diverse technologies have been developed to study protein–protein interactions (PPIs) in a cellular context including a range of biological and computational methods. The latter derive from techniques originally published and applied almost half a century ago and are based on interdisciplinary knowledge from the nexus of the fields of biology, chemistry, and physics about protein sequences, structures, and their folding. Protein–protein docking, the main protagonist of this chapter, is routinely treated as an integral part of protein research. Herein, we describe the basic foundations of the whole process in general terms, but step by step from protein representations through docking methods and evaluation of complexes to their final validation.

Key words Protein, protein docking, PPI, Rigid docking, Flexible docking, Docking foundations

1 A Historical Overview

A living cell is a very congested environment, where several thousand different molecules are constantly trying to interact and exchange information. These interactions play a pivotal role not only in shaping the cell's metabolic state but also in governing its development and ensuring its ongoing survival. Of all the participants in this traffic, proteins are the most active as well as the most populous. They are able to recognize and diffuse into each other in this highly crowded environment and then bind to each other in a highly specific and precise manner. Thus, they act as the main mediators of function in living systems, thereby becoming the key molecules whose functions, and especially these interactions, namely protein–protein interactions (PPIs), we need to understand if we are to understand the complex essence of life. Unfortunately, in spite of the significant scientific and technological progress, we still managed to identify only a small number of protein structures from the vast proteome, whether using the methods of X-ray

Agnieszka A. Kaczor (ed.), *Protein-Protein Docking: Methods and Protocols*, Methods in Molecular Biology, vol. 2780, https://doi.org/10.1007/978-1-0716-3985-6_1,

© The Author(s), under exclusive license to Springer Science+Business Media, LLC, part of Springer Nature 2024

crystallography, NMR, or cryo-electron microscopy (Cryo-EM), which in recent years has climbed to the top of the ladder of protein structure determination [1, 2], not to mention possible protein–protein or protein–nucleic acid complexes (X-ray crystallography methods are mainly helpful here) [3, 4]. However, despite the development of methods for the experimental determination of protein structures, the dissonance between the number of known sequences and their solved structures is still enormous. Given this and the fact that high-resolution protein structures are generally not needed for interface (binding site) prediction alone [5], it is acceptable (and commonly used) to use protein homology models [6].

To fill complex structures gap, diverse technologies have been developed to study PPIs in a cellular context including a range of biological and computational methods [7–9]. The latter, which are the keynote of this book, derive from techniques originally published and applied almost half a century ago and are based on interdisciplinary knowledge from the nexus of the fields of biology, chemistry, and (quantum) physics about protein sequences, structures, and their folding. Independently of the methods used, it helps the scientists to not only answer the question of whether two (or more) given proteins would interact together, but if they do, what would be the strength and the spatial configuration they might adopt in their bound state, or if one of them mutates, would they bind as well. Answering those questions might be crucial for understanding not only life itself but its pathological background on numerous diseases [10]. Today, computational methods of protein structure predictions as well as PPIs using docking are routinely treated as an integral part of protein research.

Docking is usually defined as the structural prediction of a molecular complex based on the resolved structures of its members. However, the first docking methods did not deal with protein–protein complexes per se, but rather with interactions of proteins with other ligands at specific binding sites [11, 12]. Nevertheless, early on these methods were extremely sophisticated, allowing, for example for flexible docking, taking into account the internal coordinates of not only the ligand but in some cases also the receptor. Such a task is not only time consuming but also computationally expensive even with current computational capabilities. As a result, it tends to be replaced by methods that rely on substantial approximations. Soon after, the pioneering work of Wodak et al. on the use of computer methods to predict protein–protein interactions appeared: hemoglobin in sickle cell fibers [13] as well as the studies on the trypsin–BPT complex [14]. Calculations were performed using simplified representations of proteins—one interaction center for each amino acid residue. Favorable electrostatic interactions, including hydrogen bonds, were identified manually and the evaluation of the resulting complexes was based on a scoring function that favored large interface areas and pairs of molecules in contact but not occupying the same space [15]. In the following years, with the improvement of computational power and its general availability, as well as the emergence of the first bioinformatics tools, the focus was shifted toward the development of methods that would, on the one hand, reduce computational costs and, on the other hand, predict protein–protein complexes with an accuracy close to the native one. Undoubted merit in this respect is attributed to the algorithm using the fast Fourier transform coarse method published in 1992, extended by coarse electrostatics 5 years later [16, 17], and which forms the basis of classical docking algorithms to this day and will be briefly discussed in the following section.

2 Structural Representations of Proteins

Analogous to classical molecular docking, a preprocessing step is required before the actual docking can take place, including cleaning of the input structures of the proteins, adding missing atoms or residues, abstracting surface information, and finding a suitable representation scheme for the input proteins. Especially the latter is incredibly important, for the sake of computational cost. Given that eukaryotic proteins have an average size of ~470 aa (compared to, e.g., bacterial proteins ~320 aa) [18] and consist of at least 4 atoms, assuming the considered 6 degrees of freedom (3 rotational and 3 translational), an all-atom representation would force the use of very high power computing resources for a very long time. Such a move seems both: bold (close to insane) and expensive. Therefore, several techniques have been introduced over the years to simplify the surface representation of proteins, moving from all-atom methods to the level of residue representation. A classic example is the coarse-grained approach (also used in molecular dynamics simulations), which preserves the accuracy of the calculations while improving their performance [19].

Most recent methods reduce the number of representative atoms from four ($C\alpha$ & $C\beta$ atoms and two pseudoatoms: one representing the center of the two $C\alpha$ atoms and the other representing the center of mass of the side chain atoms; CABS [20, 21], also used in protein folding prediction), three (considering $C\alpha$, $C\beta$, and an oxygen atom; AWSEM [22]), to even two (an atom representing the backbone atom and the center of mass of the side chain atoms; UNRES method used in UNRES-DOCK [23–25]). Methods using pseudoatoms alone, whose number depends on the size of the amino acid (but does not represent glycine), such as the Zacharias model, are also known. This method allowed for obtaining complexes similar to those solved experimentally [26].

3 Docking Process in a Pill

Despite the variety of methods and algorithms, protein–protein docking strategies are usually based on a two-step approach, that in most cases take place simultaneously. First, a set of possible orientations and conformations that individual proteins can adopt in three-dimensional space is generated, looking for putative binding sites. This phase, depending on the source, is called either the "search/sampling" or "docking" phase (*global search*). In the next stage, "scoring" or "ranking" phase, the formed complexes are evaluated usually on the basis of complementarity of shape and physicochemical features for each model, taking into account the highest scored model and distribution of docking results, with the hope that the complex closest to the native one will be scored if not the highest, then at least very high and not be overlooked. Most often, for a limited number of selected complexes, a final refinement (*local search*) is performed [4, 9, 27].

3.1 Docking Phase

In fact, the first phase—the generation of poses—is practically the most relevant, and in the case of an ideal algorithm, should not allow the exclusion of poses close to the native ones. The search can therefore be performed systematically (exhaustively) or stochastically. While the first option considers all possible alignments of ligand and receptor at a given time and requires considerable time and computational costs, the second option can also lead to the prediction of structures close to the native ones using algorithms known for example from conformational analysis of small ligands. Nevertheless, due to the huge number of possible degrees of freedom, the aforementioned computational cost of search algorithms is the biggest problem in PPI prediction. Due to the huge number of possible orientations to be explored, it is simply computationally inefficient to evaluate all binding complexes using the standard search method. In order to reduce costs and thus optimize the translational and rotational orientation of the docked protein at the docking stage (and the dihedral angles of specific residues at the refining stage), many different search algorithms are used, which can roughly be divided into three main categories: exhaustive global search (fast Fourier transform (FFT) and derivatives, direct search in Cartesian space), local shape feature matching (geometric hashing, local shape feature-matching algorithms, genetic algorithms), and randomized search (Monte Carlo-based methods, particle swarm and genetic algorithms) [9, 27–29].

FFT-based algorithms, a milestone in exhaustive searching methods, originally described by Katchalski-Katzir et al. [16], represent protein surfaces in a Cartesian grid model, favoring close contacts and penalizing core overlap. Rigid-body conformation searches are typically performed in a few hours, using search accelerations in translational space, proceeding to discrete space through several FFT calculations. Since the FFT algorithm does not accelerate the rotation space, it must be sampled explicitly and the ligand is rediscretized after each rotation. This method, based solely on geometric complementarity, is one of the more commonly used and modified (mostly involving mapping and adjusting potentials for different atoms' grips, and selecting potentials as well as scoring methods to describe the interactions) methods in rigid docking, e.g., ZDOCK [30, 31], PIPER [32], PyDock [33], MEGADOCK [34], or recent 5D rotation search algorithms [35]. The direct search in Cartesian space on the contrary, unlike FFT-like algorithms, has simpler values on the grid and Boolean operators and heuristic rules are used to streamline the calculation. Despite the lower search efficiency for this type of approach than the FFT, it has the main advantage of being able to account for protein flexibility, causing the PPI prediction to move to a higher level than rigid docking (e.g., SOFTDOCK [36], SKEDOCK [37]).

Local shape feature matching algorithms consider Connolly surfaces and then search for matches between two proteins with high local shape complementarity. Thus, the algorithm can generate an abundance of binding orientations in minutes. Due to the nature of the method itself, complexes with steric clashes are filtered out first. In addition, a post-clustering step is also necessary to remove possible redundancies. Furthermore, local-shape feature matching algorithms do not search the six-degree-of-freedom landscape—the three translations and three rotations are implicitly included in the transformation matrix for matching and are only computed in the case of a match. Interestingly, this algorithm is successfully used for protein–protein docking using the classic ligand–protein docking program DOCK [38]. Next, geometric hashing is a technique originally developed in computer vision for matching geometric features against a database of such features and was first developed by Wolfson and co-workers [39]. It allows the molecular surface of interacting molecules to be divided into fragments of interactions and to match triplets or pairs of critical points with associated normals, which should be aligned in roughly opposite directions. Consistency in the representation of the data points indicates matching regions. This process called complementary spatial pattern detection usually consists of three components: (1) molecular shape representation (2), surface patch matching (concave patches to convex patches or flat patches to any patch), and (3) filtering and scoring. Algorithms based on geometric hashing are used, e.g., in PatchDock [40], SnapDock [41], or the recently published LZeRD [42].

An interesting solution to simplify the possible energy landscape of two docked protein structures is the use of non-deterministic, meta-heuristic algorithms based on the laws of genetics. They allow us to describe the attraction/repulsion forces of the complexes (electrostatics, desolvation energy, van der Waals interaction) and model the docking process as a natural evolution. These algorithms balance exploration and exploitation of possible solution space, using operations of reproduction, mutation, recombination, and selection. The first of these, genetic algorithms (GA), developed by Holland in the 1970s [43], are mainly used to determine the maxima/minima of energetic functions. First, random sets of chromosomes are generated to form the initial population and each set represents ligand degrees of freedom in six dimensions [44]. As a result of the algorithm/evolution, the population is improved in each generation, through crossover and mutation. The possible population explosion, on the other hand, is regulated by the controlled survival of specific "best" chromosomes. Particle Swarm Optimization (PSO) metaheuristics represent ligands as a swarm orbiting a receptor and each particle is assigned a position and velocity, updated according to the best position reached by any particle in the swarm (e.g., SwarmDOCK) [45, 46]. Thus, it allows for the exploration of the flexibility of interacting partners after bonding. This method has also seen many modifications and optimizations (e.g., recent Glowworm algorithm, GSO implemented in LightDock [47]). In nature, circulating swarms usually help plants to pass on their genetic information. Analogically, in docking procedures, the flower pollination algorithm (FPA) was proposed by Sunny et al. in FPDock [48]. Pollen, the agents in the algorithm, are quaternions representing transformation parameters. In each iteration, they try to improve their score by combining with the best or random pollen. The iterations are followed analogously as above by the removal of unhealthy pollen to avoid population explosion and to preserve the quality of the population. The result of pollen binding is influenced by its electrostatic potential and van der Waals potential, which form the basis for identifying the best pollen.

Last but not least, Monte Carlo (MC) methods are mainly used to refine the resulting complexes, but can also be used to randomly generate the translational and rotational orientations of the docked protein. Nevertheless, the randomized search does not exhaust the entire 6D space. The parameters of each binding orientation corresponding to the six degrees of freedom are obtained from its initial distribution and adjusted by a subsequent optimization process, such as in RosettaDock [49, 50], where, e.g., MC is used to generate and select high probability conformations with optimization of all side chains. Another classic example of randomized search is HADDOCK [51], using simulated annealing in torsion angle space. What seems most important, the introduction of randomized methods has made it possible to dispense with traditional, time-consuming, and most importantly computationally expensive molecular dynamics simulations in Cartesian space in order to study the flexibility of the interacting partners after binding. In other words, it was a significant step forward to take into account the fact that proteins are not rigid when modeling protein–protein interactions. Due to the size of the proteins and the complex nature of the docking algorithms, the flexibility of the proteins has challenged modelers for years.

Most of the methods presented so far have focused on rigid docking (with the exception of some direct Cartesian search and meta-heuristic algorithms). However, like most entities of nature, proteins are mobile forms and, upon binding, might and usually change their conformation. While the global flexibility of proteins can be tackled with an initial calculation/prediction of the conformational ensemble for the interacting molecules and subsequent docking of separate conformers, local fit remains a challenge [52].

Protein flexibility can be treated implicitly, through surface softening (soft docking) or, e.g., the use of multiple frames with MD (ensemble docking—the docking process is repeated from different combinations of starting structures); however, the computational cost in this case is not the least either. On the other hand, in many docking methods, the refinement step involves short MD simulations, thus allowing limited flexibility to improve the energy and remove possible steering failures. Recent years have brought several modifications to the "classical" MD leading to improved algorithms per se and, more importantly, a more accurate selection of native-like structures compared to classical MD methods: steered molecular dynamics (SMD) using external force constraints, Markov state models (MSM) and the Hamiltonian replica exchange MD protocol (H-REMD), thus accelerating exploration of the surfaces so as to identify possible binding sites [53, 54]. Yet still, this is not a holistic approach; for example, due to the discretization methods used, the implicit approach is not able to consider large conformational changes, such as backbone rearrangements and domain movements.

In the case of explicit representation of molecules, it is possible to take into account the flexibility of both, the side chains (relatively simpler) and the backbone (definitely more difficult). An example of allowing both is the previously mentioned HADDOCK (which, however, uses simulated MD annealing). Side-chain flexibility might be considered, for example, on the basis of rotamer libraries in conjunction with replica-exchange methods (ATTRACT [55]), Monte Carlo optimization after prior docking against a soft grid (ICM-CISCO [56]), low-resolution search and subsequent side-chain repacking, or side-chain conformation sampling outside the rotamer (RosettaDock) [50].

Last but not least, recent computational trends, not only in science but in computer science more generally, include the use of machine learning and deep learning algorithms for protein–protein docking and scoring [57]. These algorithms found their way to the top from usage in identifying the protein interfaces to flexible protein docking (e.g., Bayesian network-based algorithms) and are discussed in detail in Chapter 7.

3.2 Scoring Stage

Assessing the quality of the predicted protein–protein complex is crucial in accurately predicting structures. Modern scoring functions can be classified into five groups: force-field-based, knowledge-based, empirical, consensus, and machine learning-based. Force-field-based functions make use of atomic physical interactions such as van der Waals and electrostatic interactions and (in some cases) desolvation energy and bond stretching, bending, and torsion forces. The latter are not taken into account for rigid-body methods. Knowledge-based functions use existing knowledge of the interacting proteins and statistical observations, while empirical functions estimate the binding affinity of a complex from a set of weighted energy coefficients and are significantly faster than force-field-based functions. Finally, consensus-based and AI-based scoring methods take into account a range of interactions and parameter combinations. In many cases, an additional refinement of obtained complexes is also performed after scoring [58–62].

4 Evaluation Methods

Regardless of the algorithms used, the performance evaluation of different docking techniques should be carried out on the same basis. Benchmarks—ready-made sets of experimentally generated structures, available in bound and unbound form—provide the first aid in this regard. By definition: to avoid bias, the benchmark should be diverse and include different types of proteins; to avoid false-positive and false-negative results, the structures should be solved experimentally (and not with the help of homology modelling); to reflect the actual conformational changes after binding, it should include both bound and unbound structures. The first of these was the Protein–Protein Docking Benchmark proposed by Chen et al. in 2003. This set, which contains proteins of varying degrees of difficulty, has been refined several times and reached its last release in 2015 (Protein–Protein Docking Benchmark 5.5), which includes more than 160 rigid bodies, 60 medium targets, and 35 difficult ones [63, 64]. The most numerous benchmark, however, is PPI4Dock, containing more than 1400 targets, but (for purists) not meeting one of the triads of conditions—the structures were obtained by homology modeling but based on an unbound template [65].

However, benchmarks alone are not sufficient for a factual evaluation of the complexes obtained. For numerical evaluation, three measures, originally introduced in the fifth CAPRI experiment evaluation in 2013, are used: L_RMSD, I_RMSD, and f_nat [66]. L_RMSD is the root mean square deviation (RMSD) between the ligand protein and the corresponding structure docked after the overlay of the receptor protein. I_RMSD is the

RMSD of the interface after fitting the backbone-only prediction to the target interface residues. The last, f_nat, is the fraction of interface native residues that are mapped to the predicted structure. These parameters are used together, or separately for individual docking programs, and allow the generated structures to be classified as high, medium, acceptable, or incorrect.

References

1. Assaiya A, Burada AP, Dhingra S, Kumar J (2021) An overview of the recent advances in cryo-electron microscopy for life sciences. Emerg Top Life Sci 5:151–168
2. Chua EYD, Mendez JH, Rapp M, Ilca SL, Tan YZ, Maruthi K, Kuang H, Zimanyi CM, Cheng A, Eng ET, Noble AJ, Potter CS, Carragher B (2022) Better, faster, cheaper: recent advances in cryo–electron microscopy. Annu Rev Biochem 91:1–32. https://doi.org/10.1146/annurev-biochem-032620-110705
3. Cukuroglu E, Gursoy A, Nussinov R, Keskin O (2014) Non-redundant unique interface structures as templates for modeling protein interactions. PLoS One 9. https://doi.org/10.1371/journal.pone.0086738
4. Rodrigues JPGLM, Bonvin AMJJ (2014) Integrative computational modeling of protein interactions. FEBS J 281:1988–2003. https://doi.org/10.1111/febs.12771
5. Vakser IA, Matar OG, Lam CF (1999) A systematic study of low-resolution recognition in protein-protein complexes. Proc Natl Acad Sci USA 96:8477–8482. https://doi.org/10.1073/pnas.96.15.8477
6. Tovchigrechko A, Wells CA, Vakser IA (2002) Docking of protein models. Protein Sci 11: 1888–1896. https://doi.org/10.1110/ps.4730102
7. Titeca K, Lemmens I, Tavernier J, Eyckerman S (2019) Discovering cellular protein-protein interactions: technological strategies and opportunities. Mass Spectrom Rev 38:79–111. https://doi.org/10.1002/mas.21574
8. Elhabashy H, Merino F, Alva V, Kohlbacher O, Lupas AN (2022) Exploring protein-protein interactions at the proteome level. Structure 30:462–475. https://doi.org/10.1016/j.str.2022.02.004
9. Sunny S, Jayaraj PB (2022) Protein–protein docking: past, present, and future. Protein J 41:1–26. https://doi.org/10.1007/s10930-021-10031-8
10. Wodak SJ, Vlasblom J, Turinsky AL, Pu S (2013) Protein-protein interaction networks. Curr Opin Struct Biol 23:941–953. https://doi.org/10.1016/j.sbi.2013.08.002
11. Platzer KE, Momany FA, Scheraga HA (1972) Conformational energy calculations of enzyme-substrate interactions. Int J Pept Protein Res 4:201–219
12. Pincus MR, Zimmerman SS, Scheraga HA (1976) Prediction of three dimensional structures of enzyme substrate and enzyme inhibitor complexes of lysozyme. Proc Natl Acad Sci USA 73:4261–4265. https://doi.org/10.1073/pnas.73.12.4261
13. Levinthal C, Wodak SJ, Kahn P, Dadivanian AK (1975) Hemoglobin interaction in sickle cell fibers I: theoretical approaches to the molecular contacts. Proc Natl Acad Sci USA 72:1330–1334. https://doi.org/10.1073/pnas.72.4.1330
14. Wodak SJ, Janin J (1978) Computer analysis of protein-protein interaction. J Mol Biol 124: 323–342. https://doi.org/10.1016/0022-2836(78)90302-9
15. Wodak SJ, De Crombrugghe M, Janin J (1987) Computer studies of interactions between macromolecules. Prog Biophys Mol Biol 49: 29–63. https://doi.org/10.1016/0079-6107(87)90008-3
16. Katchalski-Katzir E, Shariv I, Eisenstein M, Friesem AA, Aflalo C, Vakser IA (1992) Molecular surface recognition: determination of geometric fit between proteins and their ligands by correlation techniques. Proc Natl Acad Sci USA 89:2195–2199. https://doi.org/10.1073/pnas.89.6.2195
17. Gabb HA, Jackson RM, Sternberg MJE (1997) Modelling protein docking using shape complementarity, electrostatics and biochemical information. J Mol Biol 272:106–120. https://doi.org/10.1006/jmbi.1997.1203
18. Tiessen A, Pérez-Rodríguez P, Delaye-Arredondo L (2012) Mathematical modeling and comparison of protein size distribution in different plant, animal, fungal and microbial species reveals a negative correlation between protein size and protein number, thus

providing insight into the evolution of proteomes. BMC Res Notes 5:85. https://doi.org/10.1186/1756-0500-5-85

19. Kmiecik S, Gront D, Kolinski M, Wieteska L, Dawid AE, Kolinski A (2016) Coarse-grained protein models and their applications. Chem Rev 116:7898–7936. https://doi.org/10.1021/acs.chemrev.6b00163
20. Kolinski A (2004) Protein modeling and structure prediction with a reduced representation. Acta Biochim Pol 51:349–371. https://doi.org/10.18388/abp.2004_3575
21. Kurcinski M, Jamroz M, Blaszczyk M, Kolinski A, Kmiecik S (2015) CABS-dock web server for the flexible docking of peptides to proteins without prior knowledge of the binding site. Nucleic Acids Res 43:W419–W424. https://doi.org/10.1093/nar/gkv456
22. Davtyan A, Schafer NP, Zheng W, Clementi C, Wolynes PG, Papoian GA (2012) AWSEM-MD: protein structure prediction using coarse-grained physical potentials and bioinformatically based local structure biasing. J Phys Chem B 116:8494–8503. https://doi.org/10.1021/jp212541y
23. Krupa P, Karczyńska AS, Mozolewska MA, Liwo A, Czaplewski C (2021) UNRES-Dock-protein-protein and peptide-protein docking by coarse-grained replica-exchange MD simulations. Bioinformatics 37:1613–1615. https://doi.org/10.1093/bioinformatics/btaa897
24. Khalili M, Liwo A, Rakowski F, Grochowski P, Scheraga HA (2005) Molecular dynamics with the united-residue model of polypeptide chains. I. Lagrange equations of motion and tests of numerical stability in the microcanonical mode. J Phys Chem B 109:13785–13797
25. Khalili M, Liwo A, Jagielska A, Scheraga HA (2005) Molecular dynamics with the united-residue model of polypeptide chains. II. Langevin and Berendsen-bath dynamics and tests on model a-helical systems. J Phys Chem B 109:13798–13810. https://doi.org/10.1021/jp058007w
26. Zacharias M (2010) Accounting for conformational changes during protein-protein docking. Curr Opin Struct Biol 20:180–186. https://doi.org/10.1016/j.sbi.2010.02.001
27. Zhang Q, Ting F, Xu L, Sun H (2016) Recent advances in protein-protein docking. Curr Drug Targets 17:1586–1594. https://doi.org/10.2174/1389450117666160112111
28. Matsuzaki Y, Uchikoga N, Ohue M, Akiyama Y (2017) Rigid-docking approaches to explore protein–protein interaction space. Adv Biochem Eng Biotechnol 160:33–55. https://doi.org/10.1007/10_2016_41
29. Huang SY (2014) Search strategies and evaluation in protein-protein docking: principles, advances and challenges. Drug Discov Today 19:1081–1096. https://doi.org/10.1016/j.drudis.2014.02.005
30. Chen R, Weng Z (2002) Docking unbound proteins using shape complementarity, desolvation, and electrostatics. Proteins Struct Funct Genet 47:281–294. https://doi.org/10.1002/prot.10092
31. Chen R, Li L, Weng Z (2003) ZDOCK: an initial-stage protein-docking algorithm. Proteins 87:1–8. https://doi.org/10.1002/prot.10389
32. Kozakov D, Brenke R, Comeau SR, Vajda S (2006) PIPER: an FFT-based protein docking program with pairwise potentials. Proteins: Struct Funct Bioinf 65:392–406. https://doi.org/10.1002/prot.21117
33. Jiménez-García B, Pons C, Fernández-Recio J (2013) pyDockWEB: a web server for rigid-body protein-protein docking using electrostatics and desolvation scoring. Bioinformatics 29:1698–1699. https://doi.org/10.1093/bioinformatics/btt262
34. Ohue M, Matsuzaki Y, Uchikoga N, Ishida T, Akiyama Y (2014) MEGADOCK: an all-to-all protein-protein interaction prediction system using tertiary structure data. Protein Pept Lett 21:766–778. https://doi.org/10.2174/09298665113209990050
35. Padhorny D, Kazennov A, Zerbe BS, Porter KA, Xia B, Mottarella SE, Kholodov Y, Ritchie DW, Vajda S, Kozakov D (2016) Protein-protein docking by fast generalized Fourier transforms on 5D rotational manifolds. Proc Natl Acad Sci USA 113:E4286–E4293. https://doi.org/10.1073/pnas.1603929113
36. Jiang FKS-H (1991) "Soft docking": matching of molecular surface cubes. J Mol Biol 219:79–102. https://doi.org/10.1002/prot
37. Terashi G, Takeda-Shitaka M, Kanou K, Iwadate M, Takaya D, Umeyama H (2007) The SKE-DOCK server and human teams based on a combined method of shape complementarity and free energy estimation. Proteins: Struct Funct Bioinf 69:866–872. https://doi.org/10.1016/0022-2836(91)90859-5
38. Kuntz ID, Blaney JM, Oatley SJ, Langridge R, Ferrin TE (1982) A geometric approach to macromolecule-ligand interactions. J Mol Biol 161:269–288. https://doi.org/10.1016/0022-2836(82)90153-x

39. Wolfson HJ, Rigoutsos I (1997) Geometric hashing: an overview. IEEE Comput Sci Eng 4:10–21. https://doi.org/10.1109/99.641604
40. Schneidman-Duhovny D, Inbar Y, Nussinov R, Wolfson HJ (2005) PatchDock and SymmDock: servers for rigid and symmetric docking. Nucleic Acids Res 33:363–367. https://doi.org/10.1093/nar/gki481
41. Estrin M, Wolfson HJ (2017) SnapDock-template-based docking by Geometric Hashing. Bioinformatics 33:i30–i36. https://doi.org/10.1093/bioinformatics/btx233
42. Christoffer C, Chen S, Bharadwaj V, Aderinwale T, Kumar V, Hormati M, Kihara D (2021) LZerD webserver for pairwise and multiple protein-protein docking. Nucleic Acids Res 49:W359–W365. https://doi.org/10.1093/nar/gkab336
43. Holland JH (1975) Adaptation in natural and artificial systems. The University of Michigan Press, Ann Arbor
44. Gardiner EJ, Willett P, Artymiuk PJ (2001) Protein docking using a genetic algorithm. Proteins Struct Funct Genet 44:44–56. https://doi.org/10.1002/prot.1070
45. Kennedy J, Eberhardt R (1995) Particle swarm optimisation. In: Proc ICNN'95- IEEE Int Conf Neural Netw, vol 4, pp 1942–1948. https://doi.org/10.1007/978-3-030-61111-8_2
46. Moal IH, Bates PA (2010) SwarmDock and the use of normal modes in protein-protein Docking. Int J Mol Sci 11:3623–3648. https://doi.org/10.3390/ijms11103623
47. Jiménez-García B, Roel-Touris J, Romero-Durana M, Vidal M, Jiménez-González D, Fernández-Recio J (2018) LightDock: a new multi-scale approach to protein-protein docking. Bioinformatics 34:49–55. https://doi.org/10.1093/bioinformatics/btx555
48. Sunny S, Jayaraj PB (2021) FPDock: protein–protein docking using flower pollination algorithm. Comput Biol Chem 93:107518. https://doi.org/10.1016/j.compbiolchem.2021.107518
49. Gray JJ, Moughon S, Wang C, Schueler-Furman O, Kuhlman B, Rohl CA, Baker D (2003) Protein-protein docking with simultaneous optimization of rigid-body displacement and side-chain conformations. J Mol Biol 331:281–299. https://doi.org/10.1016/S0022-2836(03)00670-3
50. Marze NA, Roy Burman SS, Sheffler W, Gray JJ (2018) Efficient flexible backbone protein-protein docking for challenging targets. Bioinformatics 34:3461–3469. https://doi.org/10.1093/bioinformatics/bty355
51. Dominguez C, Boelens R, Bonvin AMJJ (2003) HADDOCK: a protein-protein docking approach based on biochemical or biophysical information. J Am Chem Soc 125:1731–1737. https://doi.org/10.1021/ja026939x
52. Kurkcuoglu Z, Bonvin AMJJ (2020) Pre- and post-docking sampling of conformational changes using ClustENM and HADDOCK for protein-protein and protein-DNA systems. Proteins: Struct Funct Bioinf 88:292–306. https://doi.org/10.1002/prot.25802
53. Ostermeir K, Zacharias M (2017) Accelerated flexible protein-ligand docking using Hamiltonian replica exchange with a repulsive biasing potential. PLoS One 12. https://doi.org/10.1371/journal.pone.0172072
54. Król M, Chaleil RAG, Tournier AL, Bates PA (2007) Implicit flexibility in protein docking: cross-docking and local refinement. Proteins: Struct Funct Bioinf 69:750–757
55. Zacharias M (2005) ATTRACT: protein-protein docking in CAPRI using a reduced protein model. Proteins Struct Funct Genet 60:252–256. https://doi.org/10.1002/prot.20566
56. Fernández-Recio J, Totrov M, Abagyan R (2003) ICM-DISCO docking by global energy optimization with fully flexible side-chains. Proteins Struct Funct Genet 52:113–117. https://doi.org/10.1002/prot.10383
57. Harmalkar A, Gray JJ (2021) Advances to tackle backbone flexibility in protein docking. Curr Opin Struct Biol 67:178–186. https://doi.org/10.1016/j.sbi.2020.11.011
58. Huang SY, Zou X (2008) An iterative knowledge-based scoring function for protein-protein recognition. Proteins Struct Funct Genet 72:557–579. https://doi.org/10.1002/prot.21949
59. Xue LC, Jordan RA, Yasser EM, Dobbs D, Honavar V (2014) DockRank: ranking docked conformations using partner-specific sequence homology-based protein interface prediction. Proteins: Struct Funct Bioinf 82:250–267. https://doi.org/10.1002/prot.24370
60. Chermak E, Petta A, Serra L, Vangone A, Scarano V, Cavallo L, Oliva R (2015) CONSRANK: a server for the analysis, comparison and ranking of docking models based on inter-residue contacts. Bioinformatics 31:1481–1483. https://doi.org/10.1093/bioinformatics/btu837
61. Feng T, Chen F, Kang Y, Sun H, Liu H, Li D, Zhu F, Hou T (2017) HawkRank: a new

scoring function for protein-protein docking based on weighted energy terms. J Cheminf 9:1–15. https://doi.org/10.1186/s13321-017-0254-7

62. Li H, Sze KH, Lu G, Ballester PJ (2020) Machine-learning scoring functions for structure-based drug lead optimization. Wiley Interdiscip Rev Comput Mol Sci 10:1–20. https://doi.org/10.1002/wcms.1465
63. Chen R, Mintseris J, Weng Z (2003) A protein – protein docking benchmark. Proteins: Struct Funct Bioinf 91:88–91
64. Vreven T, Moal IH, Vangone A, Pierce BG, Kastritis PL, Torchala M, Chaleil R, Jiménez-García B, Bates PA, Fernandez-Recio J, Bonvin AMJJ, Weng Z (2015) Updates to the integrated protein-protein interaction benchmarks: docking benchmark version 5 and affinity benchmark version 2. J Mol Biol 427:3031–3041. https://doi.org/10.1016/j.jmb.2015.07.016
65. Yu J, Guerois R (2016) PPI4Dock: large scale assessment of the use of homology models in free docking over more than 1000 realistic targets. Bioinformatics 32:3760–3767. https://doi.org/10.1093/bioinformatics/btw533
66. Lensink MF, Wodak SJ (2013) Docking, scoring, and affinity prediction in CAPRI. Proteins: Struct Funct Bioinf 81:2082–2095. https://doi.org/10.1002/prot.24428

Chapter 2

Sampling and Scoring in Protein–Protein Docking

Agata Zięba and Dariusz Matosiuk

Abstract

Protein–protein docking is considered one of the most important techniques supporting experimental proteomics. Recent developments in the field of computer science helped to improve this computational technique so that it better handles the complexity of protein nature. Sampling algorithms are responsible for the generation of numerous protein–protein ensembles. Unfortunately, a primary docking output comprises a set of both near-native poses and decoys. Application of the efficient scoring function helps to differentiate poses with the most favorable properties from those that are very unlikely to represent a natural state of the complex. This chapter explains the importance of sampling and scoring in the process of protein–protein docking. Moreover, it summarizes advances in the field.

Key words Molecular docking, Protein, protein docking, Conformational sampling, Scoring function

1 Introduction

The technological advances we have been witnessing in the past decades have influenced fields of computational chemistry and biology. They have led to the development of numerous experimental, theoretical, and computational approaches aiming to determine the structure of biologically active macromolecules. Such data support the process of understanding the physiological and pathological processes occurring in organisms. More specifically, interactions observed between each of the complex components give insight into its kinetics, thermodynamics, and organization, as well as the disease itself. Moreover, information on the complexes improve our understanding of the pharmacological action of therapeutic agents and enable us to learn more about structural features underlying their effectiveness or ineffectiveness [1].

Proteins are macromolecules responsible for the maintenance of cell functions. Studies on complexes formed between two proteins or proteins and a molecule of DNA or RNA help to understand the cascade of reactions underlying a plethora of physiological and pathological processes occurring in organisms [2]. Research on

Agnieszka A. Kaczor (ed.), *Protein-Protein Docking: Methods and Protocols*, Methods in Molecular Biology, vol. 2780, https://doi.org/10.1007/978-1-0716-3985-6_2,

© The Author(s), under exclusive license to Springer Science+Business Media, LLC, part of Springer Nature 2024

this category of macromolecules remains a hot topic and an important task for many scientists. Despite the fact that advances in proteomics provided numerous data improving our understanding in this field, some biologically important interactions occur only in transient complexes and experimental evaluation of such complexes remains a challenging task. Luckily, available in silico approaches can be applied to this problem, and in many cases, they deliver accurate results. In this chapter, a special emphasis will be put on a particular technique termed *molecular docking*. This technique aims to deliver the prediction of the structure of the complex with the use of its individual components [3]. Since not only high-resolution protein structures can be used for this procedure, docking approaches have to be designed in a way to compensate for these inaccuracies and reflect conformational changes observed between bound components [4].

To date, numerous tools have been developed to perform molecular docking. Thanks to advances in the field of computer science, it is now possible to perform an exhaustive search of the full docking space and include numerous restriction criteria such as (**i**) the geometrical shape of molecules; (**ii**) hydrophobicity; (**iii**) binding affinity; and (**iv**) flexibility of protein backbones, enabling the retrieval of possible ensembles of protein complexes [5]. Despite this remarkable over-the-year technological evolution of docking algorithms, this area is far from being a trouble-free zone. Most of the challenges identified in this field evolve from the complex nature of proteins, which corresponds with the complexity of the connections they create. A successful in silico simulation should be able to address the issue adequately. When it is necessary, it should include information on different folding patterns identified in considered proteins that reflect the influence of certain factors commonly identified in the natural environment of a protein. Thus, the prediction of the three-dimensional structure of such a complex remains one of the most challenging tasks in computational biology [6, 7]. Protein–protein docking remains an active area of research because a proper resolution of this problem could help generate approximately 65% of all interactions identified in human interactome [8]. The importance of the problem is also highlighted by the fact that the Critical Assessment of Predicted Interaction (CAPRI) contest, founded in 2001, continues to test available solutions for the prediction of the protein–protein complexes and stimulates the research on the development of novel tools, potentially valuable in this field [9].

Regardless of which docking protocol is applied to solve a certain problem, a typical computational workflow comprises at least two phases: rapid generation of numerous docking models (involving the application of certain algorithms for sampling conformational and configurational degrees of freedom) and application of more sophisticated scoring algorithms, thanks to which it is possible to choose the closest-to-native state combination [10, 11]. Since the topic of molecular docking is a very broad discipline, this chapter will be exclusively devoted to a closer understanding of the problems of sampling and scoring identified in in silico experiments involving two protein structures. A short overview of the recent advances in the field, along with some practical information, will be provided to ease the understanding of this topic and suggest some troubleshooting strategies.

2 Materials

Currently, a variety of web tools are available for protein–protein docking. These programs offer numerous advanced and easy-to-modify options to match the requirements of the experiment. Many of these tools exist as regularly updated web servers easily accessible to the user. In this section, we provide a brief overview of a few such tools with a special emphasis on some tools classified in the last CASP-14 CAPRI experiment [12].

2.1 MDockPP

The MDockPP Online server (available at: https://zougrouptoolkit.missouri.edu/MDockPP) provides solutions for the prediction of the structure of protein–protein complexes, protein–DNA, and protein–RNA complexes. In addition, this server can handle homo-oligomeric docking for complexes of cyclic and dihedral symmetry [12, 13]. The submitted molecule can comprise multiple proteins or nucleic acid monomers. The utilized program is based on:

1. Docking algorithm using the Fast Fourier Transform (FFT).
2. ITScorePP function for protein–protein interaction evaluation, ITScorePR for protein–RNA interaction, and ITScorePD for protein–DNA interaction.

FFT-based docking algorithm takes advantage of the reduced protein model, representing side chains on the protein surface by their center of mass for the receptor and the ligand (a smaller protein involved in the formation of the complex). The receptor protein is fixed in space, while the ligand is rotated in the rotational plane. The complementarity of shapes is evaluated with the use of FFT [13]. The distance-dependent knowledge-based scoring function aims to improve the interatomic pair potentials with the use of data on protein–protein complexes available in databases such as RCSB PDB. This results in the adjustment of pair potentials by iteration, which is done up until the scoring function can distinguish native-like models from decoys [14].

2.2 LZerD

LZerD (accessible via: https://kiharalab.org/proteindocking) is a server that provides online access for the LZerD pairwise docking and Multi-LZerD multiple chain docking protocols.

The first docking algorithm takes two input proteins and returns docking conformations, samples interface regions and angles of interactions. Complexes with too many atom clashes, little interaction area, or insufficient complementarity at the interface region are being rejected from further steps. Later, models are clustered with user-defined parameters. The remaining poses are ranked by the sum of ranks(ranksum) from three utilized scoring functions: DFIRE, GOAP, and ITScorePro.

The Multi-LZerD algorithm was designed to handle problems associated with the docking of three or more proteins. It creates reliable assembles out of provided structures. In the beginning, it utilizes a coarse-grained conformation search on pairs of docking candidates. In the following steps, Multi-LZerD builds models of multimeric proteins by iteratively generating components through pairwise docking using a genetic algorithm (different combinations of pairwise models are constantly and iteratively generated and selected) [12, 15–17]. Created models are clustered with cutoffs defined by the user in groups of personalized populations. Resulting models are scored with ransum, and models considered as the best ones among a scored population are presented on the result page.

Another interesting feature of this server is that it provides access to a solution called AttentiveDist, useful for the de novo structure prediction of single-chain proteins. The procedure does not use any references to template structures; nonetheless, a combination of specific features, e.g., the amino-acid sequence or a position-specific scoring matrix is passed into a deep neural network to generate pairwise distances between the residues. For full-atom models, coordinates are generated with PyRosetta and optimized to fulfill the requirements of predicted contacts. Final models are also ranked by presented above rank-sum and returned on the output page [17].

2.3 ClusPro

This server (accessible via: https://cluspro.org) was primarily introduced in 2004. Ever since then, it is maintained and regularly updated [18, 19]. It solves the problem of protein–protein docking with a three-step workflow:

1. Rigid-body docking with a PIPER program.
2. RMSD-based clustering.
3. CHARMM minimization of the previously selected complexes.

The abovementioned PIPER algorithm is a docking program applied for rigid-body docking. It makes use of a correlation approach termed the Fast Fourier Transform, which, as it was introduced, revolutionized the field of molecular docking. During the application, one protein is placed on a fixed grid at the origin of the coordinate system, while the second is moving on another grid. The interaction energy of a complex is written in the form of a correlation function. Thanks to the application of FFTs, an exhaustive search of docking space and evaluation of energy at each grid point is available [19, 20]. Later, a number of conformations are selected for further optimization, in a process involving a pairwise RMSD calculation and identification of highly populated clusters. Further complex minimization takes advantage of CHARMM to remove steric clashes. Up to this date, the abovementioned server was applied to solve various computational problems, e.g., docking X-ray or NMR structures of proteins, modeling antibody–antigen interactions, or constructing large multidomain proteins [21–24].

3 Methods

3.1 Sampling in Protein–Protein Docking

Sampling is a term widely recognized in many fields of life. In statistics, it refers to a selection of individuals from within a statistical population to estimate the characteristics of the entire population. The main aim of the procedure is to collect samples of various types that adequately describe the analyzed environment. On the other hand, in the field of protein–protein docking, sampling provides a suitable solution for the process of acceptable pose generation and their placement in the active site of another component of the complex [25] (*see* **Note 1**). To determine the active site of an analyzed protein, one can refer to the experimental data, a place of a dimer interface, or another plane of interest. Advances in the use of sampling algorithms were strongly correlated with rapid developments in the field of computer science and technology in general. The inclusion of additional degrees of freedom and the transition from approaches that consider parts of the analyzed complex as rigid to those that take into account the flexibility of both proteins have led to the development of numerous techniques, which can be divided into the following categories:

1. Approaches considering protein–protein complexes as rigid:
   - Shape matching.
   - Fast Fourier Transform.
2. Approaches considering protein–protein complexes as flexible:
   - Systematic search.
   - Stochastic methods.
   - Deterministic methods.

3.1.1 Representation of Proteins as Rigid Bodies

Primarily protein–protein complexes were considered as rigid. It means that certain aspects referring to their nature were neglected and their position in space was understood as the location of the center of mass and their altitude. Their movement in space was represented by three rotational and three translational parameters, termed degrees of freedom. In rigid docking algorithms, the main emphasis is put on the identification of the geometry match between analyzed molecules [26].

Shape matching techniques lead to the identification of the optimal transformation (understood as rotation and translation) that needs to be applied to one or two of the considered objects, to fit it well into the available cavity [27]. In practice, it means that complementary patches of a receptor are matched with similar regions of the second of considered macromolecules to estimate the shape complementarity of the examined fragments [27]. The main advantage of the technique is related to its computational efficiency. Unfortunately, in many cases, during the shape matching process, a ligand (smaller protein) conformation is fixed. Thus, if this technique is applied to flexible complexes, it is usually performed along with the docking of the pre-generated ligand conformations into the protein, followed by the merging/re-ranking of docked poses from different docking runs according to the energy values [28]. Similarly, Fast Fourier Transform and other grid-based methods were initially utilized for the relatively rigid protein–protein docking simulations. These grid models scanned the entire space by rotating and translating one component onto another. In the end, a proper surface match was identified by the use of the correlation function favoring close contacts and penalizing surface clashes. Theoretically, the FFT method does not need any post-processing steps to remove poses with steric clashes. Unfortunately, the scope of the considered space makes FFT techniques slower than geometric hashing-based ones [29]. More recently, the FFT-docking methods were suited to handle flexible docking by implementing, e.g., docking of multiple peptide conformations drawn from the library of PDB-derived fragments followed by the short Monte Carlo Simulation minimization [30].

3.1.2 Representation of Proteins as Flexible Bodies

As the research on proteins continued and more information on their nature was available, it became clear that rigid approaches are far from accurately describing these biological molecules. Therefore, a new category of methods with the flexible representation of docked components was introduced. At first, these methods considered a ligand (smaller protein) as a flexible one and took advantage of the "induced-fit" paradigm. Later, certain approaches considering the flexibility of a receptor were also introduced [26]. A brief description of these techniques is presented below.

In systematic search techniques, all possible binding poses of a ligand are created and it is obtained by the exploration of all six degrees of freedom of the ligand. The least complex algorithms within this category use an exhaustive search approach, which aims to accommodate the ligand in the receptor binding pocket by systematically rotating all possible bonds at a given interval. In many cases, geometric or chemical constraints are applied in this procedure to narrow down the number of possible pose combinations. More advanced sampling techniques that are based on fragmentation divide ligands into different parts and re-build accurate binding conformation of the macromolecule inside the binding pocket of the receptor. The conformational ensemble methods make use of pre-generated ligand conformations and dock these to the receptor. The ligand binding modes are collected and ranked due to their binding energy scores [26, 31].

In stochastic methods, random changes are applied to the ligand at each step, and the final decision on whether the change will be accepted or rejected is based on probabilistic criteria. One of the most commonly recognized representatives of these techniques is the Monte Carlo method, which searches for the global energy minimum of the conformers [32]. This method generates random changes in the ligand conformation and analyzes changes in its thermodynamics. For further steps, only conformations with favorable energy outcomes are taken into consideration [33]. Other stochastic methods such as simulated annealing make use of temperature changes to obtain a detailed characterization of a conformational state of a complex.

The third category of flexible-protein methods considers certain degree of flexibility in both components of the ligand–receptor complex and is termed deterministic. Thanks to the application of molecular simulations, it is capable of delivering information on the movement of individual particles as a function of time. AMBER and CHARMM are examples of tools, which take advantage of molecular simulation techniques for the process of pose generation [34–36].

Although considering proteins as flexible macromolecules makes docking predictions more reliably depict the natural state, it brings a lot of chaos since numerous additional factors need to be considered to make these predictions accurate. In particular, using unbound protein structures for such simulations creates the problem of side-chain flexibility, which influences the formation of the tight fit between docked proteins. Some ways to handle this difficulty involve implementation of the “soft” surface representation, to enable a small volume of a steric clash between docked components or the use of NMR-derived data of alternative models to consider multiple backbone conformations [37].

3.2 Scoring in Protein–Protein Docking

Since the appearance of the scoring functions (SFs) in the early 1990s, the main goal of the application of these algorithms is to ease the process of scraping through hundreds or thousands of binding possesses and yield those the most probable to exist in nature. Although a significant improvement has been made in this field, selecting the conformation in the best way representing a native state of a protein complex remains a great challenge. One of the most palpable reasons for such a state is the fact that a substantial imbalance, between the number of accurate near-native models and incorrect decoys in most protein–protein docking results is observed. Therefore, a certain strategy to examine docking results and sort obtained poses to highlight those of a special value is needed [3].

Despite the fact that the definition of the problem seems to be pretty easy to understand, the main question—how to apply the most suitable scoring function from the perspective of problem complexity—remains a mystery. In this context, the main difficulty associated with docking scoring is predicting benchmarks for experimental ΔG and $\Delta\Delta G$ values. Usually, only the poses characterized by the lowest energy are considered the closest to the correct docking conformation.

In a benchmark study published by Moal H.I. et al., all out of 115 examined scoring functions attempted to identify the complexes characterized by the lowest binding energy. Moreover, three set of theoretical approaches were applied to identify pairs of scoring techniques, which are capable of properly discriminating analyzed complexes. Interestingly, all docking-specific scoring functions performed well. Contrastingly, simple residue contact potential and novel statistical potential techniques comprised a cluster of the poorest performers [10]. The first models published for ΔG calculation took advantage of molecular mechanics force fields and statistical potentials. These algorithms performed well; however, after an extensive evaluation, they turned out to be biased toward rigid proteins. Unfortunately, the more diverse complex was taken into consideration, the less effective energy functions were [11]. In the next years, special emphasis was put on the attempts aiming to identify a method enabling an accurate prediction of a binding affinity of more flexible complexes, and it proved that conformational energy played a vital role in the calculation of ΔG energy. Moreover, it became clear that a good scoring function should accommodate small changes in spatial conformation and steric clashes, and not let slight deviations of these parameters greatly affect the verdict [27]. Some of the currently available scoring approaches utilize smaller datasets, for testing. This makes them efficient in solving only a narrow range of problems, and rather ineffective for the general docking problem [10]. Other methodologies make use of (**i**) various clustering and refining approaches; (**ii**) information on constraints derived from the interaction interface; and (**iii**) contact-prediction algorithms, to obtain specific pairs of residues from various chains located within a threshold distance [7, 38].

The primary classification of scoring functions divided this set into three major groups: force field, empirical, and knowledge-based [39]. However, as the research on protein and its interactions advanced, it became clear that ranking algorithms applied to docking poses should be a bit more diverse. Current classification focuses on the nature of examined data. Thus, the main categories of currently applied scoring functions can be divided into the following groups: physics-based SF, empirical SF, knowledge-based SF, and machine learning SF [40]. One of the great reviews on this topic has been published by Li J. et al. [39]. It describes the advantages and disadvantages of each scoring algorithm and provides information on examples of tools using each technique. Although currently known scoring functions have numerous limitations, which make their use less effective, combining several functions in one program (hybrid SF) is considered a very promising direction. Numerous strategies have been introduced to make the most of such a combination, and those include, e.g., linear combinations, rank-by-number, or vote-by-number approaches [39]. A significant drawback of such a strategy is extended calculation time. Therefore, currently, the main research focus has been placed on the application of machine learning-based scoring functions, since delivered outcomes can outperform conventional SF and deliver higher predictive accuracy. Unfortunately, ML-based techniques rely on training data accurately depicting only features correlated to the examined problem. Thus, their application to more diverse problems is yet impossible [39] (*see* **Note 2**).

Despite the undeniable advances in the effectiveness of scoring, many algorithms often fail to deliver a universal solution easily applicable to all protein–protein docking experiments. Therefore, the work on the development of an ideal scoring function is considered as far from over. A large number of structural information creates a high demand for an advanced solution capable of delivering precise and computationally efficient results.

4 Notes

1. Sampling remains an inevitable part of a docking protocol. Thanks to the sampling algorithms, it is possible to generate hundreds or even thousands of poses representing different conformations for the protein–protein complex. Some sampling methods such as shape matching consider both components of a complex as rigid bodies. On the other hand, techniques such as systematic methods (incremental

construction) and stochastic methods (Monte Carlo approach) consider additional criteria of rotational, translational, and conformational degrees of freedom of biological molecules.

2. Works on the development of an ideal scoring function for protein–protein docking are far from being over. A growing number of structural data provides new insight into the complex world of proteins and their interactions. The more we know in this field, the more features we want to consider during the actual scoring. Current attempts in this field focus on delivering versatile solutions for solving numerous docking problems of various complexity. There are numerous web servers for protein–protein docking available, enabling the user to specify various parameters to fully match the requirements of the performed in silico experiment.

References

1. Du X, Li Y, Xia Y-L et al (2016) Insights into protein–ligand interactions: mechanisms, models, and methods. Int J Mol Sci 17:144. https://doi.org/10.3390/ijms17020144
2. Keskin O, Gursoy A, Ma B, Nussinov R (2008) Principles of protein–protein interactions: what are the preferred ways for proteins to interact? Chem Rev 108:1225–1244. https://doi.org/10.1021/cr040409x
3. Meng X-Y, Zhang H-X, Mezei M, Cui M (2011) Molecular docking: a powerful approach for structure-based drug discovery. Curr Comput Aided Drug Des 7:146–157
4. Vakser IA (2014) Protein-protein docking: from interaction to Interactome. Biophys J 107:1785–1793. https://doi.org/10.1016/j.bpj.2014.08.033
5. Tradigo G, Rondinelli F, Pollastri G (2019) Algorithms for structure comparison and analysis: docking. In: Ranganathan S, Gribskov M, Nakai K, Schönbach C (eds) Encyclopedia of bioinformatics and computational biology. Academic, Oxford, pp 77–80
6. Maleki M, Aziz MDM, Rueda L (2011) Analysis of relevant physicochemical properties in obligate and non-obligate protein-protein interactions. In: 2011 IEEE international conference on bioinformatics and biomedicine workshops (BIBMW), pp 345–351
7. Pozzati G, Kundrotas P, Elofsson A (2022) Scoring of protein–protein docking models utilizing predicted interface residues. Protein Struct Funct Bioinformatics 90:1493–1505. https://doi.org/10.1002/prot.26330
8. Mosca R, Céol A, Aloy P (2013) Interactome3D: adding structural details to protein networks. Nat Methods 10:47–53. https://doi.org/10.1038/nmeth.2289
9. Janin J (2002) Welcome to CAPRI: a critical assessment of PRedicted interactions. Protein Struct Funct Bioinformatics 47:257–257. https://doi.org/10.1002/prot.10111
10. Moal IH, Torchala M, Bates PA, Fernández-Recio J (2013) The scoring of poses in protein-protein docking: current capabilities and future directions. BMC Bioinformatics 14:286. https://doi.org/10.1186/1471-2105-14-286
11. Kastritis PL, Bonvin AMJJ (2010) Are scoring functions in protein–protein docking ready to predict Interactomes? Clues from a novel binding affinity benchmark. J Proteome Res 9: 2216–2225. https://doi.org/10.1021/pr9009854
12. Prediction of protein assemblies, the next frontier: the CASP14-CAPRI experiment. https://doi.org/10.1002/prot.26222
13. Huang S-Y, Zou X (2010) MDockPP: a hierarchical approach for protein-protein docking and its application to CAPRI rounds 15–19. Proteins 78:3096–3103. https://doi.org/10.1002/prot.22797
14. Huang S-Y, Zou X (2008) An iterative knowledge-based scoring function for protein–protein recognition. Protein Struct Funct Bioinformatics 72:557–579. https://doi.org/10.1002/prot.21949
15. Venkatraman V, Yang YD, Sael L, Kihara D (2009) Protein-protein docking using region-based 3D Zernike descriptors. BMC Bioinformatics 10:407. https://doi.org/10.1186/1471-2105-10-407

16. Esquivel-Rodríguez J, Yang YD, Kihara D (2012) Multi-LZerD: multiple protein docking for asymmetric complexes. Proteins 80: 1818–1833. https://doi.org/10.1002/prot.24079
17. Christoffer C, Chen S, Bharadwaj V et al (2021) LZerD webserver for pairwise and multiple protein-protein docking. Nucleic Acids Res 49:W359–W365. https://doi.org/10.1093/nar/gkab336
18. Desta IT, Porter KA, Xia B et al (2020) Performance and its limits in rigid body protein-protein docking. Structure 28:1071–1081.e3. https://doi.org/10.1016/j.str.2020.06.006
19. Kozakov D, Hall DR, Xia B et al (2017) The ClusPro web server for protein–protein docking. Nat Protoc 12:255–278. https://doi.org/10.1038/nprot.2016.169
20. Kozakov D, Brenke R, Comeau SR, Vajda S (2006) PIPER: an FFT-based protein docking program with pairwise potentials. Protein Struct Funct Bioinformatics 65:392–406. https://doi.org/10.1002/prot.21117
21. Kamal JKA, Benchaar SA, Takamoto K et al (2007) Three-dimensional structure of cofilin bound to monomeric actin derived by structural mass spectrometry data. Proc Natl Acad Sci 104:7910–7915. https://doi.org/10.1073/pnas.0611283104
22. Luxán G, Casanova JC, Martínez-Poveda B et al (2013) Mutations in the NOTCH pathway regulator MIB1 cause left ventricular noncompaction cardiomyopathy. Nat Med 19: 193–201. https://doi.org/10.1038/nm.3046
23. Tran K, Poulsen C, Guenaga J et al (2014) Vaccine-elicited primate antibodies use a distinct approach to the HIV-1 primary receptor binding site informing vaccine redesign. Proc Natl Acad Sci 111:E738–E747. https://doi.org/10.1073/pnas.1319512111
24. Shams N, Khoshtinat Nikkhoi S, Gu Z, Rahbarizadeh F (2022) Isolation and characterization of human anti-CD20 single-chain variable fragment (scFv) from a naive human scFv library. Med Oncol 39:177. https://doi.org/10.1007/s12032-022-01757-1
25. Elokely KM, Doerksen RJ (2013) Docking challenge: protein sampling and molecular docking performance. J Chem Inf Model 53: 1934–1945. https://doi.org/10.1021/ci400040d
26. Novič M, Tibaut T, Anderluh M (2016) The comparison of docking search algorithms and scoring functions: an overview and Case studies. Methods and algorithms for molecular docking-based drug design and discovery
27. Voruganti HK, Dasgupta B (2018) A novel volumetric criterion for optimal shape matching of surfaces for protein-protein docking. J Comput Des Eng 5:180–190. https://doi.org/10.1016/j.jcde.2017.10.003
28. Lorber DM, Shoichet BK (2005) Hierarchical docking of databases of multiple ligand conformations. Curr Top Med Chem 5:739–749
29. Padhorny D, Kazennov A, Zerbe BS et al (2016) Protein–protein docking by fast generalized Fourier transforms on 5D rotational manifolds. Proc Natl Acad Sci 113: E4286–E4293. https://doi.org/10.1073/pnas.1603929113
30. Padhorny D, Hall DR, Mirzaei H et al (2018) Protein – ligand docking using FFT based sampling: D3R case study. J Comput Aided Mol Des 32:225–230. https://doi.org/10.1007/s10822-017-0069-7
31. Huang S-Y, Zou X (2010) Advances and challenges in protein-ligand docking. Int J Mol Sci 11:3016–3034. https://doi.org/10.3390/ijms11083016
32. Jagiello K, Grzonkowska M, Swirog M et al (2016) Advantages and limitations of classic and 3D QSAR approaches in nano-QSAR studies based on biological activity of fullerene derivatives. J Nanopart Res 18:256. https://doi.org/10.1007/s11051-016-3564-1
33. Caflisch A, Niederer P, Anliker M (1992) Monte carlo minimization with thermalization for global optimization of polypeptide conformations in cartesian coordinate space. Protein Struct Funct Bioinformatics 14:102–109. https://doi.org/10.1002/prot.340140111
34. Brooks BR, Bruccoleri RE, Olafson BD et al (1983) CHARMM: a program for macromolecular energy, minimization, and dynamics calculations. J Comput Chem 4:187–217. https://doi.org/10.1002/jcc.540040211
35. Salomon-Ferrer R, Case DA, Walker RC (2013) An overview of the Amber biomolecular simulation package. WIREs Comput Mol Sci 3:198–210. https://doi.org/10.1002/wcms.1121
36. Wingbermühle S, Schäfer LV (2020) Capturing the flexibility of a protein-ligand complex: binding free energies from different enhanced sampling techniques. https://doi.org/10.26434/chemrxiv.11106806.v3
37. Francis-Lyon P, Gu S, Hass J et al (2010) Sampling the conformation of protein surface residues for flexible protein docking. BMC Bioinformatics 11:575. https://doi.org/10.1186/1471-2105-11-575

38. Schneider S, Zacharias M (2012) Scoring optimisation of unbound protein–protein docking including protein binding site predictions. J Mol Recognit 25:15–23. https://doi.org/10.1002/jmr.1165
39. Li J, Fu A, Zhang L (2019) An overview of scoring functions used for protein–ligand interactions in molecular docking. Interdiscip Sci Comput Life Sci 11:320–328. https://doi.org/10.1007/s12539-019-00327-w
40. Wang X, Flannery ST, Kihara D (2021) Protein docking model evaluation by graph neural networks. Front Mol Biosci 8

Chapter 3

Free-Docking and Template-Based Docking: Physics Versus Knowledge-Based Docking

Magdalena A. Krupa and Paweł Krupa

Abstract

Docking methods can be used to predict the orientations of two or more molecules with respect of each other using a plethora of various algorithms, which can be based on the physics of interactions or can use information from databases and templates. The usability of these approaches depends on the type and size of the molecules, whose relative orientation will be estimated. The two most important limitations are (i) the computational cost of the prediction and (ii) the availability of the structural information for similar complexes. In general, if there is enough information about similar systems, knowledge-based and template-based methods can significantly reduce the computational cost while providing high accuracy of the prediction. However, if the information about the system topology and interactions between its partners is scarce, physics-based methods are more reliable or even the only choice. In this chapter, knowledge-, template-, and physics-based methods will be compared and briefly discussed providing examples of their usability with a special emphasis on physics-based protein–protein, protein–peptide, and protein–fullerene docking in the UNRES coarse-grained model.

Key words Physics-based docking methods, Knowledge-based docking methods, Template-based docking methods, Force fields, Molecular dynamics, Genetic algorithms, Proteins and peptides, Carbon nanoparticles, Fullerenes, UNRES

1 Introduction

Docking algorithms contain two most important elements: (i) search of the conformational space and (ii) selection of the best docking poses (*see* **Notes 1–3**). These two elements usually are connected with each other, as the search of the optimal orientation of two molecules is performed in a way to optimize the local and global geometry of the system and find local minima of the free energy (scoring) function, which can be reflecting physics of interactions, or be derived from statistical analysis of databases or machine learning methods. Then, generated conformations (which number can easily overcome millions) have to be limited (by excluding some of them and clustering the remaining ones) and

Agnieszka A. Kaczor (ed.), *Protein-Protein Docking: Methods and Protocols*, Methods in Molecular Biology, vol. 2780,
https://doi.org/10.1007/978-1-0716-3985-6_3,
© The Author(s), under exclusive license to Springer Science+Business Media, LLC, part of Springer Nature 2024

sorted and the best candidates have to be selected using a scoring method.

Docking methods can be used to study many various systems, such as protein–protein, protein–peptide, protein–ligand, protein–nanoparticle, protein–nucleic acid, and protein–sugar complexes using similar approaches. One of the most popular applications is to study the interaction of the protein with potential drug compounds (usually a ligand or peptide) and to predict whether the ligand will bind to the pocket-binding (or other region of the protein) and if it works as it was designed for. Some docking tools include flexibility (or it can be studied after docking using predicted complex structure as an initial model in Molecular Dynamics (MD) or Monte Carlo (MC) simulations) allowing one to observe the mechanism of opening and closing the binding pocket. Therefore, if the binding mechanism is known, one can try to design new and better potential drug compounds. The use of computational docking procedures can not only save money and time by limiting the number of in vitro and in vivo experiments but also allow the determination of binding processes at the atomistic level and easily predict impact of modifications and mutations of both receptor and the ligand.

Molecular docking gives us the possibility to pair different molecules with each other allowing the search for novel ligands using virtual screening or dock already existing ligands or known drugs. Such studies on already existing drugs were performed during the pandemic of SARS-CoV-2 to search for a potential cure [1] or on the interaction between remdesivir and ribavirin with SARS-CoV-2 [2] and Hepatitis E virus where AutoDock Vina and LeDock software were used [3]. Protein–protein structure prediction is important, but proteins and their complexes tend to interact also with other molecules, such as DNA and RNA [4, 5] and carbohydrates [6]. There are also studies on the implementation of new particles into existing force fields and docking algorithms such as nanotubes [7] and fullerenes.

Many proteins do not exist in the human body as separate units but can form complexes, most often with the same (homomeric) or other (heteromeric) proteins [8]. Many human genetic diseases are caused not by altering the conformation of a single protein, but by impacting complex structure, dynamics, and stability [9]; therefore, it is crucial to be able to predict protein–protein structure and interactions with high accuracy. Contrary to the protein–ligand docking methods, in which the latter molecule can be easily treated with the inclusion of its flexibility due to its small size, in protein–protein docking, it is much more difficult and computationally expensive. The docking procedure often allows us to freeze part or the whole molecule, which is very useful when the simulated system is big or there is some information about the interaction site, which allows us to limit the search and required flexibility. This type of docking is known as semi-rigid or rigid docking and it is used for systems, which exhibit a "lock and key" binding scheme [9, 10]. On the other hand, flexible docking, in which fragments of both systems have to be able to move, is necessary to properly study some systems with an "induced fit" binding mechanism [11] (*see* **Notes 2** and **3**).

Molecular docking tools can use different approaches to perform the docking procedure based on the energy and scoring functions. Two most popular approaches are: (i) knowledge-based (KB), and (ii) physics-based (PB) (*see* **Note 1**). Knowledge-based approach is based on a statistical interpretation of information from existing databases, e.g., of an atom–pair interactions from protein–protein, protein–peptide and protein–ligand complexes with already solved structures by experiments such as X-ray crystallography [12], NMR (nuclear magnetic resonance) spectroscopy [13] and Cryo-EM (Cryo-electron microscopy) [14] and such data is then converted into pseudopotentials [15]. However, while the number of single-molecule structures rises rapidly, the number of structures of complexes is far behind, especially taking into account that a single protein can be involved in multiple different complexes [16]. This trend is caused by the fact that protein–protein complexes are larger and more difficult to solve by NMR and often more complicated to crystallize to perform X-ray studies. Moreover, crystal complexes may not have the same binding interface as in physiological conditions [17]. On the other hand, physics-based approaches are based on the potentials describing physical interactions between atoms (in all-atom models) or interaction centers (beads; in coarse-grained models). Scoring functions in physics-based methods are based on molecular mechanical calculations, which in most cases are calculated based on desolvation energies, Van der Waals interactions, and electrostatic interactions. These values can be obtained from ab initio simulations [18] and experimental data. Due to simplified representation of the systems during docking and simplifications and inaccuracies of data used to derive the parameters for PB potentials, they are often less accurate than KB potentials for typical systems leading to lower average accuracy. However, KB potentials often completely fail for non-typical systems, for which PB potentials should provide consistent performance. The number of KB scoring functions grows more rapidly [19, 20] than PBs [21, 22]; however, in general, hybrid KB-PB scoring functions tend to perform better [23, 24], showing importance to develop both types of methods.

In general, most popular tools used for docking do not emphasize the use of purely physics-based or knowledge-based approaches, as in most cases, the combination of them provides the best results [24]. Therefore, there is a very limited number of tools, which can be called truly physics-based. PMFF is one of the recent examples that physics-based methods can be effectively used for docking purposes [25], although mostly in protein–ligand scheme (*see* **Note 1**). On the other hand, there are methods based purely on the analysis of data from databases using machine learning techniques, such as AlphaFold [26], which has been highly successful in protein structure prediction. AlphaFold can also be used for protein–peptide docking [27] and can be extended for protein–protein docking [28]. However, such tools cannot be used to provide any other information about studied systems, such as their dynamics or impact of individual fragments of binding.

Docking procedures can be run freely (without any additional information) achieving satisfactory accuracy or with some restraints based on available experimental information or templates [29], the latter approach being very successful, if reliable similar models are available [30]. There are also some tools, like InterEvDock3, which can combine both approaches to achieve higher accuracy of the prediction [31], while others, like LZerD [32], use de novo protein structure prediction to enhance the quality of the results. There are some reports showing that template-free docking is generally more reliable; however, template-based prediction of protein–protein complexes works better if partner or partners exhibit larger conformational changes [33]. However, in most cases, only scarce information about the protein–protein system is available and then free docking is not a choice, but a necessity.

Even more complicated case is when no available experimental structures of one or more proteins forming a complex are available. Such cases should be treated with tools suitable for models, which may have larger imperfections than regular experimental structures [34] and should be followed by the additional step, such as molecular dynamics simulations, to confirm obtained data and allow for the larger flexibility of the partners [35, 36]. Despite the fact that docking is more successful if higher quality models are used in rigid-body docking, in general, with high-quality template-based models, satisfactory quality of protein–protein complex prediction [37] can be achieved.

It should be noted that the protein–protein docking problem is very similar in nature to the prediction of the orientation of protein domains with restriction that linker between the domains limits the conformational space search. And as TBM methods can predict single-domain proteins with high accuracy, prediction of the multi-domain proteins and protein–protein complexes, due to uncertainty of their orientation, is still problematic, despite the tremendous development of methods [38]. An interesting example illustrating the limitations of knowledge-based and template-based modeling methods occurs during the CASP10 experiment, specifically with target protein T0663 (PDB code: 4EXR). In this case, all of the TBM methods failed to predict the correct orientation of the domains. It was only the physics-based UNRES force field method that successfully predicted it with satisfactory quality

[39]. However, it should be noted that only with the addition of the conformational information about each of the domains, prediction resulted in high-quality structure of the whole protein [40].

2 Materials and Methods

Major characteristic properties of popular docking tools are summarized in Table 1, while their brief description is presented in further sentences (*see* **Notes 1–3**). One of the most popular physics-based tools used for docking is AutoDock4 [41], which utilizes a standard all-atom force field energy function. It simplifies the process by running predictions on a precalculated grid, making it two orders of magnitude faster than using pairwise calculations. However, a drawback is that the receptor must be completely rigid. Contrary to standard all-atom simulations, this tool uses the Lamarckian genetic algorithm. The knowledge-based extension of AutoDock4 is AutoDockVina [42], which is much faster but not that accurate in prediction of binding affinities [43]. However, with the recent developments, AutoDock Vina can be run in a mode similar to AutoDock 4 [44] making it a hybrid PB/KB approach. Due to high speed of calculations, several open-source software can be used to perform computational and virtual screening such as AutoDock Vina (for virtual screening), Raccoon2 (graphical tool which is used for virtual screening and to perform analysis) [45], AutoDockTools (graphical interface for preparation and analysis) [41], and AutoLigand (for predicting binding sites of ligand on receptors) [46]. Although these tools are designed for docking of protein–ligand complexes, they can also be used for prediction of protein–peptide complexes or with some modifications even for rigid-body docking of two proteins, if their size is relatively small [36].

Thanks to the fact that knowledge-based protein–protein docking in a rigid scheme is fast and requires low computational cost, some of the tools are available as servers. One of the examples of such a tool is ZDOCK [47, 48], which is probably the most popular protein–protein server based on fast Fourier transform (FFT) used for rigid-body docking of proteins and prediction of protein multimers using pdb files as input data with Atomic Contact Energy score [49]. HADDOCK [50] (High Ambiguity Driven protein–protein DOCKing) is a very popular method allowing semi-rigid docking of protein–protein complexes based on a mixture of physics- and knowledge-based terms. HDOCK, which allows users to predict binding sites in protein–protein docking and DNA/RNA–protein docking, is a web-server using a hybrid algorithm containing free docking and template-based modeling to assure constant performance for both highly and lowly homologous complexes. Thanks to the TBM function, HDOCK allows users to provide as

Table 1
Comparison of popular computational tools for docking of macromolecules

Tool	PB, TBM, KB	Search function	Scoring function	Type of molecules	Flexibility
UNRES-dock	PB	Semi-random placement and coarse-grained MD and REMD simulations	WHAM and clustering	Protein–protein, protein–peptide, protein–fullerene	Both semi-rigid and flexible
CABS-dock	KB	Monte Carlo simulations	Filtering and energy scoring	Protein–peptide	Limited
ATTRACT	KB	Energy minimization in a simplified model	Clustering of lowest energy complexes	Protein–protein	Full/limited
AutoDock 4	PB	Lamarckian Genetic Algorithm	All-atom force field parameters	Protein-ligand, protein–peptide, protein–protein (with modifications)	Limited or rigid
AutoDock Vina	KB	Iterated Local Search global optimizer	KB scoring function	Protein-ligand, protein–peptide	Limited or rigid
ZDOCK	PB/KB	Fast Fourier transform search	Atomic Contact Energy score	Protein–protein	Rigid
HADDOCK	PB/KB	Minimization of randomized positions followed by semirigid simulated annealing and refinement	Clustering and scoring based on average interaction energies and their average buried surface area	Protein–protein	Semi-rigid
HDOCK	KB	Fast Fourier transform search	KB scoring function	Protein–protein, protein–peptide, protein–nucleic acid	Rigid
SwarmDock	KB	Normal mode search with particle swarm optimization	Scoring function	Protein–protein	Full/limited
pyDockWeb	KB	Fast Fourier transform search	Electrostatics and desolvation scoring	Protein–protein	Rigid
LZerD	KB	Geometric hashing algorithm	Clustering and Ranksum (combination of DFIRE, GOAP, and ITScorePro)	Protein–protein	Rigid

RosettaDock	KB	Broad local search of pregenerated orientations by Monte Carlo with minimization	Motif Dock Score	Protein–protein	Limited
ClusPro	KB	Fast Fourier Transform search by PIPER	Various scoring functions to choose from	Protein–protein	Rigid
InteEvDock3	KB/ TBM	Template-based and free modeling conformational search	Combination of information including covariation-inferred contact maps and evolutionary information	Protein–protein	Rigid

PB physics-based method, *TBM* template-based method, *KB* knowledge-based method

an input both protein sequence or structure [51, 52]. Another server is SwarmDock, which allows to perform protein–protein docking in global docking mode or with narrowing down binding sites by providing some particular residues that should or should not form inter-chain interactions [53, 54]. Another server is pyDockWEB, which uses adjusted grid size for optimal FFT calculations along with a new version of pyDock. This combination allows to speed up the calculation while maintaining high accuracy [55]. Another server is LZreD where the AttentiveDist [56] deep-learning method for inter-chain distance prediction is used, which allows to combine the structures deposited in the PDB database with the de novo predictions [32]. RosettaDock server uses Monte Carlo (MC) docking algorithm where different resolutions and grains of structures are used to determine the position of main chain and side chains [32, 57–59]. ClusPro is yet another server that requires two protein structures from PDB (Protein DATA Bank) as input data utilizing PIPER [60] as the FFT search engine. This server stands out from others by providing the availability of advanced options like the removal of protein regions with undetermined structure or including small angle X-ray scattering (SAXS) data, additionally providing choice from six different energy functions depending on the system [61]. InterEvDock3 is a server in which both protein sequence and structure (in the form of a PDB file) can be used as input files for docking due to the TBM capabilities of the method. Obtained TBM models from a large variety of homologs are compared with data obtained by other implemented methods predicting inter-molecule contacts with the option to include or not deep learning data [31].

As the majority of the methods are based on knowledge-based energy functions, below we will describe probably the only available tool based purely on physics of interactions, without the need to use any information from databases or templates to perform efficient protein–protein docking.

2.1 UNRES-Dock

UNRES-dock is an extension of the coarse-grained physics-based UNRES model and force field for prediction of protein–peptide, protein–protein [62], and protein–fullerene structures [63]. It is available both as a standalone code allowing to freely modify parameters, such as flexibility of the molecules and web-server (https://unres-server.chem.ug.edu.pl). In the UNRES model, every amino-acid residue is represented by two centers of interaction: peptide group placed between two Cα atoms and side-chain (SC), while fullerenes can be represented as implicit or explicit models. The implicit model is based on Kihara potential [7], while the explicit model [63] utilizes alanine SCs for C_{60} representation, similar to the coarse-grained representation used in the MARTINI model [64]. In addition to 20 standard amino-acid residues, the UNRES model includes some non-standard residues and their D-variants

[65] and supports the dynamic formation and disruptions of disulfide bonds (including non-native) [66, 67], which are often involved in the formation of protein–protein complexes [68] having a significant impact on protein–protein docking poses and stability of obtained orientations [69]. In general, the performance of UNRES-Dock was better than in the case of CABS-Dock in the prediction of protein–peptide complexes, however, requiring larger computational resources [62]. For protein–protein docking, UNRES-dock on average provided worse performance than ATTRACT; however, the latter tool completely failed to predict protein–protein complexes outside of regular benchmark databases (e.g., randomly chosen homodimers and heterodimers taking from PDB database [70]) proving again that PB potentials cannot reach the accuracy of KB ones for well-known targets but are much more versatile and their reliability is consistently high even in the case of non-typical systems [62]. The performance of protein-C_{60} docking was tested on a group of systems, previously suspected of being able to bind carbon nanoparticles [71–73], including Human Serum Albumin (HSA), which was also investigated for the role of disulfide-bonds using dynamic disulfide breaking and formation potential [63].

The UNRES-Dock docking procedure (Fig. 1) is based on the Replica Exchange Molecular Dynamics (REMD) [74] simulations run at different temperatures starting from a pre-generated set of most different molecule orientations in periodic boundary conditions [75]. Ensemble structures are then analyzed using the Weighted Histogram Analysis Method (WHAM) [76]. Structures obtained from the analysis with the lowest free energy at a given temperature are clustered. Subsequently, cluster centroids are determined and their all-atom models are reconstructed using PULCHRA [77] and SCWRL [78] tools. Obtained conformations are treated as predicted model candidates.

Both conformational search and scoring are based on the force field version with physics-based potentials calibrated using the maximum-likelihood method [79, 80] and are frequently improved as the standard UNRES model develops [81]. However, as the whole method is based on the physics of interactions and not on knowledge-based potentials, it should be comparably accurate for systems similar to the one in which structures are known and the one with new types of fold and protein–protein orientations. The three docking variants are based on the same methodology, but there are a few differences between them: in protein–protein docking by default, there are weak restraints placed on the separate proteins to prohibit them from making too big conformational changes at high-temperature replicas in REMD scheme, while in protein–peptide variant the latter molecule is unrestrained (with the option to generate the initial conformation from sequence) to adequately represent high flexibility of the short peptide chain. To

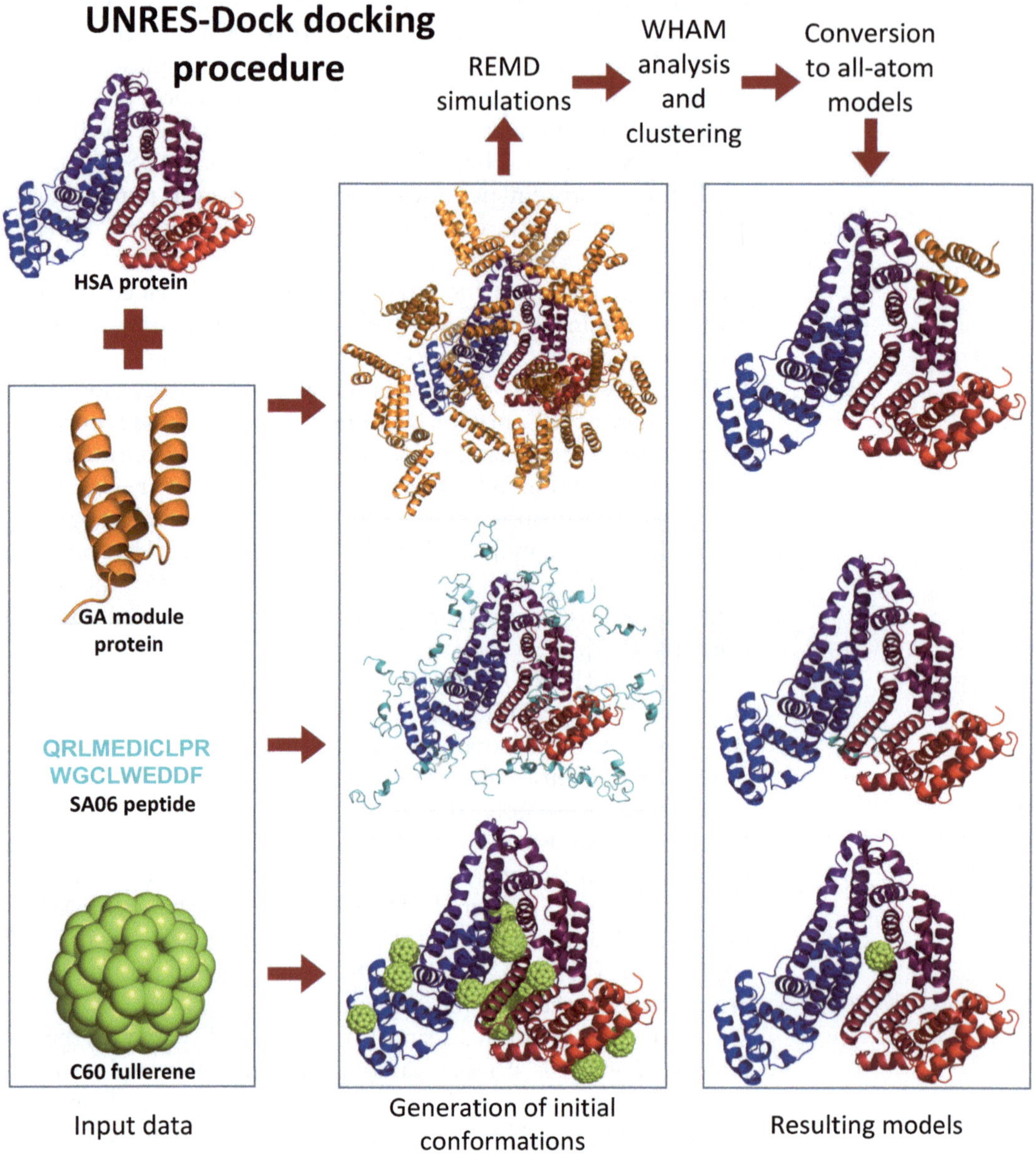

Fig. 1 Schematic of the UNRES-Dock docking procedure, in which protein structures, peptide structures or sequences, and carbon nanoparticles, such as fullerenes, can be used as input data. Subsequently, geometrically different initial conformations are generated and used to conduct REMD simulations in a coarse-grained representation. After the simulations are concluded, WHAM and cluster analysis are performed to generate the most probable orientations of molecules, which are then converted into all-atom models

avoid initial protein–protein and protein–peptide systems being stuck in the local minima, the initial conformation generation is set up to produce as different as possible orientations, which are not docked, but only weakly interact with the receptor. This approach however is not optimal for protein–fullerene docking, as the second molecule can be treated as fully rigid, and in this case the pre-docked conformations with stronger interactions are generated. In all variants, multiple initial orientations (24 by default) are used to run REMD simulations to improve sampling even in short simulations.

The docking scheme in UNRES-Dock is presented on the example of Human Serum Albumin (HSA) (Fig. 1), which is the most abundant protein in human plasma and can bind other proteins, peptides, and fullerenes [82]. For the purpose of this example, we selected bacterial albumin-binding GA module [83], SA06 peptide [84], and C60 fullerene [85], which are known HSA ligands. Not only all of them can be docked to HSA protein with satisfactory accuracy, but due to the physics-based character of the method, it can also predict the dynamics and thermodynamics of the complex and the impact of the disulfide bonds on the binding and stability. In the near future, it will be possible to also include other molecules interacting with proteins, such as lipids, which will significantly increase predicting capabilities for proteins embedded in lipid bilayers.

3 Notes

1. There are not many tools with purely physics-based scoring functions. Those that do exist are designed to perform mostly protein–ligand docking. Examples of such tools include Dock [86, 87], GalaxyDock [88, 89], GoldScore [90–92], and an early version of AutoDock 2 [93].
2. To perform docking with larger molecules not just ligands to protein, the physics-based force field can be used. Although all-atom force fields can be used for protein–ligand docking [94] and protein–peptide [95], they often require applying simplifications, such as assuming grid search or rigid-body docking.
3. As protein–protein docking is even more computationally expensive, molecular dynamics simulations with all-atom force fields are rarely used for that purpose. One solution is to use coarse-grained force fields as the main engine for conformational searches, such as ATTRACT [96–98], CABS (CABS-Dock) [99–102], or UNRES (UNRES-Dock) [62, 103, 104].

Acknowledgments

This work was supported by grants UMO-2015/17/N/ST4/03935 and 2019/35/D/ST4/03156 from the National Science Centre (Poland).

References

1. Elfiky AA (2020) Ribavirin, Remdesivir, Sofosbuvir, Galidesivir, and Tenofovir against SARS-CoV-2 RNA dependent RNA polymerase (RdRp): a molecular docking study. Life Sci 253:117592
2. Fadlalla M, Ahmed M, Ali M et al (2022) Molecular docking as a potential approach in repurposing drugs against COVID-19: a systematic review and novel pharmacophore models. Curr Pharmacol Rep 8:212–226
3. Cancela F, Rendon-Marin S, Quintero-Gil C et al (2021) Modelling of Hepatitis E virus RNA-dependent RNA polymerase genotype 3 from a chronic patient and interaction analysis by molecular docking with Ribavirin. J Biomol Struct Dyn:1–17
4. Setny P, Bahadur RP, Zacharias M (2012) Protein-DNA docking with a coarse-grained force field. BMC Bioinformatics 13:228
5. Tuszynska I, Magnus M, Jonak K et al (2015) NPDock: a web server for protein-nucleic acid docking. Nucleic Acids Res 43:W425–W430
6. Kerzmann A, Fuhrmann J, Kohlbacher O, Neumann D (2008) BALLDock/SLICK: a new method for protein-carbohydrate docking. J Chem Inf Model 48:1616–1625
7. Sieradzan AK, Mozolewska MA (2018) Extension of coarse-grained UNRES force field to treat carbon nanotubes. J Mol Model 24:121
8. Marsh JA, Teichmann SA (2015) Structure, dynamics, assembly, and evolution of protein complexes. Annu Rev Biochem 84:551–575
9. Bergendahl LT, Therese Bergendahl L, Gerasimavicius L et al (2019) The role of protein complexes in human genetic disease. Protein Sci 28:1400–1411
10. Tripathi A, Bankaitis VA (2017) Molecular docking: from lock and key to combination lock. J Mol Med Clin Appl 2(1)
11. Koshland DE (1995) The key–lock theory and the induced fit theory. Angew Chem Int Ed Engl 33:2375–2378
12. Tutton AEH (1915) X-rays and crystal structure. Nature 95:198–199
13. Wagner G (1993) Prospects for NMR of large proteins. J Biomol NMR 3:375–385
14. Adrian M, Dubochet J, Lepault J, McDowall AW (1984) Cryo-electron microscopy of viruses. Nature 308:32–36
15. Velec HFG, Gohlke H, Klebe G (2005) DrugScore(CSD)-knowledge-based scoring function derived from small molecule crystal data with superior recognition rate of near-native ligand poses and better affinity prediction. J Med Chem 48:6296–6303
16. Hopf TA, Schärfe CPI, Rodrigues JPGLM et al (2014) Sequence co-evolution gives 3D contacts and structures of protein complexes. elife 3:e03430
17. Kobe B, Guncar G, Buchholz R et al (2008) Crystallography and protein-protein interactions: biological interfaces and crystal contacts. Biochem Soc Trans 36:1438–1441
18. Monticelli L, Peter Tieleman D (2013) Force fields for classical molecular dynamics. Methods Mol Biol 924:197–213
19. Sasse A, de Vries SJ, Schindler CEM et al (2017) Rapid design of knowledge-based scoring potentials for enrichment of near-native geometries in protein-protein docking. PLoS One 12:e0170625
20. Huang S-Y, Zou X (2008) An iterative knowledge-based scoring function for protein-protein recognition. Proteins 72: 557–579
21. Bryce RA (2011) Physics-based scoring of protein-ligand interactions: explicit polarizability, quantum mechanics and free energies. Future Med Chem 3:683–698
22. Faraggi E, Krupa P, Mozolewska MA et al (2018) Reoptimized UNRES potential for protein model quality assessment. Genes 9(12):601
23. Baek M, Shin W-H, Chung HW, Seok C (2017) GalaxyDock BP2 score: a hybrid scoring function for accurate protein–ligand docking. J Comput Aided Mol Des 31:653–666
24. Guedes IA, Barreto AMS, Marinho D et al (2021) New machine learning and physics-based scoring functions for drug discovery. Sci Rep 11:3198
25. Hwang SB, Lee CJ, Lee S et al (2020) PMFF: development of a physics-based molecular force field for protein simulation and ligand docking. J Phys Chem B 124:974–989
26. Jumper J, Evans R, Pritzel A et al (2021) Highly accurate protein structure prediction with AlphaFold. Nature 596:583–589
27. Tsaban T, Varga JK, Avraham O et al (2022) Harnessing protein folding neural networks for peptide-protein docking. Nat Commun 13:176
28. Jones G, Jindal A, Ghani U et al (2022) Elucidation of protein function using computational docking and hotspot analysis by ClusPro and FTMap. Acta Crystallogr D Struct Biol 78:690–697

29. Xue LC, Rodrigues JPGLM, Dobbs D et al (2017) Template-based protein-protein docking exploiting pairwise interfacial residue restraints. Brief Bioinform 18:458–466
30. Porter KA, Desta I, Kozakov D, Vajda S (2019) What method to use for protein–protein docking? Curr Opin Struct Biol 55:1–7
31. Quignot C, Postic G, Bret H et al (2021) InterEvDock3: a combined template-based and free docking server with increased performance through explicit modeling of complex homologs and integration of covariation-based contact maps. Nucleic Acids Res 49: W277–W284
32. Christoffer C, Bharadwaj V, Luu R, Kihara D (2021) LZerD protein-protein docking webserver enhanced with de novo structure prediction. Front Mol Biosci 8:724947
33. Vreven T, Hwang H, Pierce BG, Weng Z (2014) Evaluating template-based and template-free protein-protein complex structure prediction. Brief Bioinform 15:169–176
34. Bohnuud T, Luo L, Wodak SJ et al (2017) A benchmark testing ground for integrating homology modeling and protein docking. Proteins 85:10–16
35. Shen J, Zhang W, Fang H et al (2013) Homology modeling, molecular docking, and molecular dynamics simulations elucidated α-fetoprotein binding modes. BMC Bioinformatics 14:S6
36. Mozolewska MA, Krupa P, Scheraga HA, Liwo A (2015) Molecular modeling of the binding modes of the iron-sulfur protein to the Jac1 co-chaperone from *Saccharomyces cerevisiae* by all-atom and coarse-grained approaches. Proteins 83:1414–1426
37. Yu J, Guerois R (2016) PPI4DOCK: large scale assessment of the use of homology models in free docking over more than 1000 realistic targets. Bioinformatics 32:3760–3767
38. Ozden B, Kryshtafovych A, Karaca E (2021) Assessment of the CASP14 assembly predictions. Proteins 89:1787–1799
39. He Y, Mozolewska MA, Krupa P et al (2013) Lessons from application of the UNRES force field to predictions of structures of CASP10 targets. Proc Natl Acad Sci USA 110:14936–14941
40. Krupa P, Mozolewska MA, Joo K et al (2015) Prediction of protein structure by template-based modeling combined with the UNRES force field. J Chem Inf Model 55:1271–1281
41. Morris GM, Huey R, Lindstrom W et al (2009) AutoDock4 and AutoDockTools4: automated docking with selective receptor flexibility. J Comput Chem 30:2785–2791
42. Trott O, Olson AJ (2010) AutoDock Vina: improving the speed and accuracy of docking with a new scoring function, efficient optimization, and multithreading. J Comput Chem 31:455–461
43. Nguyen NT, Nguyen TH, Pham TNH et al (2020) Autodock Vina adopts more accurate binding poses but Autodock4 forms better binding affinity. J Chem Inf Model 60:204–211
44. Eberhardt J, Santos-Martins D, Tillack A, Forli S (2021) AutoDock Vina 1.2.0: new docking methods, expanded force field, and python bindings. J Chem Inf Model 61(8): 3891–3898
45. Forli S, Huey R, Pique ME et al (2016) Computational protein-ligand docking and virtual drug screening with the AutoDock suite. Nat Protoc 11:905–919
46. Harris R, Olson AJ, Goodsell DS (2008) Automated prediction of ligand-binding sites in proteins. Proteins 70:1506–1517
47. Chen R, Weng Z (2002) Docking unbound proteins using shape complementarity, desolvation, and electrostatics. Proteins 47:281–294
48. Pierce BG, Wiehe K, Hwang H et al (2014) ZDOCK server: interactive docking prediction of protein-protein complexes and symmetric multimers. Bioinformatics 30:1771–1773
49. Zhang C, Vasmatzis G, Cornette JL, DeLisi C (1997) Determination of atomic desolvation energies from the structures of crystallized proteins. J Mol Biol 267:707–726
50. Dominguez C, Boelens R, Bonvin AMJ (2003) HADDOCK: a protein–protein docking approach based on biochemical or biophysical information. J Am Chem Soc 125:1731–1737
51. Yan Y, Zhang D, Zhou P et al (2017) HDOCK: a web server for protein-protein and protein-DNA/RNA docking based on a hybrid strategy. Nucleic Acids Res 45:W365–W373
52. Yan Y, Tao H, He J, Huang S-Y (2020) The HDOCK server for integrated protein–protein docking. Nat Protoc 15:1829–1852
53. Moal IH, Bates PA (2010) SwarmDock and the use of normal modes in protein-protein docking. Int J Mol Sci 11:3623–3648
54. Torchala M, Moal IH, Chaleil RAG et al (2013) SwarmDock: a server for flexible protein–protein docking. Bioinformatics 29: 807–809
55. Jiménez-García B, Pons C, Fernández-Recio J (2013) pyDockWEB: a web server for rigid-

body protein-protein docking using electrostatics and desolvation scoring. Bioinformatics 29:1698–1699
56. Jain A, Terashi G, Kagaya Y et al (2021) Analyzing effect of quadruple multiple sequence alignments on deep learning based protein inter-residue distance prediction. Sci Rep 11: 7574
57. Gray JJ, Moughon S, Wang C et al (2003) Protein-protein docking with simultaneous optimization of rigid-body displacement and side-chain conformations. J Mol Biol 331: 281–299
58. Lyskov S, Gray JJ (2008) The RosettaDock server for local protein-protein docking. Nucleic Acids Res 36:W233–W238
59. Marze NA, Roy Burman SS, Sheffler W, Gray JJ (2018) Efficient flexible backbone protein–protein docking for challenging targets. Bioinformatics 34:3461–3469
60. Kozakov D, Brenke R, Comeau SR, Vajda S (2006) PIPER: an FFT-based protein docking program with pairwise potentials. Proteins 65:392–406
61. Kozakov D, Hall DR, Xia B et al (2017) The ClusPro web server for protein-protein docking. Nat Protoc 12:255–278
62. Krupa P, Karczyńska AS, Mozolewska MA et al (2021) UNRES-dock-protein-protein and peptide-protein docking by coarse-grained replica-exchange MD simulations. Bioinformatics 37:1613–1615
63. Rogoża NH, Krupa MA, Krupa P, Sieradzan AK (2024) Integrating explicit and implicit fullerene models into UNRES force field for protein interaction studies. Molecules 29(9): 1919
64. Souza PCT, Alessandri R, Barnoud J et al (2021) Martini 3: a general purpose force field for coarse-grained molecular dynamics. Nat Methods 18:382–388
65. Sieradzan AK, Hansmann UHE, Scheraga HA, Liwo A (2012) Extension of UNRES force field to treat polypeptide chains with d-amino acid residues. J Chem Theory Comput 8:4746–4757
66. Chinchio M, Czaplewski C, Liwo A et al (2007) Dynamic formation and breaking of disulfide bonds in molecular dynamics simulations with the UNRES force field. J Chem Theory Comput 3:1236–1248
67. Krupa P, Sieradzan AK, Mozolewska MA et al (2017) Dynamics of disulfide-bond disruption and formation in the thermal unfolding of ribonuclease a. J Chem Theory Comput 13:5721–5730
68. Oka OBV, Bulleid NJ (2013) Forming disulfides in the endoplasmic reticulum. Biochim Biophys Acta 1833:2425–2429
69. Inaba K, Murakami S, Suzuki M et al (2006) Crystal structure of the DsbB-DsbA complex reveals a mechanism of disulfide bond generation. Cell 127:789–801
70. Berman HM, Westbrook J, Feng Z et al (2000) The protein data bank. Nucleic Acids Res 28:235–242
71. Calvaresi M, Zerbetto F (2010) Baiting proteins with C60. ACS Nano 4:2283–2299
72. Giełdoń A, Witt MM, Gajewicz A, Puzyn T (2017) Rapid insight into C60 influence on biological functions of proteins. Struct Chem 28:1775–1788
73. Ahmed L, Rasulev B, Kar S et al (2017) Inhibitors or toxins? Large library target-specific screening of fullerene-based nanoparticles for drug design purpose. Nanoscale 9:10263–10276
74. Czaplewski C, Kalinowski S, Liwo A, Scheraga HA (2009) Application of multiplexed replica exchange molecular dynamics to the UNRES force field: tests with alpha and alpha+beta proteins. J Chem Theory Comput 5:627–640
75. Sieradzan AK (2015) Introduction of periodic boundary conditions into UNRES force field. J Comput Chem 36:940–946
76. Kumar S, Rosenberg JM, Bouzida D et al (1992) THE weighted histogram analysis method for free-energy calculations on biomolecules. I. The method. J Comput Chem 13:1011–1021
77. Rotkiewicz P, Skolnick J (2008) Fast procedure for reconstruction of full-atom protein models from reduced representations. J Comput Chem 29:1460–1465
78. Krivov GG, Shapovalov MV, Dunbrack RL Jr (2009) Improved prediction of protein side-chain conformations with SCWRL4. Proteins 77:778–795
79. Zaborowski B, Jagieła D, Czaplewski C et al (2015) A maximum-likelihood approach to force-field calibration. J Chem Inf Model 55: 2050–2070
80. Krupa P, Hałabis A, Żmudzińska W et al (2017) Maximum likelihood calibration of the UNRES force field for simulation of protein structure and dynamics. J Chem Inf Model 57:2364–2377
81. Antoniak A, Biskupek I, Bojarski KK et al (2021) Modeling protein structures with the coarse-grained UNRES force field in the CASP14 experiment. J Mol Graph Model 108:108008

82. Fanali G, di Masi A, Trezza V et al (2012) Human serum albumin: from bench to bedside. Mol Asp Med 33:209–290
83. Lejon S, Cramer JF, Nordberg P (2008) Structural basis for the binding of naproxen to human serum albumin in the presence of fatty acids and the GA module. Acta Crystallogr Sect F Struct Biol Cryst Commun 64:64–69
84. Hou J, Peng J, Yu Y et al (2017) Allosteric modulation of human serum albumin induced by peptide ligand. Chin J Chem 35:1270–1277
85. Song M, Liu S, Yin J, Wang H (2011) Interaction of human serum album and C60 aggregates in solution. Int J Mol Sci 12:4964–4974
86. DesJarlais RL, Sheridan RP, Seibel GL et al (1988) Using shape complementarity as an initial screen in designing ligands for a receptor binding site of known three-dimensional structure. J Med Chem 31:722–729
87. Allen WJ, Balius TE, Mukherjee S et al (2015) DOCK 6: impact of new features and current docking performance. J Comput Chem 36: 1132–1156
88. Shin W-H, Seok C (2012) GalaxyDock: protein-ligand docking with flexible protein side-chains. J Chem Inf Model 52:3225–3232
89. Shin W-H, Kim J-K, Kim D-S, Seok C (2013) GalaxyDock2: protein-ligand docking using beta-complex and global optimization. J Comput Chem 34:2647–2656
90. Jones G, Willett P, Glen RC (1995) Molecular recognition of receptor sites using a genetic algorithm with a description of desolvation. J Mol Biol 245:43–53
91. Jones G, Willett P, Glen RC et al (1997) Development and validation of a genetic algorithm for flexible docking. J Mol Biol 267: 727–748
92. Verdonk ML, Cole JC, Hartshorn MJ et al (2003) Improved protein-ligand docking using GOLD. Proteins 52:609–623
93. Goodsell DS, Morris GM, Olson AJ (1996) Automated docking of flexible ligands: applications of AutoDock. J Mol Recognit 9:1–5
94. Wu G, Robertson DH, Brooks CL 3rd, Vieth M (2003) Detailed analysis of grid-based molecular docking: a case study of CDOCKER-A CHARMm-based MD docking algorithm. J Comput Chem 24:1549–1562
95. Wang J, Alekseenko A, Kozakov D, Miao Y (2019) Improved modeling of peptide-protein binding through global docking and accelerated molecular dynamics simulations. Front Mol Biosci 6:112
96. Zacharias M (2005) ATTRACT: protein-protein docking in CAPRI using a reduced protein model. Proteins 60:252–256
97. de Vries SJ, Schindler CEM, Chauvot de Beauchêne I, Zacharias M (2015) A web interface for easy flexible protein-protein docking with ATTRACT. Biophys J 108: 462–465
98. Chéron J-B, Zacharias M, Antonczak S, Fiorucci S (2017) Update of the ATTRACT force field for the prediction of protein-protein binding affinity. J Comput Chem 38:1887–1890
99. Kolinski A (2004) Protein modeling and structure prediction with a reduced representation. Acta Biochim Pol 51:349–371
100. Kurcinski M, Ciemny MP, Oleniecki T et al (2019) CABS-dock standalone: a toolbox for flexible protein–peptide docking. Bioinformatics 35:4170–4172
101. Kurcinski M, Jamroz M, Blaszczyk M et al (2015) CABS-dock web server for the flexible docking of peptides to proteins without prior knowledge of the binding site. Nucleic Acids Res 43:W419–W424
102. Kurcinski M, Badaczewska-Dawid A, Kolinski M et al (2020) Flexible docking of peptides to proteins using CABS-dock. Protein Sci 29: 211–222
103. Liwo A, Oldziej S, Pincus MR et al (1997) A united-residue force field for off-lattice protein-structure simulations. I. Functional forms and parameters of long-range side-chain interaction potentials from protein crystal data. J Comput Chem 18:849–873
104. Liwo A, Baranowski M, Czaplewski C et al (2014) A unified coarse-grained model of biological macromolecules based on mean-field multipole-multipole interactions. J Mol Model 20:2306

Part II

Novel Methodological Developments in Protein–Protein Docking

Chapter 4

Approaches to Backbone Flexibility in Protein–Protein Docking

Ayesha Asim

Abstract

Proteins are the fundamental organic macromolecules in living systems that play a key role in a variety of biological functions including immunological detection, intracellular trafficking, and signal transduction. The docking of proteins has greatly advanced during recent decades and has become a crucial complement to experimental methods. Protein–protein docking is a helpful method for simulating protein complexes whose structures have not yet been solved experimentally. This chapter focuses on major search tactics along with various docking programs used in protein–protein docking algorithms, which include: direct search, exhaustive global search, local shape feature matching, randomized search, and broad category of post-docking approaches. As backbone flexibility predictions and interactions in high-resolution protein–protein docking remain important issues in the overall optimization context, we have put forward several methods and solutions used to handle backbone flexibility. In addition, various docking methods that are utilized for flexible backbone docking, including ATTRACT, FlexDock, FLIPDock, HADDOCK, RosettaDock, FiberDock, etc., along with their scoring functions, algorithms, advantages, and limitations are discussed. Moreover, what progress in search technology is expected, including not only the creation of new search algorithms but also the enhancement of existing ones, has been debated. As conformational flexibility is one of the most crucial factors affecting docking success, more work should be put into evaluating the conformational flexibility upon binding for a particular case in addition to developing new algorithms to replace the rigid body docking and scoring approach.

Key words Protein, protein interaction (PPI), Protein, protein docking, Backbone flexibility, Confirmations, Algorithms, CAPRI benchmark, Monte Carlo

1 Introduction

Proteins are the fundamental organic macromolecules in living systems that play a key role in a variety of biological functions including immunological detection, intracellular trafficking, and signal transduction. In addition, protein interactions are essential for gene expression, signal transmission, control of metabolic pathways, catalysis, homeostasis of enzymes, and other vital biological processes. Proteins frequently interact with other proteins to form

Agnieszka A. Kaczor (ed.), *Protein-Protein Docking: Methods and Protocols*, Methods in Molecular Biology, vol. 2780,
https://doi.org/10.1007/978-1-0716-3985-6_4,

© The Author(s), under exclusive license to Springer Science+Business Media, LLC, part of Springer Nature 2024

complexes, in order to carry out their specific tasks in living organisms [1]. As nearly all biological processes in human health and illness primarily include interactions between proteins, a thorough understanding of protein–protein interactions is essential in medical and pharmaceutical sciences. Therefore, discovering binding modes of PPI holds great significance in the diagnostics and therapeutics world [2].

The docking of proteins has greatly advanced during recent decades and has become a crucial complement to experimental methods [3]. Molecular docking plays a significant role in predicting interactions among proteins, enlightening the mechanism of PPI and pinpointing the "hot-spot" residues for drug discovery [4]. Even when complex analytically defined structures are not present, protein–protein docking aids in understanding how molecular processes are carried out in the living cell [5]. Furthermore, when there is no prior structural information available, the docking of one protein to another protein aims at structure prediction of the protein complex that begins with 3D coordinates of the unbound structures and identifying key binding residues in PPI [6].

1.1 Protein–Protein Docking

The process that utilizes physicochemical and steric complementarity at the interface of proteins to computationally determine which relative transformation and conformation of two proteins result in a stable compound is known as protein–protein docking [7]. It involves the estimation of conformational changes between bound and unbound structures [8].

Proteins are treated as rigid bodies (three dimensions of translational and rotational freedom) at the starting stage [9]. The proteins contact each other repeatedly and the partners create native complexes after the development of new interactions and conformational alterations at the interface regions with a large steric interface complementarity [10].

1.2 Search Techniques for Protein–Protein Docking

Following are the fundamental categories to classify the search tactics used in protein–protein docking algorithms:

- Direct search.
- Exhaustive global search.
- Local shape feature matching.
- Randomized search.
- Broad category of post-docking approaches.

Putative binding modes are effectively sampled using direct search techniques in the physical world, or Cartesian space. Depending on the availability of binding site information, the sampling process may be local or global [11].

Due to the lack of knowledge concerning protein binding sites for binding orientations with more than six degrees of freedom, the docking of proteins needs a global approach (3D translational and rotational). The computing cost of conducting a comprehensive global search is therefore O(N6) in theory, where O(N3) is the cost of translational 3D search and O(N3) is the cost of the rotational 3D search. Usually, one protein is fixed known as "static molecule" in an actual docking, while another protein is placed around a stationary protein known as "moving molecule." The mobile protein molecule is initially rotated by an Euler angle in three-dimensional rotational space, which distinguishes searches greater than three degrees of freedom in translation and over three degrees of freedom in rotation. This is followed by a thorough search for the transient protein throughout the complete 3D translational space with regard to the rotation of the static protein. The preceding procedure continues until all of the three-dimensional rotating space has been sampled. Originally introduced by Katchalski-Katzir, FFT-based docking is a grid-based technique that improves the search for binding modes in 3D translational space. As a result, in global sampling, the computational time was reduced from O (N6) to O(N3 log(N3)) [12]. In the last 10 years, a number of FFT-based protein–protein docking algorithms were created due to its quick global search and they have had significant success in CAPRI (Critical Assessment of Prediction of Interactions) trials conducted by the entire community.

Proteins are modeled by molecular forms in local shape feature-matching methods, which include Connolly surface, in which a probe molecule is used for computing and is frequently referred to as the "solvent" excluded accessible surface [4]. Afterward, by using algorithms, it is possible to identify those matches between two proteins that have decent local shape complementarity.

In molecular docking, both local and global searches frequently employ randomized searches. Despite the fact that a grid depiction of the receptor or simplified protein models could be employed to expedite the search, randomized search algorithms frequently conduct their searches at the atomic level [1].

Post-docking methods are intended to enhance the quality and ranking of proper binding modes by enhancing the potential modes derived from different sampling techniques [13]. The post-docking process has been of standard practice during protein–protein docking operations and achieved remarkable accomplishments. Biological information and protein flexibility may be simply integrated into a limited number of docking solutions. All docking methods must have a scoring function to rank and assess the sampled conformations. The most fundamental part of the scoring function, shape complementarity, is crucial for finding potential binding poses and evaluating the sampled poses.

Table 1 displays the search methods utilized by the protein–protein docking algorithms that are currently in use.

Table 1
The search methods utilized by the protein–protein docking algorithms that are currently in use

Search algorithms		Description	Docking programs	References
Exhaustive global search	FFT-based search	The FFT-based technique shortens the calculation time from the standard O(N3) to O(log(N3)), enhancing the search process across three degrees of freedom in translation	GRAMM, SDOCK, ZDOCK, FTDock, F2DOCK, ASPDock, PIPER, DOT, MolFit, Cell-Dock	[4]
	Spherical Fourier transform-based search	A straightforward Cartesian grid model of protein topology is used in Fourier correlation methods, which automatically penalizes steric conflicts and favors tight interactions	FRODO, HEX	[4]
	Direct search in Cartesian space	With the use of some acceleration techniques, these methods identify similarities in 3D Cartesian grid space of two proteins	SOFTDOCK, SKE-DOCK, BIGGER	[4]
Local shape feature matching	Distance geometry algorithm	The molecular surface of receptor protein is calculated using distance geometry technique. The sphere points are created for depiction of receptor protein's binding site's shape. By aligning the atoms of the ligand–protein with the sphere sites on the receptor protein, the next potential complexes are built	DOCK	[14]
	Geometric hashing	By locating the local matches (surface patches) of shape analysis between proteins, geometric hashing algorithms carry out protein–protein docking search globally	SymmDock, PatchDock	[15, 16]
	Genetic algorithm	An established theory of natural selection and biological evolution serves as the foundation for a search algorithm known as a genetic algorithm (GA). It employs the "survival of the fittest" approach when dealing with a population of solutions	GAPDOCK	[17]

Randomized search	Monte Carlo search	Determining the global minimum matching to the native binding state through the investigation of the binding energy landscape	HADDOCK, RosettaDock, ATTRACT, ICM-DISCO	[18]
	Particle swarm optimization	A computational method for problem optimization that involves repeated attempts to improve the quality of a proposed solution	SwarmDock	[19]
	Genetic algorithm	Search algorithm focusing on the principle of "survival of the fittest"	AutoDock	[20]
Post-docking approach	Using advanced scoring functions	To precisely assess the interaction between the associated proteins, scoring (or energy) functions are used (i.e., binding affinity). In general, near-native selection and conformational sampling, coupled with refinement, are the two uses for scoring functions	RPScore, EMPIRE, DECK, PyDock, SIPPER, DARS, PIE, ZRANK	[4]

2 Challenges in Protein–Protein Docking

Despite the latest advancements in computational docking techniques, the structure prediction of protein complexes from parts remains challenging [21]. Major complications include the requirement for docking of various types of molecules, appropriate binding accounting for conformational flexibility, novel techniques based on the prediction of affinity and residue co-evolution, deep learning, and the development of fully automated docking servers [22].

Recent improvements in rigid-body scoring functions and sampling techniques enable the creation of slightly elevated docking models that are ideal for functional and biological annotations, at least in some circumstances. However, docking prediction still faces significant obstacles:

- Computational biology research is greatly hampered by the flexibility of protein docking. The docking problem is significantly more challenging as a result of the backbone flexibility's enormous degree of freedom added to the search space.
- One of the obvious shortcomings of the present rigid-body docking techniques is conformational flexibility upon binding. Fully unconstrained rigid-body docking methods are insufficient in circumstances with greater mobility involving multi-domain proteins and should be combined with domain–domain linker residue constraints, binding site predictions, or evolutionary information.
- Other difficult situations include weak or brief contacts, which include proteins playing a role in electrons transferring in which the landscape of binding energies is complicated by the availability of encounter complexes and numerous binding orientations [21].
- The models used for interacting subunits during the course of simulations are another significant docking challenge.

The backbone flexibility predictions and interactions mediated by water in high-resolution protein–protein docking remain important issues in the overall optimization context. High-resolution docking in contrast to low-resolution docking traverses the conformational space with atomic precision to produce predictions with minute features. It enables solutions to more complex docking issues including conformational changes brought on by binding [23].

3 Solution for Handling Backbone Flexibility in Protein–Protein Docking

It remained a struggle to deal with backbone flexibility in protein–protein docking. The four kinds of docking procedures can be used to treat backbone flexibility. The first method involves docking with soft interfaces and permits a few steric conflicts among complicated models. The second method executes ensemble docking using potential conformations of proteins that were previously produced. The third kind focuses on hinge bending motions and the fourth group explores a large conformational search space in search of energetically preferred conformations heuristically [24].

3.1 Soft Interface

Soft interface docking techniques carry out rigid-body docking very quickly, allowing certain steric conflicts (penetration). Three main categories can be used to classify these techniques:

- Brute force methods [25].
- Randomized methods [17].
- Shape complementarity methods.

Although the exact alterations are not explicit, modeled proteins are thought to be able to make necessary conformational adjustments to prevent penetrations. Since steric conflicts frequently appear in the outcomes of this soft docking, they need to be fixed using a later refinement stage.

3.2 Ensemble Docking

The ensemble docking method samples an ensemble of various viable conformations preceding docking, avoiding the necessity to conduct a comprehensive, flexible conformational space search for two proteins while docking or refining. It is then followed by the docking simulations of the entire ensemble. The many conformations can be docked collectively using various techniques, such as the mean-field approach or individually using cross-docking, which greatly lengthens calculation time. The protein's NMR conformers and several crystal structures might be included in the ensemble. Using computational sampling techniques established from the flexibility of protein studies (molecular dynamics, essential dynamics, loop modeling, normal modes, etc.), other structures can also be determined. Additionally, feasible structures can be chosen at random using techniques like Monte Carlo as well as genetic algorithms. The mean field technique can be used during the docking process to look for an ideal loop conformation [24].

3.3 Modeling Hinge Motion

When creating complicated systems, hinge-bending actions are frequent. The relatively inflexible portions of proteins such as domains or subdomains are separated by hinges, which are flexible segments. The algorithm Sandak proposed deals with this kind of adaptability. Multiple hinge locations can be specified by the user and are supported by the algorithm and just a single interacting protein that can have hinges, for example, ligand. Using a geometric hashing method, the algorithm simultaneously docks each of the flexible ligand's rigid parts [26].

The FlexDock algorithm, which allows flexibility for hinge bending in one of the partnering proteins, represents a more advanced docking technique [27]. The HingeProt algorithm automatically determines where the hinges are. The intricacy of the running time is unaffected by and is not limited by the number of hinges. However, a linear chain must be formed by the subdomains that the hinges have divided causing hinges to impose a chain-type topology [28]. The flexible protein's hinge sites are where the algorithm divides it into subdomains. The PatchDock method docks these subdomains independently to the second protein followed by the construction of an assembly graph. The docking score determines the weight of each node in the graph, and each node reflects the outcome of a subdomain rigid docking (transformation). Between nodes that indicate consistent answers to successive rigid subdomains, edges are added. The score of complementarity among structures between two subdomains represented by two nodes translates to an edge weight. The docking findings of the individual subdomains are then integrated to provide fully consistent results for the complex by means of an efficient dynamic programming method that locates top-scoring paths in the graph.

3.4 Treating Backbone Flexibility Via Refinement and Minimization Methods

Since the additional flexibility improves the docking outcomes, rigid-body docking is performed by HADDOCK accompanied by MD designed to simulate refinement of the degrees of freedom for the side chain and backbone during annealing [29]. HADDOCK expressly permits flexibility of backbone throughout a rotational and translational stage of an MD-simulated annealing refinement. The explicit addition of flexibility of both the side and back chain boosted many viable options and enhanced the grading of complexes generated for the accurate predictions made using HADDOCK [30].

A new refining technique for docking solutions was developed by Lindahl and Delarue that reduces the energy of interaction in complexity in the low-frequency normal mode directions, along axes 5–10 [31]. The search space's degrees of freedom are provided by the normal mode amplitudes. The quasi-Newtonian method is employed for the minimization of energy. Using techniques that had been verified on complexes such as protein–DNA and protein–ligand, there might be a 0.3–3.6 Å reduction in the RMSD of the docking model and actual complex.

3.5 Identifying Putative Binding Sites

The models widely utilized for global search to limit degrees of freedom and to simplify the enormous conformation space of flexible proteins are coarse-grained. In methods like ClusPro and ZDOCK, protein binding partners are projected onto a discretized 3D-grid using the fast Fourier transform (FFT) correlation. Traditional FFT methods only speed up translational space sampling where each rotation requires a new FFT [2]. Geometric hashing is a different shape-based method that uses point sets or curves as indices to reveal geometric properties when transformed in various ways, such as rotations, scaling, and translations. One of the most effective techniques in CAPRI known as local 3D Zernike descriptor-based docking (LZerD) involves projections of 3D surfaces onto spheres to effectively record protein surface complementarity. To further enhance the differentiation of created structures, several rigid body techniques incorporate data from small-angle X-ray scattering (SAXS) or cross-linking chemical research. These techniques were employed by numerous predictors in previous CAPRI rounds as the initial step in the discovery of putative binding patches, and they were then complemented by other refining methods to capture the flexibility of the backbone of the protein.

4 Methods Accounting Backbone Flexibility

4.1 Molecular Dynamics

One method that is usually utilized for refinement following grid-search or template-based techniques is molecular dynamics (MD). A time-resolved high-resolution microscopic model of PPI can be produced via impartial, all-atom MD simulations. MD simulates protein attachment and dissociation events by computing Newtonian trajectories with energy levels based on physics [32].

Steered molecular dynamics (SMD), which uses external force limitations, and Markov sampling, which divides lengthy MD simulations into numerous small paths, are two new approaches that have emerged in the last 10 years to capture conformational changes [33]. A Hamiltonian replica exchange MD procedure (H-REMD) was created by Ostermeir et al. to hasten the detachment of protein partners at less-than-ideal binding sites for docking proteins. Biassing potentials in H-REMD are estimated based on the atoms of protein partners that are closest to one another (known as "ambiguity restrictions"). In one replica, related partner proteins are compelled to dissociate whereas in another the biassing potential and linked ambiguity restraints differ between replicates [34]. For a benchmark set of five objectives, Pan et al. simulated lengthy timescales in a global search space on a special-purpose computer. Scaling of soft-core van der Waals intermolecular potential takes place in their "tempered binding" technique to increase the frequency of long-lived states dissociating and hence increase the sampling efficiency. In addition, it was discovered by Pan et al.

that the pattern of repetitive association and dissociation is followed by proteins instead of searching for local binding sites by scanning the surface continuously. There have not yet been any reports of docking of proteins that use MD methods to move greater than 2.2 Å RMSD after binding.

4.2 Monte Carlo Methods

Unlike MD procedures, which use Newtonian dynamics to seek flexibility, Monte Carlo (MC) methods sample by random motions are accompanied generally by minimization reduction (MCM). MC permits a large variety of conformational motion types to sample different conformations. The key–lock and other kinetic binding models such as induced-fit (IF) mechanisms and conformer selection (CS) have been imitated by MC algorithms [35]. The advantage of this method is that one partner's conformational can be docked at a time because the CS model selects the backbones of protein from an ensemble that has already been created. However, if the ensemble lacks native-like backbone conformations, CS docking may not succeed [35]. Zhang et al. developed 40 structures ensemble to perform MC-based docking for targets with RMSDBU closer to 2.5 Å [36]. This ensemble docking method makes use of the coarse-grained protein ATTRACT model as well as replica exchange (RE) to explore rigid body and backbone space. Despite the fact that the ensemble rarely contains bound-like protein conformations, their REMC-ensemble docking method nonetheless delivers higher-quality structures than REMC and MCM approaches [37]. To manage 100-structure ensembles for every partner protein, RosettaDock4.0, which is a conformer choice MCM method, controls backbone swaps by modulating each conformer's sampling rates. The procedure creates monomer structures using normal modes, backrub movements, and all-atom backbone refinement to diversify backbone conformations.

4.3 Using Normal Modes for Sampling Backbone Conformations

In some docking techniques, harmonic dynamics are used to record backbone motions of proteins because intrinsic protein variations contribute to conformational change [10]. Ordinary vibration modes illustrate a protein's internal motions by using a Hookean potential among neighboring residues.

There have been new advancements in the past few years wherein normal mode analysis (NMA) is included in numerous docking techniques. When NMA-derived harmonic potentials were being created, residues at the interface were shifted in Cartesian coordinate space. Schindler et al. introduced an attraction force to simulate induced fit. iATTRACT functioned as a stage of refinement and increased the estimated percentage of native contacts by 70%. For targets having an unbound-to-bound interface, iATTRACT can build high-quality models with RMSD values greater than 4 Å [38].

NMA has also been used in population-based techniques like particle swarm optimization (PSO). PSO is a statistical method that employs several different systems to optimize many degrees of freedom. To sample variations in molecular complexes, the complex NMA was introduced in docking [39]. This method narrows the search space's dimensions while concentrating on the binding region by calculating the modes of an encounter complex [40]. Higher frequency modes of NMA frequently cause protein bonds to be distorted, which is one of its issues. To get over this limitation, the internal coordinate NMA (iNMA) technique was created. It allows for movement in rotational and translational space with predetermined angles and bond lengths. A basic protein model was used to record the changes in larger conformations in an internal coordinate space from low-frequency mode eigenvectors. In their benchmark, 39% of proteins had a single domain and 45% had multiple domains. Starting with the unbound state, iNMA can create structures that are within 3 Å of the bound state.

4.4 Machine Learning Methods

Although deep learning techniques in biology have recently placed a lot of emphasis on protein folding, only a few studies have specifically addressed issues related to protein docking. A molecular region rich in information that can be explored for understanding interactions of proteins is the protein binding site [41]. Using these data to construct score functions to use with conventional docking methods is one strategy.

In the development of a scoring potential for evaluating postures of docking, Geng et al. trained a vector machine (SVM) on both non-native and native protein complex structures. They chose graph representations as an example. Additionally, iScore, which combines HADDOCK scores and GraphRank, won the top spot in the CAPRI scoring rounds (high-quality or medium structures for 9 out of 13 targets) [11]. By applying image recognition methods to protein structures, other researchers have employed deep learning approaches to locate protein interfaces.

Degiacomi utilized MD simulations in an experiment to investigate in what manner neural networks can be used and exploited for building structures showing significant backbone motion. He developed an autoencoder using conformations that condensed the protein's motion into a small latent space [42]. The autoencoder produced an intermediary closed-apo conformation from the natural state at 0.8 Å RMSD by using simulations of a target protein in both its open (apo) and closed (bound) states. The generator could only create structures that were extremely far from the closed state (above 4.2 Å), which ultimately restricted the efficacy of this strategy for blind docking when the autoencoder was programmed entirely with open conformations.

Cao and Shen created a Bayesian Active learning (BAL) model to measure ambiguity protein structure quality using a technique appropriate for blind instances. The model was expanded subsequently to flexible docking of proteins. As it samples different backbone conformations, the Bayesian framework calculates the posterior probability [39]. The flexibility is recorded using low-frequency complex NMA modes and theoretically it can be stretched to higher frequencies to record hinge motions and loops.

5 Docking Methods for Flexible Protein Backbone

The methods used for flexible protein backbone docking are ATTRACT, FlexDock, FLIPDock, HADDOCK, RosettaDock, and FiberDock and are presented in Table 2.

5.1 ATTRACT

For bioinformatics and structural biology, structure estimation of protein complexes based on the structural models of individual partners is becoming increasingly important. The systematic docking approaches that focused on the minimization of docking energies can be carried out using the ATTRACT program. It is a component of the Python and C++-written object-oriented PTools library. The library includes a number of methods for modifying protein structures, setting up docking searches, running them, and evaluating the results [38].

Table 2
Methods for flexible protein–protein docking

Method	Description	Scoring function	Algorithm	References
ATTRACT	Mean field approach	LJ-type effective potentials and electrostatics	Randomized search	[43]
FlexDock	Hinge identification method	Energy based	Geometric docking	[27]
FLIPDock	Divide and conquer approach	AutoDock v3.05	Genetic algorithm	[44]
FiberDock	NMA approach, Flexible refinement and rescoring of rigid body	Energy-based scoring function	Post docking	[45]
HADDOCK	Flexible docking	Desolvation energy, buried surface area, uncertain interaction restraint energy, electrostatic, intermolecular van der Waals	MC search algorithm, randomization search	[46, 47]
RosettaDock	Monte Carlo minimization	Linear repVdW, hydrogen bonds, rotamer probability, attrVdW, electrostatics, EEF1 (SASA)	Randomization search	[48, 49]

A versatile protein–protein docking technique named ATTRACT is centered around a randomized search procedure. Effective potentials of the Lennard-Jones type and electrostatics are implemented as the scoring function in this application. It distinguishes between side chains that are hydrophilic and hydrophobic and is rather simple. Here, the degree of surface complementarity, hydrophilic or hydrophobic nature of contracting protein sections, and residue–residue potential have been utilized in a docking experiment [43].

In the ATTRACT approach, conformational backbone flexibility is explicitly and effectively taken into account during protein–protein docking. Based on thousands of initial protein partner placements, this docking method seeks to minimize the energy of degrees of freedom in rotation and translation of one protein with respect to another protein [50] orientation based on residues used in the ATTRACT docking minimization. The depiction of each residue by up to four pseudo atoms—two for the backbone and up to two for each side chain—helps to roughly explain the double nature of side chains of a few amino acids, such as the hydrophobic and hydrophilic side chain portions. Two pseudo atoms are used to represent bigger and much more versatile side chains, while one pseudo atom is used to represent shorter and far less flexible side chains (Ala, Thr, Asn, Asp, Ser, and Val) (geometric mean of side chain heavy atoms) [51]. The related proteins' conformational relaxation in the soft collective backbone degrees of freedom determined beforehand in ATTRACT allows for the estimated incorporation of protein partners' overall flexibility during the docking process that takes place systematically [50].

Advantages

In several CAPRI rounds, the ATTRACT docking tool has been used to successfully predict complicated structures. The characteristics of ATTRACT include the following:

- Its force field has a coarse granularity.
- Protein flexibility is utilized during the docking search.
- It enables the concurrent docking of many protein entities.

The iATTRACT protocol enables all-atom flexible refining of models produced by ATTRACT. Consequently, ATTRACT can address a wide range of docking issues due to its comprehensive feature and option set.

Limitations

- There is no re-docking provision in the ATTRACT.
- The scoring system might not be precise enough to take electrostatic solvation and Coulomb interaction into account. In theory, the simplified protein model might not be as precise as an atomic resolution description.

5.2 FiberDock

With so much flexibility of the protein, the backbone expands the range of possible searches, which poses a considerable challenge for docking techniques to handle. Thus, docking algorithms run much slower. To overcome this obstacle, the FiberDock technique was created. This new technique imitates the induced fit process [52].

The first web server for refinement of docking called FiberDock optimizes the proteins' relative rigid-body orientation by managing both side chain and backbone flexibility. There are infinite amounts of standard modes that can represent the flexibility of the backbone. Previous studies have demonstrated the value of simulating induced-fit conformational changes using high-frequency normal modes. As opposed to this, other refinement techniques that have been established so far only use the first few lowest frequency normal modes. FiberDock utilized both high-frequency and low-frequency modes. As a result, it may simulate both local and global conformational changes. Twenty test scenarios where the backbone structure of one protein changes as it interacts with another were used to evaluate the procedure. The findings showed that the ranking and accuracy of protein complexes are greatly enhanced when backbone flexibility is incorporated into the refining process [53].

FiberDock simulates the flexibility of the backbone by offering an endless number of regular modes. Iteratively, the method reduces the structure of the flexible protein along the key modes. The relationship between the translation vector of each atom in the normal mode and the chemical forces acting on each one determines the importance of a mode [53]. The shape of energy funnels surrounding the native docking orientation also changes when backbone flexibility is taken into account. In a group of candidate solutions, these energy funnels find relatively close solutions.

To find the near-native models, the FiberDock technique improves and re-ranks alternatives for soft rigid-docking solutions. Backbone and side-chain flexibility are both taken into account in the refining. The method integrates a unique backbone refinement technique based on normal mode analysis (NMA) with side-chain optimization and rigid-body minimization [24]. The NMA is carried out during the pre-processing stage. At this point, the anisotropic network model is used to determine the proteins' normal modes (ANM).

One of the characteristics of the backbone refinement stage is the level of reduced frequency normal modes that will be assessed during the refining. A user can restrict the movements of the backbone globally by specifying normal modes in small numbers, for instance, 10, but normal modes that are more in number will allow the program to utilize high-frequency modes, defining movements locally [52]. Additionally, the user has control over the flexibility of the backbone. A penalty term is applied to the backbone reduction stage to stop the backbone from excessively distorting. This penalty term severity is based on how flexible the backbone is. The weight decreases as the level rises. Most of our test scenarios were found to be satisfied by default value, i.e., 0.95.

Advantages

- At the moment, the FiberDock technique is especially useful when the receptor experiences an opening conformational change brought on by steric conflicts.
- FiberDock handles relatively minor backbone conformational changes caused by binding.
- If the RMSD (recIRMSD) of the receptor interface is less than 5, it produces good refining results.
- Antibody–antigen complexes can be predicted with the aid of FiberDock. Because of the flexible CDR loops, docking in this field is challenging.

Limitations

Optimization of the proteins' relative rigid-body orientation by managing both side and backbone flexibility is challenging.

5.3 FlexDock

FlexDock is an approach for flexible protein–protein docking using hinge bends. FlexDock is an effective docking method for molecules that takes hinges into account, as well as a hinge identification approach called HingeProt [28].

The currently used geometric docking approach ensures flexibility within one of the input molecules while assuming that the second molecule is inflexible. FlexDock manages docking by bending a molecule's hinge. The flexible molecule is forced by the hinges into a topology resembling a chain. The rigid parts of a chain-type topology can be arranged so that the hinges are positioned between any two subsequent sections. Since the molecule's backbone hinges cause a chain-like topology, the algorithm is particularly well suited for managing backbone flexibility [27].

The approach makes use of HingeProt, which is centered around an elastic network input model to recognize hinges and rigid sections in the flexible molecule at first. The geometric hashing then docks rigid parts. Finally, a flexible outcome is created by assembling the partial docking orientations [54].

There are two major stages of the FlexDock algorithm. In the first stage, rigid docking of all rigid parts is performed whereas in the second stage compilation of a high-scoring multitransformation list takes place from the rigid docking findings into flexible consistent solutions. One of the molecules' numerous hinges can be handled by the algorithm.

It's vital to note that a solution that is nearly native frequently receives a high ranking even though the rigid part docking list places the near-native solutions fairly low. This is because solutions that are consistent with hinges are preferred throughout the assembling stage. The algorithm is very efficient due to several factors [27]:

- An effective rigid docking algorithm that is based on geometric hashing executes only one docking operation for each rigid part during execution.
- Consistent multitransformations are simultaneously constructed at the assembly stage using rigid docking techniques. The algorithm runs in a similar amount of time on a desktop PC to the PatchDock rigid docking algorithm or in the range of minutes.

Advantages

- Flex Dock's capacity to manage any quantity of flexible molecular hinges makes it special.
- FlexDock manages docking with motions that cause one of the molecules to flex at the hinges without adding an additional order of computational complexity.

Limitations

- The method's primary drawback is that it depends on stiff docking approaches that identify a single accurate fractional solution. A flexible loop, shear motion, or turn may accompany a hinge motion, which in many situations can also dramatically change the rigid part's backbone structure.
- It is still difficult to control hinge motion in both molecules [28].

5.4 FLIPDock

Automated docking techniques continue to face difficulties when dealing with the conformational changes that occur in biological macromolecules when they interact with ligands. Flexibility is incorporated into a new protein-ligand docking program named FLIPDock (Flexible Ligand-Protein Docking) [45]. FLIPDock is a genetic algorithm-based docking program that allows flexible ligands and flexible proteins to be docked. It employs the FlexTree (FT) data structure to express the protein–ligand combination. The FT data structure describes the receptor as a layered system of molecular fragments capable of engaging in a variety of movements, such as twist, shear, rotameric states, side chain, hinge, normal modes, and many more. One of the approach's key strengths is how it represents protein flexibility through an intuitive hierarchy of movement classification [55].

The FLIPDock approach makes use of two FlexTree data structures to reflect the flexibility of both receptor and ligand. A divide-and-conquer strategy and genetic algorithm are then used to search for the appropriate conformations throughout the docking process [56].

There are three essential components of a docking program:

1. A depiction of the receptor–ligand system.
2. The search approach.
3. The method for ranking potential solutions.

The Flexibility Tree is used in FLIPDock to characterize molecular conformational spaces (FT). While the FT is capable of representing fully flexible ligands, its original purpose was to serve as a multiresolution, hierarchical, data structure for selectively encoding the subspace confirmations of biological macromolecules. It is possible to construct these subspace confirmations to include a variety of conformations important for a protein's biological functions [44]. A range of motions can be integrated, such as rigid movement of domains moving as entities to atoms of backbone deforming normally. Additionally, throughout the process of docking, these conformational subspaces can be specified by a limited number of variables, efficiently modelling the conformational variation in a flexible receptor. For the docking studies, AutoDock v3.05 force-field.20 provides the foundation for the score function utilized. The four nonbonded interactions in this empirical force field are as follows: repulsion or dispersion, desolvation, hydrogen bonding, and electrostatics on binding. The FLIPDock score is calculated by adding the internal energies of the small molecule, the protein, and the interaction energies between the two [44]. FLIPDock differs from other docking programs in a way that it searches for receptor conformational changes during the docking process. There is no need for a pre- or post-docking procedure.

Advantages

- FLIPDock's adaptability when using FTs to describe and encode docking difficulties is a significant advantage.
- FLIPDock allows flexible ligand molecules to automatically dock into flexible receptor molecules' active sites.
- Docks adaptable ligands to receptors that demand conformational adjustments., such as mobility of backbone.
- FLIPDock's component-based architecture allows for the insertion and interchangeability of new search engines and scoring systems.

Limitations

- During docking, look for the receptor's conformation.
- Too many possible outcomes.

5.5 HADDOCK

HADDOCK is a framework for integrated modeling of biomolecular docking that uses experimental data acquired from biophysical/biochemical experiments or bioinformatics predictions to guide the docking process. Usually, this information is converted into distance restrictions that are used to control docking [57].

HADDOCK considers the flexibility of both side chains and the backbone [47]. Simulations based on molecular dynamics (MD) are used to accomplish this. During the refinement process, flexibility is gradually introduced, beginning with side chains, then extending to backbone interface residues, and finally, during the last cooling step of explicit solvent refinement, the entire system becomes flexible. Users have the option to either directly specify the semi-flexible regions or allow HADDOCK to determine them based on an analysis of contacting residues. Furthermore, totally flexible portions can be created, which are handled totally as flexible for the whole of the simulated annealing process. This may be ideal for docking tiny flexible peptides or flexible loops or ligands with protein [49].

Two working modes are included in the HDOCK server workflow: the template-free docking mode and the default hybrid docking method. Users must first input two distinct proteins—one for the ligand and the other for the receptor—each of which supports a sequence of amino acids as well as structures from PDB [58]. In order to simulate the ligand and receptor molecules using templates, the server will first search the Protein Data Bank to search the possible homologous templates based on sequences of proteins. If a complex homologous template is discovered, a new receptor–ligand structure will be created using that template regardless of whether the input is a structure or sequence; otherwise, the 3D structure will be produced if the input is a protein sequence using a monomer template. The docking pipeline will end without making any predictions if there is no template available for a given input sequence. As a result, the HDOCK server may use one of three types of individual protein structures: (i) homology models built on a complex template, (ii) users entered PDB structures, and (iii) using a monomer template to build homology models, with the primary accessible one chosen in the order of their priorities [58].

In HADDOCK, docking is carried out in three successive steps [46]:

Rigid-body docking (it0): To create a complex, the subunits are randomly distributed throughout the area, with around 25 Å distance among these subunits. Next, the energy minimization for rigid bodies subunits is used.

- Semi-flexible refinement (it1): The highest scoring models (by default 200 out of 1000) are modified in the it0 stage by using simulated-annealing (SA) in the torsion-angle space technique. During this stage, the interface is considered flexible—first side chains only, then both side chains, and finally the backbone.

- Flexible refinement in explicit solvent (itw): A light-constrained molecular dynamics simulation is run on the models from the first iteration in an obvious solvent shell in this final stage.

Advantages

- Complexes of protein–protein, protein–oligosaccharide, protein–nucleic acid, and protein–small molecule are a few of the issues to which HADDOCK has been applied.
- HADDOCK permits changes in confirmations of both the side and the backbone of the molecules during complex formation.
- HADDOCK natively supports NMR and Protein Data Bank structures with numerous models during docking. HADDOCK has been used to solve the coordinates of more than 60 biomolecular complexes. In the CAPRI blind docking experiment, HADDOCK also showed excellent performance, achieving RMSD values for 11 out of 15 complexes [59].

Limitations

- Protein amino acid sequence input is supported by HDOCK. It should be highlighted that the degree of sequence similarity between the target and template sequences has a significant impact on the correctness of 3D structures produced by homology modelling.

5.6 RosettaDock

Accounting for monomeric conformational changes that take place during binding is a difficulty in docking of proteins. The RosettaDock approach yields docking models having atomic accuracy, with changes in confirmations primarily confined to sidechains of proteins while maintaining the backbone fixed [60]. The component of the Rosetta suite software, RosettaDock, describes rigid body internal coordinate system that orients the two docking partners. The torsion angles of protein backbone are fixed but the side chains are allowed to spin and sample various conformations of rotamer [29].

One of the best approaches for computational protein–protein docking has been RosettaDock. Numerous stiff complexes (58%) in the Docking Benchmark 3.0 set were successfully docked by RosettaDock 3.2, which combined selection of coarse-grained conformations and fit caused by complete atoms [37]. Rosetta, a tool for modeling and designing structures, has addressed backbone flexibility through numerous methods such as the following:

- Synthesis of new backbones of proteins from fragments of peptides that successfully predicted the structure of the protein using ab initio methods.
- Iterating among sequence design using Monte Carlo search followed by structural optimization through minimization.
- Utilization of a kinematic closure (KIC) approach inspired by robotics, demonstrated to describe loop conformations with sub-Angstrom precision.
- Benchmarking the Backrub algorithm using recaptures of well-known sequences [48].

The majority of flexible backbone design techniques iterates between designing the sequence on backbone that is fixed and optimizing the structure on fixed sequence, thus decoupling changes to the sequence from having an impact on the backbone structure. Forward engineering has effectively utilized Rosetta's flexible backbone design, although accuracy comparisons between various approaches have not been done using standard benchmark datasets [48].

Advantages

In situations where experimental structures were unavailable and crystallization posed difficulties, RosettaDock aids in the prediction of antibody–antigen structures of therapeutic interest to offer hypotheses on a pharmacological mechanism and insights into affinity maturation for complexes. Application targets for RosettaDock include interactions between antibodies, calcium channel proteins, and malaria proteins [61]. RosettaDock can be used alone or in conjunction with other docking services, with the option to improve recommended docking places using local searches. RosettaDock may theoretically be used to improve potential solutions from any global docking technique.

Limitations

The limitations of RosettaDock are the following:

- Using the command-line interface can be challenging.
- Model generation takes a lot of calculation time, usually needing a cluster of computers [61].

6 Future Developments Regarding Protein–Protein Docking Methods' Backbone Flexibility

Although search strategies during protein docking have garnered significant attention and have increased greatly in complexity and variety, the search algorithms, as they exist today, are still in their infancy. A search algorithm, one of the fundamental components of molecular docking, is strongly related to other elements, including scoring and flexibility during docking calculations [4]. Therefore, a significant progress in search technology is expected in the future, including not only the creation of new search algorithms but also the enhancement of existing ones. Examples of these other features include flexibility of proteins and scoring functions. The complex docking benchmark development, appropriate application of criteria for evaluating, and testing advancements by means of accurate docking experiments, including CAPRI, are possible directions in the area of research in the future.

As conformational flexibility is a key factor affecting success of the docking method, extra efforts would be put into evaluating the flexibility of confirmations when binding for a particular event in addition to developing new algorithms that can replace scoring approach and the rigid body docking [21].

In the nearest future, integrating algorithms of docking with the program for predicting protein interfaces, sequence analysis methods and structural databases would lead toward precise structural models of the essential molecular interactions inside cells and improved predictions of PPI networks [1]. The use of MD simulation methods for docking searches or the pre-selection of essential conformational transitions will become more widespread as a result of the continual increase in computer power and the use of improved sampling procedures. To quantify the flexibility of all binding partners under physiological settings, account for entropic changes, and comprehensively examine the function of water molecules at the ligand–receptor interface, MD simulations with explicit solvent could be used. The linear interaction energy approach and MD simulations could both be utilized to calculate the binding free energy from a single trajectory. Thus, increased use of MD techniques may lead to better scoring and docking of ligand–receptor complexes [57].

Acknowledgments

This research was funded under an OPUS grant from National Science Center (NCN, Poland), grant number UMO-2021/43/B/NZ7/01732.

References

1. Ritchie DW (2008) Recent progress and future directions in protein-protein docking. Curr Protein Pept Sci 9:1–15. https://doi.org/10.2174/138920308783565741
2. Harmalkar A, Gray JJ (2021) Advances to tackle backbone flexibility in protein docking. Curr Opin Struct Biol 67:178–186. https://doi.org/10.1016/j.sbi.2020.11.011
3. Yan Y, Huang SY (2019) Pushing the accuracy limit of shape complementarity for protein-protein docking. BMC Bioinformatics 20:1–10. https://doi.org/10.1186/s12859-019-3270-y
4. Huang SY (2014) Search strategies and evaluation in protein–protein docking: principles, advances and challenges. Drug Discov Today

19:1081–1096. https://doi.org/10.1016/j.drudis.2014.02.005

5. Christoffer C, Bharadwaj V, Luu R, Kihara D (2021) LZerD protein-protein docking webserver enhanced with de novo structure prediction. Front Mol Biosci 750. https://doi.org/10.3389/fmolb.2021.7249
6. Barradas-Bautista D, Rosell M, Pallara C, Fernández-Recio J (2018) Structural prediction of protein–protein interactions by docking: application to biomedical problems. Adv Protein Chem Struct Biol 110:203–249. https://doi.org/10.1016/bs.apcsb.2017.06.003
7. Banu SV, Durai S (2013) Implication of sidechain flexibility in protein-protein docking algorithm. Natl Conf Syst Des Infor Proce 46
8. Kangueane P, Nilofer C (2018) Protein-protein docking: methods and tools. In: Protein-protein and domain-domain interactions, pp 163–168. https://doi.org/10.1007/978-981-10-7347-2_14
9. Wiehe K, Peterson MW, Pierce B, Mintseris J, Weng Z (2008) Protein–protein docking: overview and performance analysis. Protein Struct Predict 413:283–314. https://doi.org/10.1007/978-1-59745-574-9_11
10. Zacharias M (2010) Accounting for conformational changes during protein–protein docking. Curr Opin Struct Biol 20:180–186. https://doi.org/10.1016/j.sbi.2010.02.001
11. Dominguez C, Boelens R, Bonvin AM (2003) HADDOCK: a protein−protein docking approach based on biochemical or biophysical information. J Am Chem Soc 125:1731–1737. https://doi.org/10.1021/ja026939x
12. Yan Y, Tao H, Huang SY (2018) HSYMDOCK: a docking web server for predicting the structure of protein homo-oligomers with Cn or Dn symmetry. Nucleic Acids Res 46:423–431. https://doi.org/10.1093/nar/gky398
13. Cheng TMK, Blundell TL, Fernandez-Recio J (2007) pyDock: electrostatics and desolvation for effective scoring of rigid-body protein–protein docking. Protein Struct Funct Bioinformatics 68:503–515. https://doi.org/10.1002/prot.21419
14. Kuntz ID, Blaney JM, Oatley SJ, Langridge R, Ferrin TE (1982) A geometric approach to macromolecule-ligand interactions. J Mol Biol 161:269–288. https://doi.org/10.1016/0022-2836(82)90153-X
15. Duhovny D, Nussinov R, Wolfson HJ (2002) Efficient unbound docking of rigid molecules. In: International workshop on algorithms in bioinformatics, vol 2452, pp 185–200. https://doi.org/10.1007/3-540-45784-4_14
16. Esquivel-Rodríguez J, Yang YD, Kihara D (2012) Multi-LZerD: multiple protein docking for asymmetric complexes. Protein Struct Funct Bioinformatics 80:1818–1833. https://doi.org/10.1002/prot.24079
17. Gardiner EJ, Willett P, Artymiuk PJ (2001) Protein docking using a genetic algorithm. Protein Struct Funct Bioinformatics 44:44–56. https://doi.org/10.1002/prot.1070
18. Gray JJ, Moughon S, Wang C, Schueler-Furman O, Kuhlman B, Rohl CA, Baker D (2003) Protein–protein docking with simultaneous optimization of rigid-body displacement and side-chain conformations. J Mol Biol 331:281–299. https://doi.org/10.1016/S0022-2836(03)00670-3
19. Moal IH, Bates PA (2010) SwarmDock and the use of normal modes in protein-protein docking. Int J Mol Sci 11:3623–3648. https://doi.org/10.3390/ijms11103623
20. Morris GM, Goodsell DS, Halliday RS, Huey R, Hart WE, Belew RK, Olson AJ (1998) Automated docking using a Lamarckian genetic algorithm and an empirical binding free energy function. J Comput Chem 19:1639–1662. https://doi.org/10.1002/(SICI)1096-987X(19981115)19:14<1639::AID-JCC10>3.0.CO;2-B
21. Pons C, Grosdidier S, Solernou A, Pérez-Cano L, Fernández-Recio J (2010) Present and future challenges and limitations in protein–protein docking. Protein Struct Funct Bioinformatics 78:95–108. https://doi.org/10.1002/prot.22564
22. Vakser IA (2020) Challenges in protein docking. Curr Opin Struct Biol 64:160–165. https://doi.org/10.1016/j.sbi.2020.07.001
23. Park H, Lee H, Seok C (2015) High-resolution protein–protein docking by global optimization: recent advances and future challenges. Curr Opin Struct Biol 35:24–31. https://doi.org/10.1016/j.sbi.2015.08.001
24. Andrusier N, Mashiach E, Nussinov R, Wolfson HJ (2008) Principles of flexible protein–protein docking. Protein Struct Funct Bioinformatics 73:271–289. https://doi.org/10.1002/prot.22170
25. Palma PN, Krippahl L, Wampler JE, Moura JJ (2000) BiGGER: a new (soft) docking algorithm for predicting protein interactions. Protein Struct Funct Bioinformatics 39:372–384. https://doi.org/10.1002/(SICI)1097-0134(20000601)39:4<372::AID-PROT100>3.0.CO;2-Q

26. Sandak B, Nussinov R, Wolfson HJ (1998) A method for biomolecular structural recognition and docking allowing conformational flexibility. J Comput Biol 5:631–654. https://doi.org/10.1089/cmb.1998.5.631
27. Schneidman-Duhovny D, Inbar Y, Nussinov R, Wolfson HJ (2005) Geometry-based flexible and symmetric protein docking. Protein Struct Funct Bioinformatics 60:224–231. https://doi.org/10.1002/prot.20562
28. Schneidman-Duhovny D, Nussinov R, Wolfson HJ (2007) Automatic prediction of protein interactions with large scale motion. Protein Struct Funct Bioinformatics 69:764–773. https://doi.org/10.1002/prot.21759
29. Wang C, Bradley P, Baker D (2007) Protein–protein docking with backbone flexibility. J Mol Biol 373:503–519. https://doi.org/10.1016/j.jmb.2007.07.050
30. Bonvin AM (2006) Flexible protein–protein docking. Curr Opin Struct Biol 16:194–200. https://doi.org/10.1016/j.sbi.2006.02.002
31. Lindahl E, Delarue M (2005) Refinement of docked protein–ligand and protein–DNA structures using low frequency normal mode amplitude optimization. Nucleic Acids Res 33:4496–4506. https://doi.org/10.1093/nar/gki730
32. Christoffer C, Terashi G, Shin WH et al (2020) Performance and enhancement of the LZerD protein assembly pipeline in CAPRI 38-46. Protein Struct Funct Bioinformatics 88:948–961. https://doi.org/10.1002/prot.25850
33. Król M, Chaleil RA, Tournier AL, Bates PA (2007) Implicit flexibility in protein docking: cross-docking and local refinement. Protein Struct Funct Bioinformatics 69:750–757. https://doi.org/10.1002/prot.21698
34. Ostermeir K, Zacharias M (2017) Accelerated flexible protein-ligand docking using Hamiltonian replica exchange with a repulsive biasing potential. PLoS One 12:e0172072. https://doi.org/10.1371/journal.pone.0172072
35. Kuroda D, Gray JJ (2016) Pushing the backbone in protein-protein docking. Structure 24: 1821–1829. https://doi.org/10.1016/j.str.2016.06.025
36. Zhang Z, Ehmann U, Zacharias M (2017) Monte Carlo replica-exchange based ensemble docking of protein conformations. Protein Struct Funct Bioinformatics 85:924–937. https://doi.org/10.1002/prot.25262
37. Marze NA, Roy Burman SS, Sheffler W, Gray JJ (2018) Efficient flexible backbone protein–protein docking for challenging targets. Bioinformatics 34:3461–3469. https://doi.org/10.1093/bioinformatics/bty355
38. Schindler CE, de Vries S, Zacharias M (2015) iATTRACT: simultaneous global and local interface optimization for protein–protein docking refinement. Protein Struct Funct Bioinformatics 83:248–258. https://doi.org/10.1002/prot.24728
39. Oliwa T, Shen Y (2015) cNMA: a framework of encounter complex-based normal mode analysis to model conformational changes in protein interactions. Bioinformatics 31:i151–i160. https://doi.org/10.1093/bioinformatics/btv252
40. Chen H, Sun Y, Shen Y (2017) Predicting protein conformational changes for unbound and homology docking: learning from intrinsic and induced flexibility. Protein Struct Funct Bioinformatics 85:544–556. https://doi.org/10.1002/prot.25212
41. Fout A, Byrd J, Shariat B, Ben-Hur A (2017) Protein interface prediction using graph convolutional networks. Adv Neural Inf Process Syst 30
42. Degiacomi MT (2019) Coupling molecular dynamics and deep learning to mine protein conformational space. Structure 27:1034–1040. https://doi.org/10.1016/j.str.2019.03.018
43. Agrawal P, Singh H, Srivastava HK, Singh S, Kishore G, Raghava GP (2019) Benchmarking of different molecular docking methods for protein-peptide docking. BMC Bioinformatics 19:105–124. https://doi.org/10.1186/s12859-018-2449-y
44. Zhao Y, Sanner MF (2007) FLIPDock: docking flexible ligands into flexible receptors. Protein Struct Funct Bioinformatics 68:726–737. https://doi.org/10.1002/prot.21423
45. Chen G, Seukep AJ, Guo M (2020) Recent advances in molecular docking for the research and discovery of potential marine drugs. Mar Drugs 18:545. https://doi.org/10.3390/md18110545
46. van Zundert GC, Bonvin AM (2014) Modeling protein–protein complexes using the HADDOCK webserver "modeling protein complexes with HADDOCK". In: Protein structure prediction, pp 163–179. https://doi.org/10.1007/978-1-4939-0366-5_12
47. Schmitz C, Melquiond AS, de Vries SJ et al (2012) Protein–protein docking with HADDOCK, pp 520–530. https://doi.org/10.1002/9783527644506.ch32
48. Loshbaugh AL, Kortemme T (2020) Comparison of Rosetta flexible-backbone computational protein design methods on binding interactions. Protein Struct Funct

Bioinformatics 88:206–226. https://doi.org/10.1002/prot.25790

49. De Vries SJ, Van Dijk M, Bonvin AM (2010) The HADDOCK web server for data-driven biomolecular docking. Nat Protoc 5:883–897. https://doi.org/10.1038/nprot.2010.32
50. May A, Zacharias M (2007) Protein–protein docking in CAPRI using ATTRACT to account for global and local flexibility. Protein Struct Funct Bioinformatics 69:774–780. https://doi.org/10.1002/prot.21735
51. Fiorucci S, Zacharias M (2010) Binding site prediction and improved scoring during flexible protein–protein docking with ATTRACT. Protein Struct Funct Bioinformatics 78:3131–3139. https://doi.org/10.1002/prot.22808
52. Mashiach E, Nussinov R, Wolfson HJ (2010) FiberDock: flexible induced-fit backbone refinement in molecular docking. Protein Struct Funct Bioinformatics 78:1503–1519. https://doi.org/10.1002/prot.22668
53. Mashiach E, Nussinov R, Wolfson HJ (2010) FiberDock: a web server for flexible induced-fit backbone refinement in molecular docking. Nucleic Acids Res 38:W457–W461. https://doi.org/10.1093/nar/gkq373
54. Grosdidier S, Fernandez-Recio J (2012) Protein-protein docking and hot-spot prediction for drug discovery. Curr Pharm Des 18:4607–4618. https://doi.org/10.2174/138161212802651599
55. Totrov M, Abagyan R (2008) Flexible ligand docking to multiple receptor conformations: a practical alternative. Curr Opin Struct Biol 18:178–184. https://doi.org/10.1016/j.sbi.2008.01.004
56. Li Z, Tianchi Z, Jing Z (2012) Optimization algorithms for flexible protein-protein docking. In: 2012 third international conference on digital manufacturing & automation, pp 571–573. https://doi.org/10.1109/ICDMA.2012.135
57. Koukos PI, Faro I, van Noort C, Bonvin AM (2018) A membrane protein complex docking benchmark. J Mol Biol 430:5246–5256. https://doi.org/10.1016/j.jmb.2018.11.005
58. Yan Y, Tao H, He J, Huang SY (2020) The HDOCK server for integrated protein–protein docking. Nat Protoc 15:1829–1852. https://doi.org/10.1038/s41596-020-0312-x
59. de Vries SJ (2009) How proteins get in touch: interface prediction and docking of protein complexes. Doctoral dissertation, Utrecht University, 9:394–406. 1389–2042
60. Chaudhury S, Sircar A, Sivasubramanian A, Berrondo M, Gray JJ (2007) Incorporating biochemical information and backbone flexibility in RosettaDock for CAPRI rounds 6–12. Protein Struct Funct Bioinformatics 69:793–800. https://doi.org/10.1002/prot.21731
61. Lyskov S, Gray JJ (2008) The RosettaDock server for local protein–protein docking. Nucleic Acids Res 36:W233–W238. https://doi.org/10.1093/nar/gkn216

Chapter 5

From Multiple Protein Docking to Protein–Protein Docking at Interactome Level

Reema Gabrani, Priyanjal Jain, Srishti Sharma, Ritu Ghildiyal, and Vijeta Prakash

Abstract

Molecular docking is used to anticipate the optimal orientation of a particular molecule to a target to form a stable complex. It makes predictions about the 3D structure of any complex based on the binding characteristics of the ligand and the target receptor usually a protein. It is an exceptionally useful tool, which is used as a model to study how ligands attach to proteins. Docking can also be used for studying the interaction of ligands and proteins to analyze inhibitory efficacy. The ligand may also be a protein, making it possible to study interactions between two different proteins using the numerous docking tools available for basic research on protein interactions. The protein–protein docking is a crucial approach to understanding the protein interactions and predicting the structure of protein complexes that have not yet been experimentally determined. Moreover, the protein–protein interactions can predict the function of target proteins and the drug-like properties of molecules. Therefore, protein docking assists in uncovering insights into protein interactions and also aids in a better understanding of molecular pathways/mechanisms. This chapter comprehends the various tools for protein–protein docking (pairwise and multiple), including their methodologies and analysis of output as results.

Key words LZerD, Molecular docking, Multiple protein docking, Structure prediction

1 Introduction

A variety of biological processes in cells depend on protein–protein interactions (PPIs), including signal transduction, regulation of the cell cycle, DNA replication, RNA transcription, and protein synthesis [1]. It is pivotal to understand the underlying PPIs of crucial biological processes during normal and diseased conditions. The knowledge paves the way to develop compounds that can interfere with PPIs with pharmaceutical significance [2]. Protein–protein docking has become a popular research topic because of its ability to predict PPIs and identify hotspot residues at the protein–protein interface [3].

Agnieszka A. Kaczor (ed.), *Protein-Protein Docking: Methods and Protocols*, Methods in Molecular Biology, vol. 2780, https://doi.org/10.1007/978-1-0716-3985-6_5,
© The Author(s), under exclusive license to Springer Science+Business Media, LLC, part of Springer Nature 2024

It is extremely challenging to sample the conformational space in protein–protein docking, despite similar principles as docking between protein and small molecule. During the interaction of two proteins that are relatively rigid, it can be difficult to explore their rotational-conformational space as they exchange mutual orientations. For protein–protein docking, the search algorithms are computationally expensive due to the number of degrees of freedom [3]. The docking algorithms of PPIs can be divided into two categories: template-based modeling and template-free docking. In template-based modeling, interacting pair proteins have at least 30% sequence identity, which indicates they will interact similarly. Hence, if an appropriate template is available, this method can be used to predict the near-native structure of protein–protein complexes. In reality, however, the number of templates available is still very limited, which makes template-free docking more popular. The scoring and sampling stages of most template-free docking algorithms are involved in the docking process [2].

Large-scale protein–protein interaction maps, electron microscopy, and electron tomography of cells have recently highlighted the importance and abundance of protein interactions and complexes. It is evident from the growing number of participants in the critical assessment of prediction of interactions (CAPRI) that many protein docking methods have been developed over the past few years. Pairwise docking, also known as pairwise assembly, is the most common method [4]. A number of studies have recently used multiple static receptor structures to represent flexible receptors in drug design applications. Using multiple protein structures for docking poses two critical challenges. The first one remains the reliability of conformational change prediction and the other challenge is an interpretation of combined conformations [5].

A few software for protein docking are described below.

1.1 LZerD

An algorithm for docking multiple proteins, Multi-LZerD, employs a multistage approach in which the initial step is a coarse-grained conformation search followed by a genetic algorithm (GA) combination of pairwise solutions. In addition to docking predictions based on pairwise matches of component proteins, Multi-LZerD is also capable of handling those protein complexes wherein some pairs of proteins do not interact with one another [4].

A multimeric protein complex with different docking topologies can be found in near-native conformations using Multi-LZerD. Multiple docking structures can be obtained by combining low-ranked pairwise predictions and handling a large number of pairwise combinations [4].

1.2 ZDock

To explore rigid-body docking positions efficiently, ZDOCK, a software program that uses grid-based representations of two proteins and prefers fast Fourier's transforms (FFTs), was developed. ZDOCK 3.0 includes a scoring function that incorporates shape complementarity electrostatics, and the atomic statistical potentials are derived from transient protein complex contact propensities in a pairwise manner. A comparison of ZDOCK 3.0 with the previous version shows vast improvements in its prediction ability, and the experiment CAPRI provides highly successful results. ZDOCK's running time and memory consumption increased significantly with improved accuracy due to the pairwise statistical capability, as seven FFTs had to be computed per docking orientation, as opposed to two in ZDOCK 2.3 [6].

1.3 ClusPro

Using ClusPro, one can directly dock two interacting proteins. A significant amount of modifications and expansions have been made to ClusPro since it was introduced in 2004. There are three computational steps that the server performs:

1. Samples billions of conformations for rigid bodies docking.
2. In order to find the largest clusters that represent the most likely model of the complex, 1000 lowest energy structures are clustered based on root mean square deviation (RMSD).
3. Energy minimization is used to refine the selected structures.

   PIPER, a program that uses FFT correlation, is used to dock rigid bodies, to optimize docking [7].

1.4 PatchDock

Protein–small ligand and protein–protein are docked very efficiently using PatchDock. A geometry-based docking algorithm called PatchDock is used to dock molecular structures. Its objective is to find docking transformations that result in good complementarity between molecular shapes. Application of such transformations reduces steric clashes and wide interface areas are created. The docked molecules have complementary characteristics covering a wide interface that also includes several matched local features. Commonly dot surface representations of the molecules are broken down into flat, convex, or concave patches by PatchDock. A candidate transformation is generated by matching complementary patches. Atomic desolvation energy and geometric fit are also considered in the scoring function for each candidate transformation. Finally, redundant solutions are discarded using an RMSD (root mean square deviation) clustering. Due to its fast transformational search, PatchDock is highly efficient because it does not rely on brute force comparisons between features but on local feature matching [8].

1.5 pyDockWEB

In order to use the protein–protein docking and scoring program pyDock, a web-based application is needed. There are three components to pyDock web: a web front-end, pyDockD, which manages pyDock executions, and a data storage system. The user-friendly front-end for pyDock allows users to send jobs for execution in simple steps. Python-based pyDockD, the controller, periodically polls the MySQL database for job requests submitted from the web front-end and submits them as pyDocks to the Slurm batch queue. The pyDockWEB uses an optimized version of pyDock version 3, along with a customized version of FTDock that is implemented using MPI (Message Passing Interface) library MPICH2, which allows docking poses to be generated on multiple processors. Here are some other optimizations that have been made. In FTDock, the two molecules are discretized onto orthogonal grids and the translational and rotational spaces are scanned globally using FFTW 2.1.5. As calculated from the single grid unit size and the protein size, the size of the transform in the FFTW scope corresponds to the number of cells in the FTDock grid [9].

2 Generation of Protein Structure

The 3D structure of proteins is retrieved from the RCSB PDB database. If the 3D structure is not available in PDB, then it is generated using Swiss-PDB Viewer and I-TASSER. These servers perform homology modeling to generate 3D protein structures. Validation of protein structures is done by RAMPAGE, which creates a Ramachandran plot. The Chiron tool is used to reduce steric clashes or for energy minimization of protein structures, which is accessible online. The protein 3D structures are now ready to be docked using various docking tools.

3 Tools Used for Multiple Protein Docking

3.1 LZerD

Materials

Server: The LZerD web server can be accessed through the Internet via the link https://kiharalab.org/proteindocking/lzerd.php. The users can access the server with or without creating an account [10].

Input: Protein structural information is required in the PDB format. It can either be taken from RCSB (https://www.rcsb.org/) or users can upload their own PDB files containing the structure [11, 12].

Fig. 1 LZerD web server home page screenshot

Methods

1. Open the LZerD web server.
2. Click on the "Server" tab displayed among many other options like Home, LZerD, Multi-LZerD, and About (Fig. 1).
3. Select "Create New Job." You will be directed to a new page.
4. Switch to Multi-LZerD for docking more than two proteins. Here, you can add multiple protein structures for docking (Fig. 2).
5. Input the structures of a minimum of 3 proteins in the PDB format (*see* **Note 1**). There are two options to input their structures; first is to enter their RCSB PDB IDs and the second is to upload your own PDB files.
6. One can also specify the chains of the protein that need to be used for docking. If this field is left empty, then all the chains of the protein are used for docking.

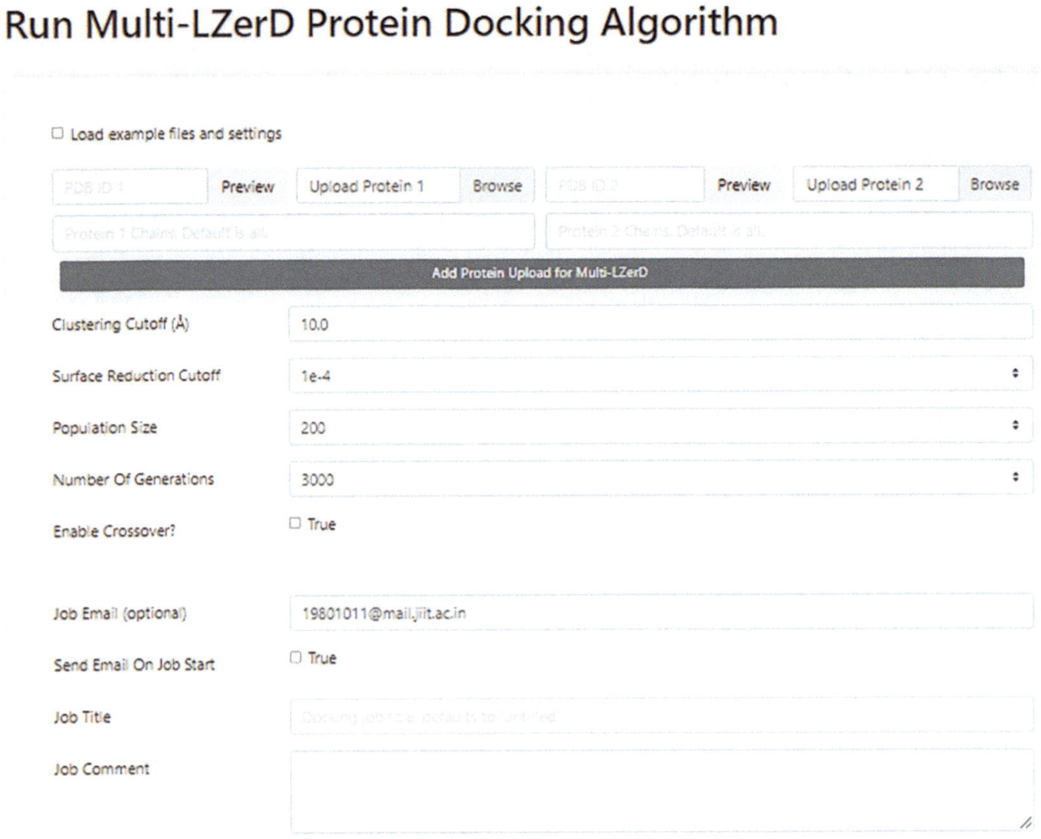

Fig. 2 Screen capture of Multi-LZerD input page

7. You can add or remove protein structures by clicking on the add protein upload and remove protein upload, respectively.
8. Fill in your email (if not registered already) and job title. Keep the rest of all settings as default.
9. In the advanced options, there are fields provided wherein you can specify overall constraints, residue–residue constraints, receptor constraints, and ligand constraints (Fig. 3). You can fill these fields if you have the required data, otherwise skip.
10. Click on "Submit."
11. You will be directed to a new page, which will show whether your job submission was successful or not.
12. On completion of docking, you will be notified via email containing a URL to view the results.
13. The result page shows centroids of the top 50 3D models, color-coded for each protein. Models are ranked according to the rank-sum score. The smaller the rank-sum score, the better the predicted model.

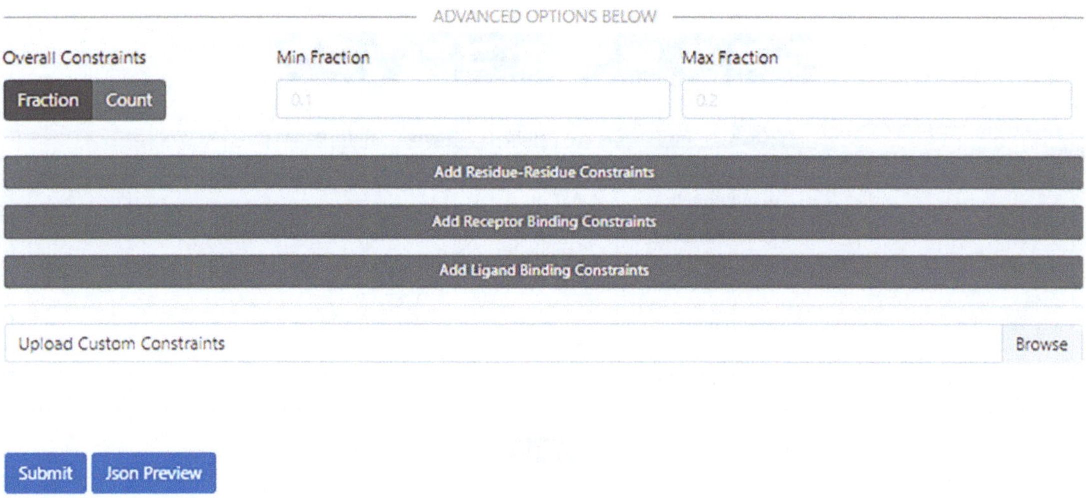

Fig. 3 Screenshot of "Advanced options" in LZERD server

14. Model scores (GOAP, DFIRE, ITScore, rank-sum) are generally listed below the 3D structure.
15. The obtained docked structures as output and all the generated model predictions can be downloaded from the results page in the form of a zip file.

4 Various Tools Used for Protein–Protein Docking

4.1 ZDOCK

Materials

Server: The ZDOCK web server can be accessed through the Internet via the link https://zdock.umassmed.edu/. Users having an academic email ID can only use the server. Java Runtime Environment is required to view the docked model predictions [13].

Input: Protein structural data must be provided in the PDB format. This can either be taken from RCSB (https://www.rcsb.org/) or users can upload their own PDB files containing the structure.

Methods

1. Open the ZDOCK server web page.
2. You will be directed to a page containing three mandatory fields and two optional fields (Fig. 4).

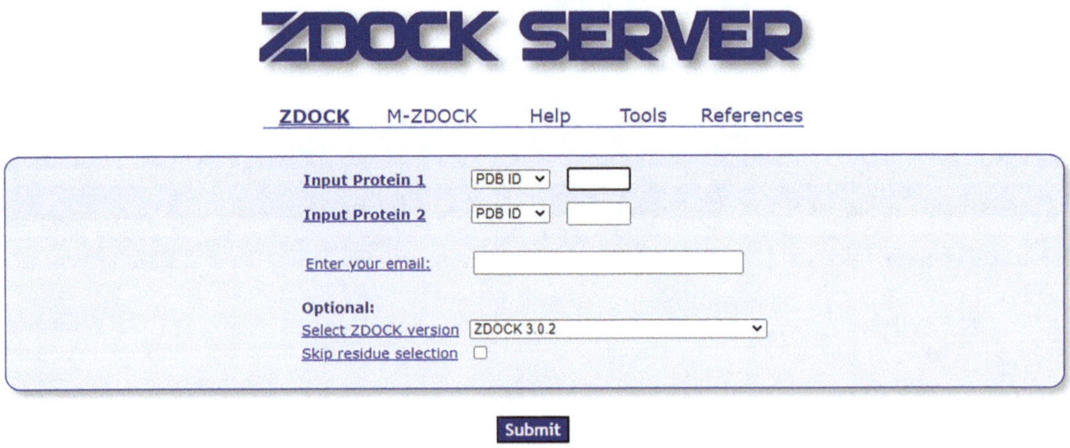

Fig. 4 Screenshot of ZDOCK server input page

3. Add the structures of protein 1 and protein 2 in the PDB format (*see* **Note 2**). There are two options to input their structures; first is to enter their RCSB PDB IDs and the second is to upload your own PDB files.
4. In the email field, enter your institutional/academic email address.
5. Selection of ZDOCK version is optional and it is by default set to the latest version.
6. Check the "Skip residue selection" field if you do not wish to select blocking and contact residues and click on submit.
7. If you do not check this field, then you will be directed to a new page that allows you to select certain residues from both the proteins that you wish to block from the binding site. It also allows specifying contact residues (if any) that are important for binding (Fig. 5) (*see* **Note 3**).
8. You will be required to confirm your selection after pressing on the submit button.
9. After confirming, click on OK.
10. A screen will appear indicating that docking is in progress and a confirmation email will also be sent to you.
11. When the job is finished, you will be notified via an email containing a URL to view the results.
12. The result page shows the top 5 model predictions (*see* **Note 4**). ZDOCK Output and top 10 model predictions can be downloaded from the results page (Fig. 6).

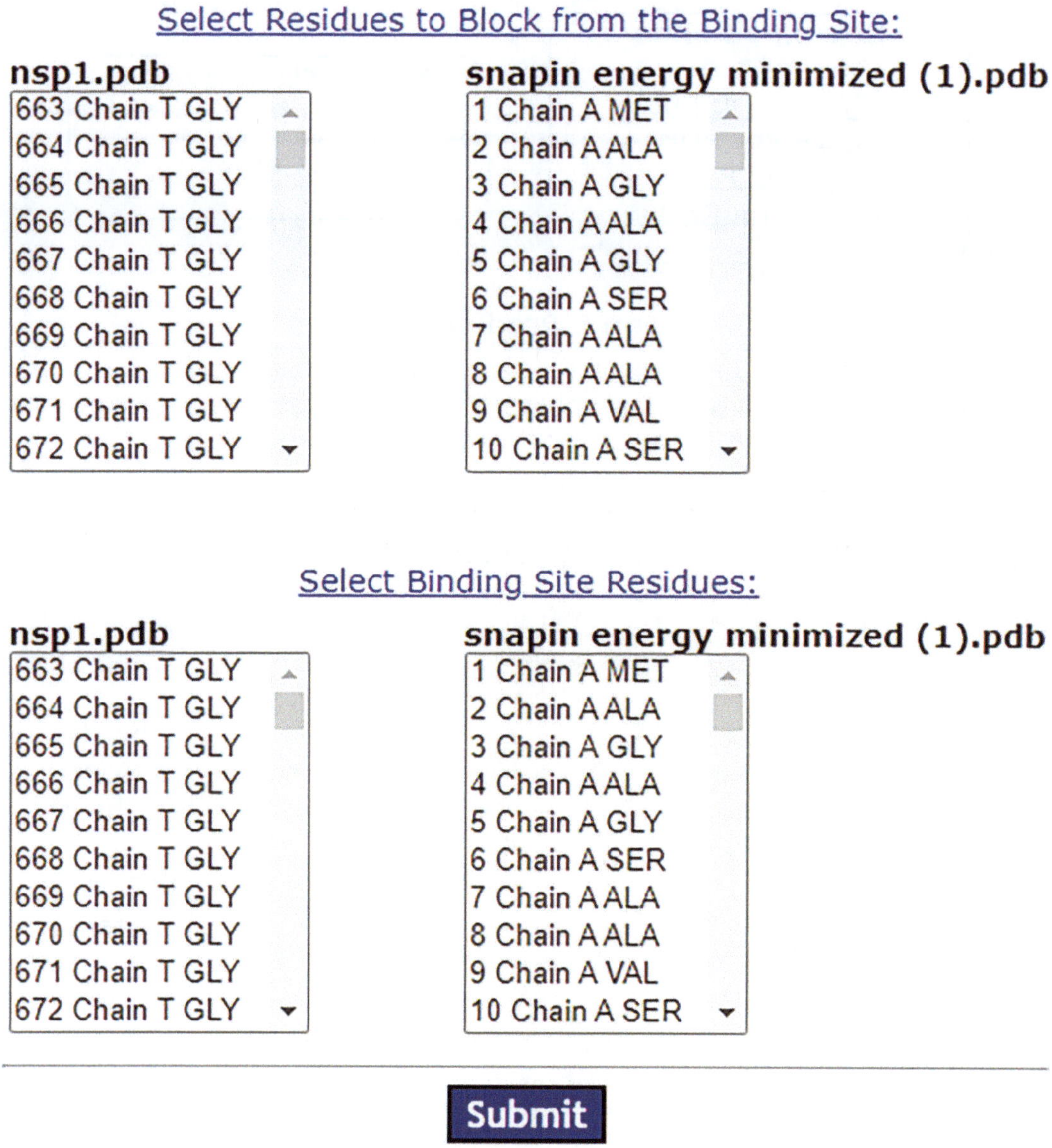

Fig. 5 Representation of ZDOCK residue selection page

4.2 ClusPro

Materials

Server: The ClusPro web server can be accessed through the Internet via the link https://cluspro.bu.edu/login.php?redir=/home.php. Users can access the server with or without creating an account (Fig. 7). However, it is beneficial to create one (*see* **Note 5**). An account can only be created if the user has an academic email ID [14, 15].

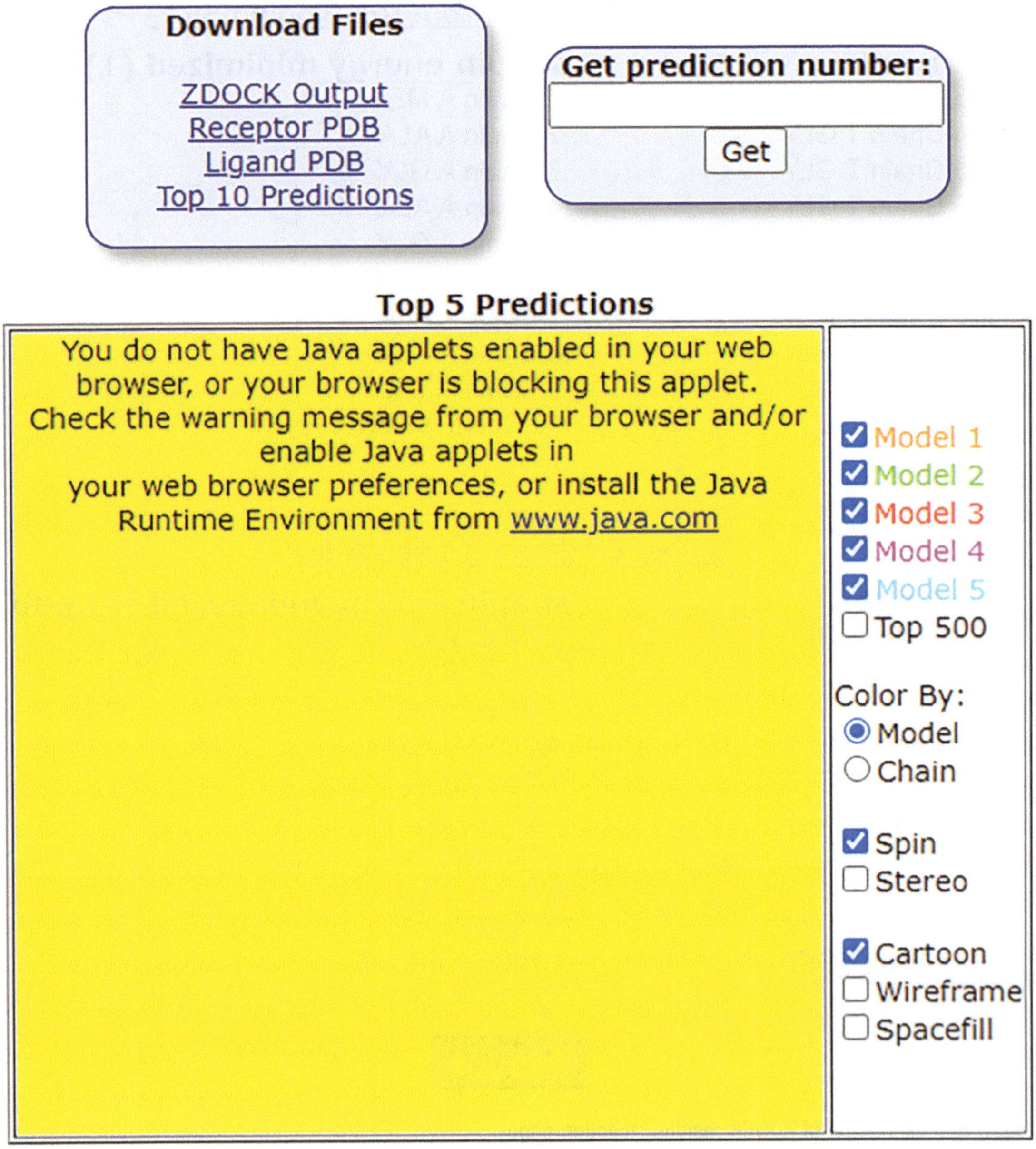

Fig. 6 Screenshot of docking results obtained from ZDOCK server

Input: ClusPro allows users to upload coordinate files for two protein structures, or they can provide the PDB codes for the corresponding structures, and ClusPro will download those files from the RCSB server (http://www.rcsb.org/pdb/) [7, 16].

Fig. 7 ClusPro web server home page screenshot

Methods

1. Open the ClusPro web server.
2. Click on the "Dock" tab displayed among many other options like Dimer Classification, Queue, Results, Preferences, Downloads, Papers, and Help.
3. A screen will be displayed containing a few empty fields (Fig. 8).
4. In the Job Name field, enter any name by which you can identify your submission. If nothing is entered in this field, the job id will be assigned as a name.
5. Then, add the structures of the receptor and ligand molecules you wish to dock in the fields provided (*see* **Note 6**). You can either enter their RCSB PDB IDs or upload your own PDB files.

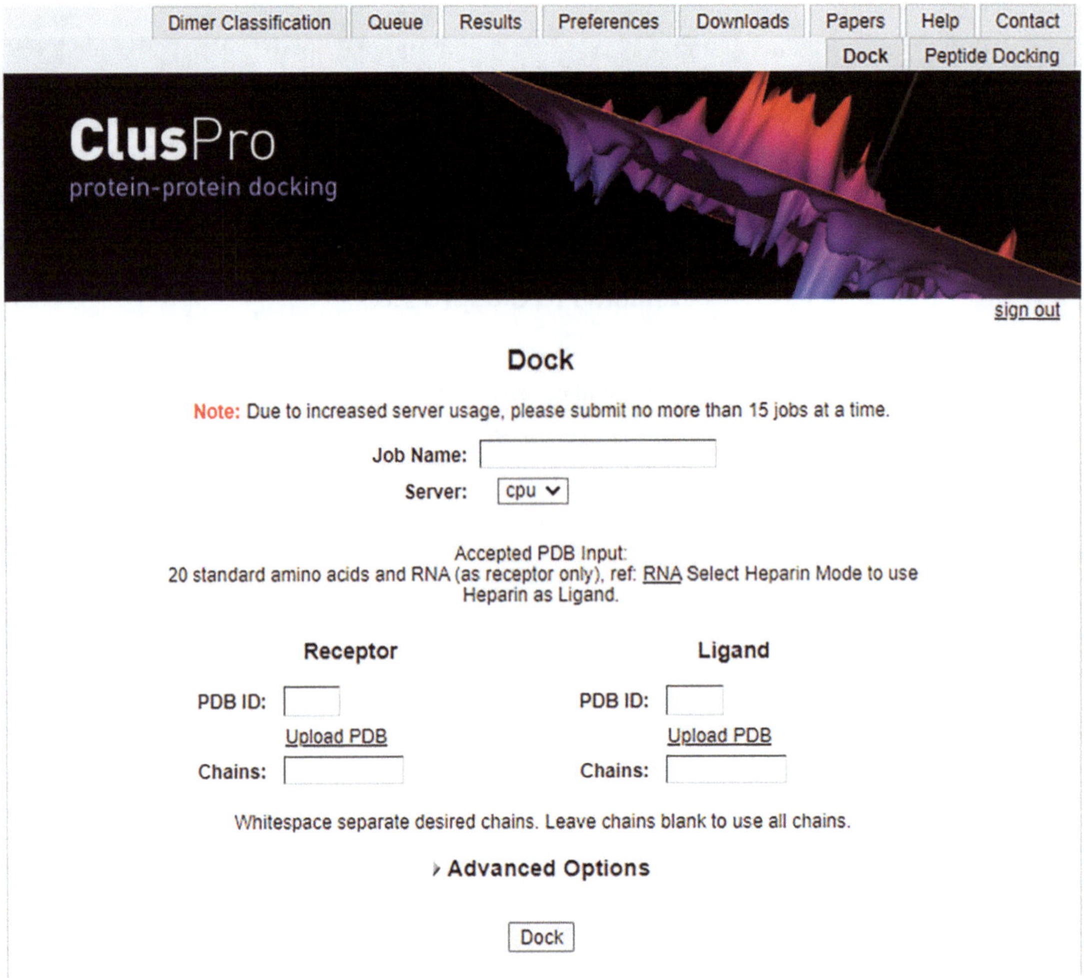

Fig. 8 ClusPro web server "Dock" page screen capture

6. One can also specify the chains, which need to be used for docking in the "Chains" field (*see* **Note 7**). If this field is left empty, then all the chains of the protein are used for docking.
7. Click on the Dock option provided at the end of the page.
8. You will be directed to a new page in which all the running jobs including yours will be displayed in the form of a queue.
9. It usually takes a few hours for docking to get completed (*see* **Note 8**). Once your job is completed, a mail will be sent to your registered email ID containing a link to view the results.
10. Alternatively, the results can also be viewed from the "Results" tab by clicking on the Job Id.
11. The server displays the top 10 models by default. This number can also be increased (Figs. 9 and 10).

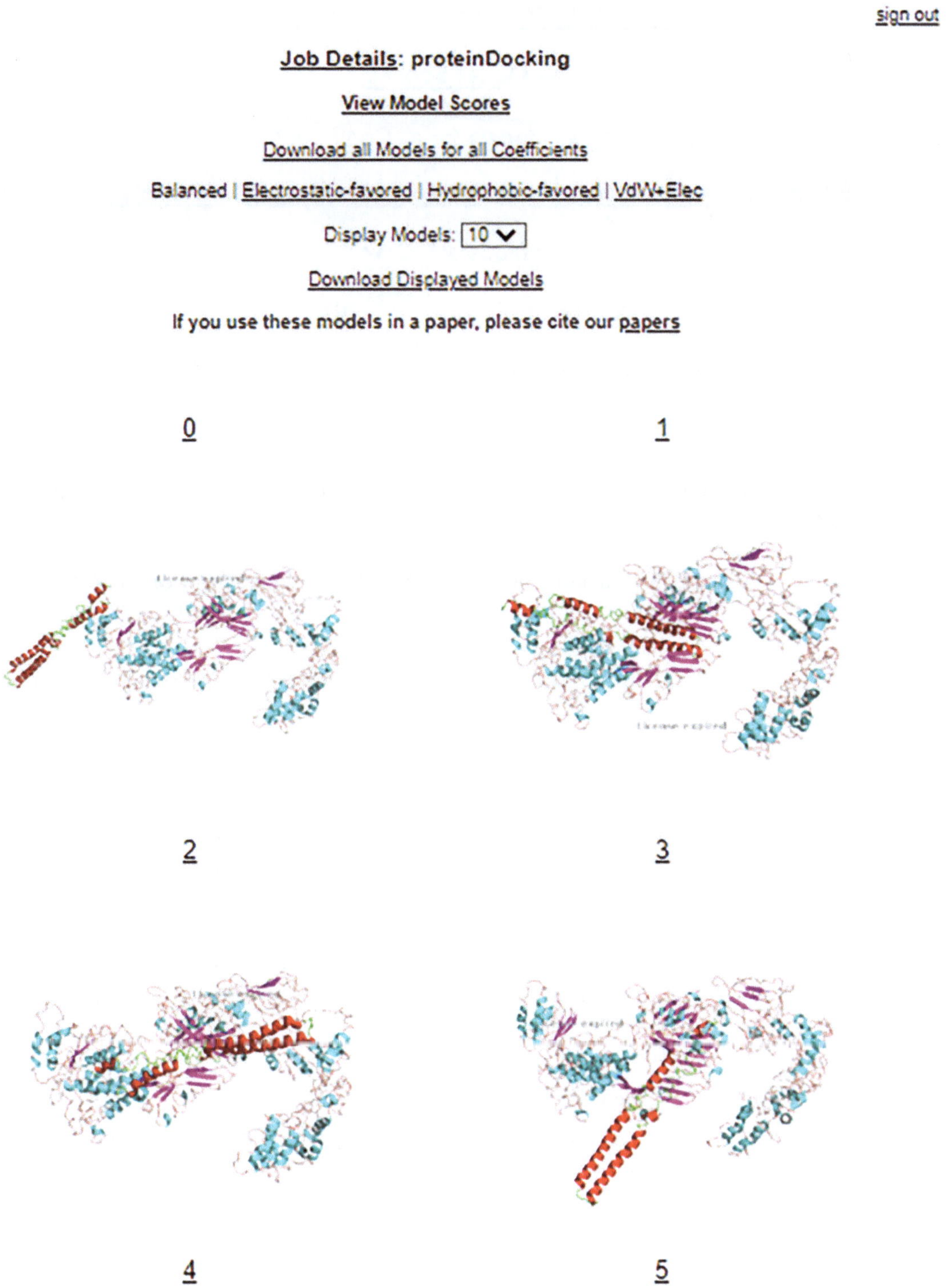

Fig. 9 Results page of ClusPro server. Screenshot showing docked models

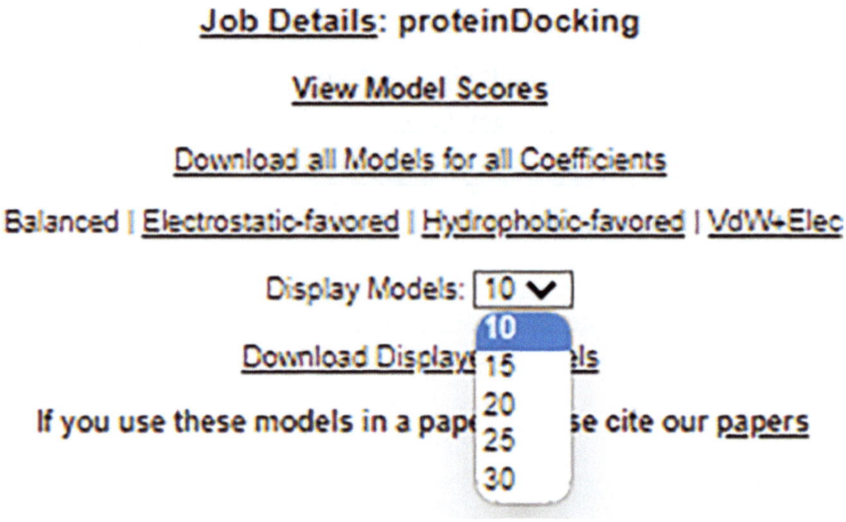

Fig. 10 The number of models displayed on the results page can be changed by the drop-down menu adjacent to the "Display Models" field

12. By clicking on the "View Model scores" option (Fig. 9), you can find the weighted score for each cluster (*see* **Notes 9** and **10**).
13. Models can be downloaded as PDB files and structures can be analyzed in Pymol (Fig. 9).

4.3 PatchDock

Materials

Server: The following URL will take the user to the PatchDock web server on the Internet: https://bioinfo3d.cs.tau.ac.il/PatchDock/php.php. Only users with academic email IDs are permitted to access the server [17].

Input: Structural information of Receptor and Ligand molecules is required in the PDB format. RCSB (https://www.rcsb.org) is one option, or users can submit their own PDB files [18].

Methods

1. Open the PatchDock web server.
2. A screen will be displayed containing a few empty fields (Fig. 11).
3. Add the structures of the receptor and ligand molecules you wish to dock in the fields provided. There are two options to input their structures; first is to enter their RCSB PDB IDs and the second is to upload your own PDB files.
4. While entering the PDB ID, you can also specify the desired chains that need to be used for docking by mentioning their chain IDs.

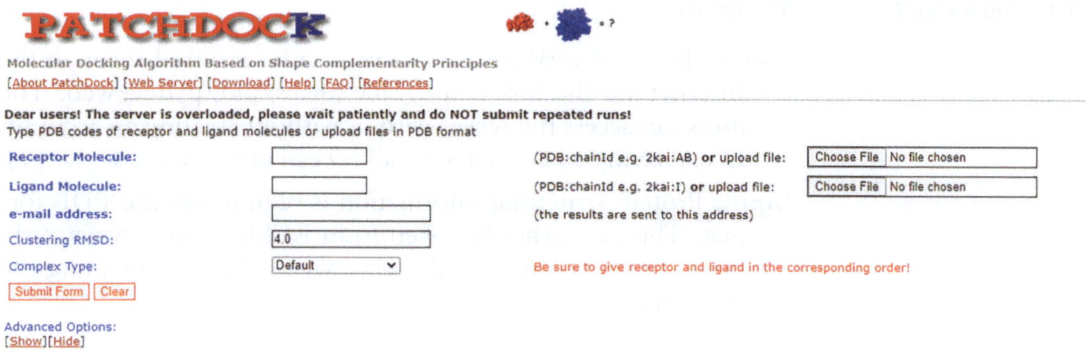

Fig. 11 Representing the PatchDock web server input page

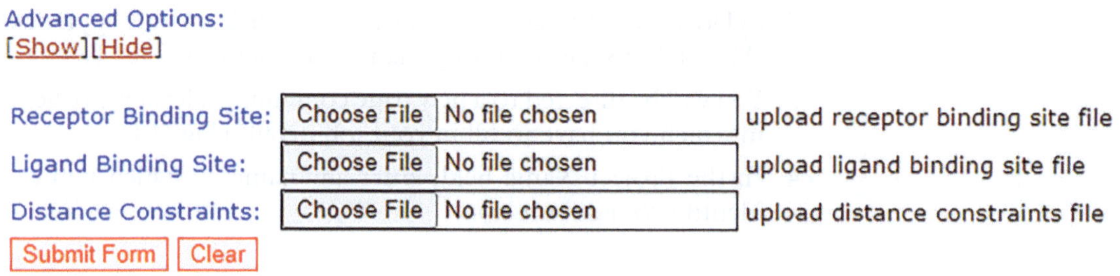

Fig. 12 Screenshot of "Advanced Options" in PatchDock

5. In the email address field, enter your institutional/academic email ID.
6. Set the "Clustering RMSD" field to 4.0 angstrom for protein–protein docking and the "Complex type" to default (*see* **Note 10**).
7. In the advanced options, there are fields to upload the files containing information about the potential receptor and ligand binding sites and distance constraints between the receptor and the ligand atoms (Fig. 12). You can fill these fields if you have the required data, otherwise skip the step.
8. Click on the "Submit Form" table.
9. A screen will appear indicating that PatchDock is running.
10. When docking is completed, you will be notified via an email containing a URL to view the results.
11. Results appear in the form of a table containing the top 20 solutions along with their score, area, Atomic Contact Energy, 3D Transformation, and PDB file of the receptor–ligand complex (*see* **Note 11**).
12. At least 100 solution complexes can be viewed and downloaded.

4.4 pyDockWEB

Materials

Server: The pyDockWEB web server can be accessed through the Internet via the link https://life.bsc.es/pid/pydockweb. The users can access the server with or without creating an account. Users having an academic email ID can only use the server [9].

Input: Protein structural information is required in the PDB format. This can either be taken from RCSB (https://www.rcsb.org/) or users can upload their own PDB files containing the structure.

Methods

1. Open the pyDockWEB server.
2. pyDock home page will be displayed. Click on the "pyDock-Web" tab displayed among many other options (Fig. 13).
3. You will be directed to a new page containing a few empty fields in which you have to fill in your job details (Fig. 14).
4. In the Project Name field, enter any name by which you can identify your submission.

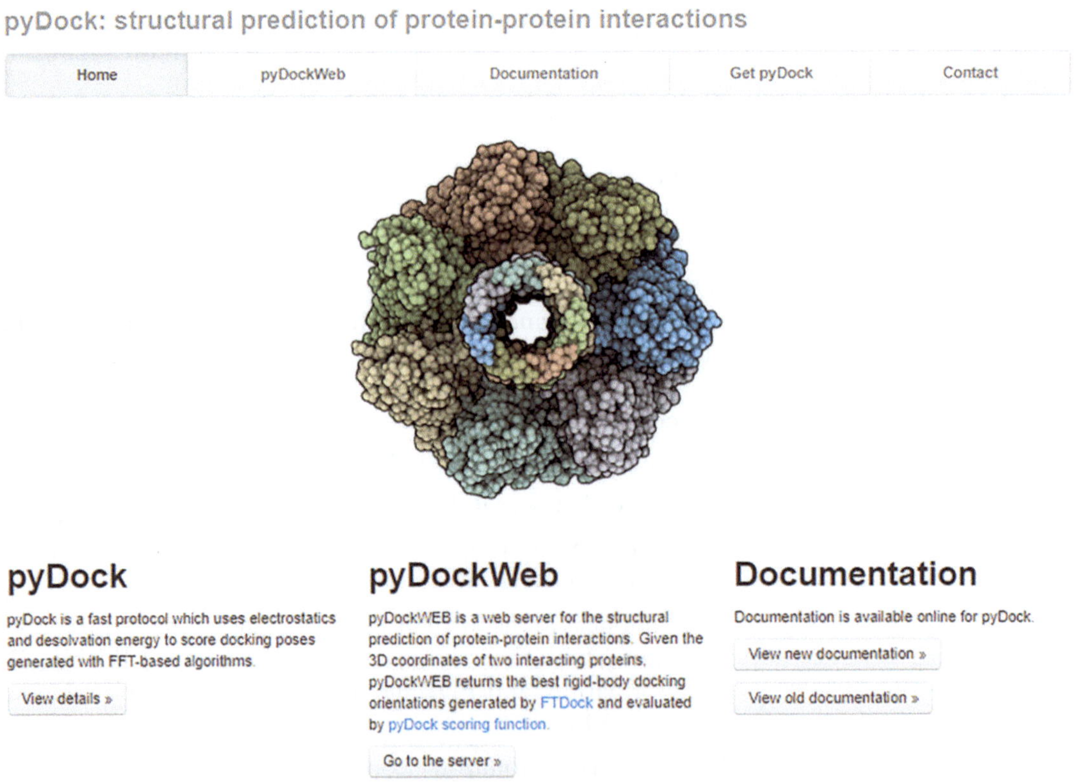

Fig. 13 Screenshot of pyDockWEB server home page

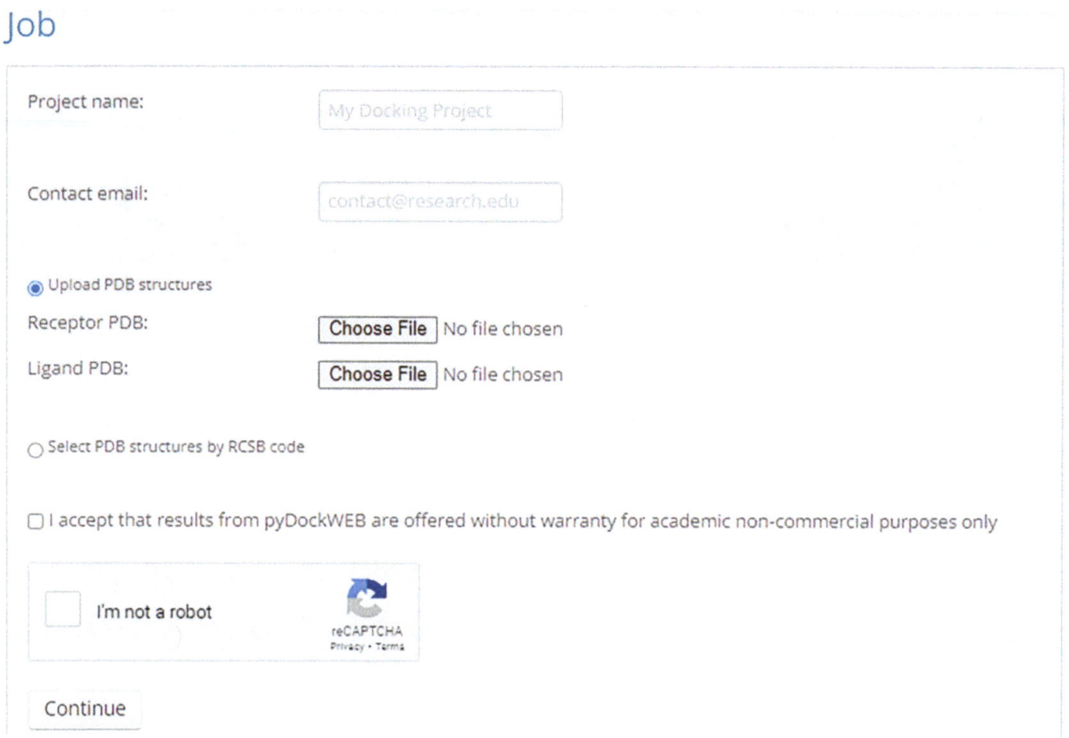

Fig. 14 Screenshot of "Job" submission page of pyDockWEB server

5. Enter your institutional/academic email ID in the Contact email field to get notifications.
6. Then, add the PDB structures of the receptor and ligand molecules you wish to dock in the fields provided (*see* **Note 12**). There are two options to input their structures; first is to upload your own PDB files and the second is to enter their RCSB PDB IDs.
7. Click on the "Continue" tab at the bottom of the screen.
8. A new page will appear containing "Structure information." Information regarding available chains in the receptor and ligand molecules is displayed on this page and you are required to select the chains for which docking needs to be performed (Fig. 15).
9. Click on "Select chains."
10. A new page will appear requiring you to specify distance restraints in the receptor and ligand residues (Fig. 16). This step is optional and can be skipped.
11. Click on "Next Step."
12. Preview of the job will appear on the screen. Check whether all the information provided by you is correct and then submit your job.

Fig. 15 Screenshot of "Structure information" page. Receptor and ligand 3D structures and their available chains are displayed

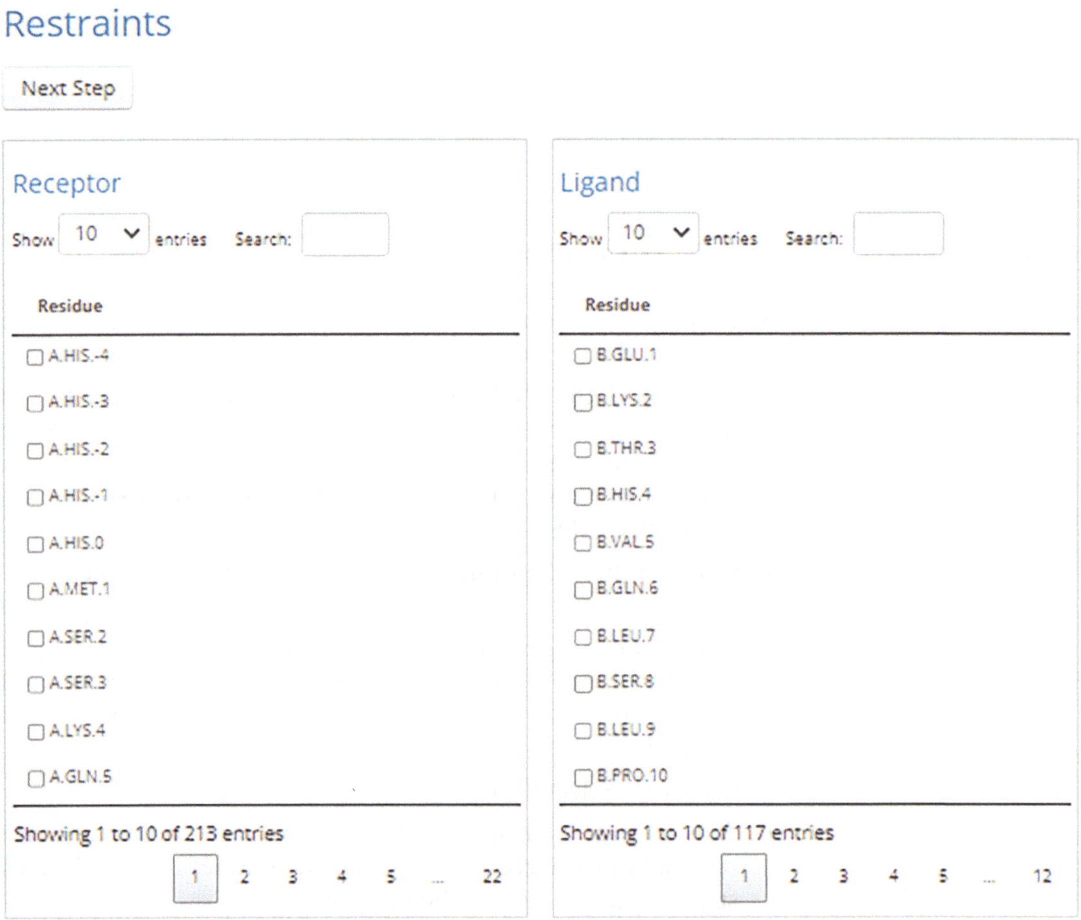

Fig. 16 Screenshot of "Restraints" page

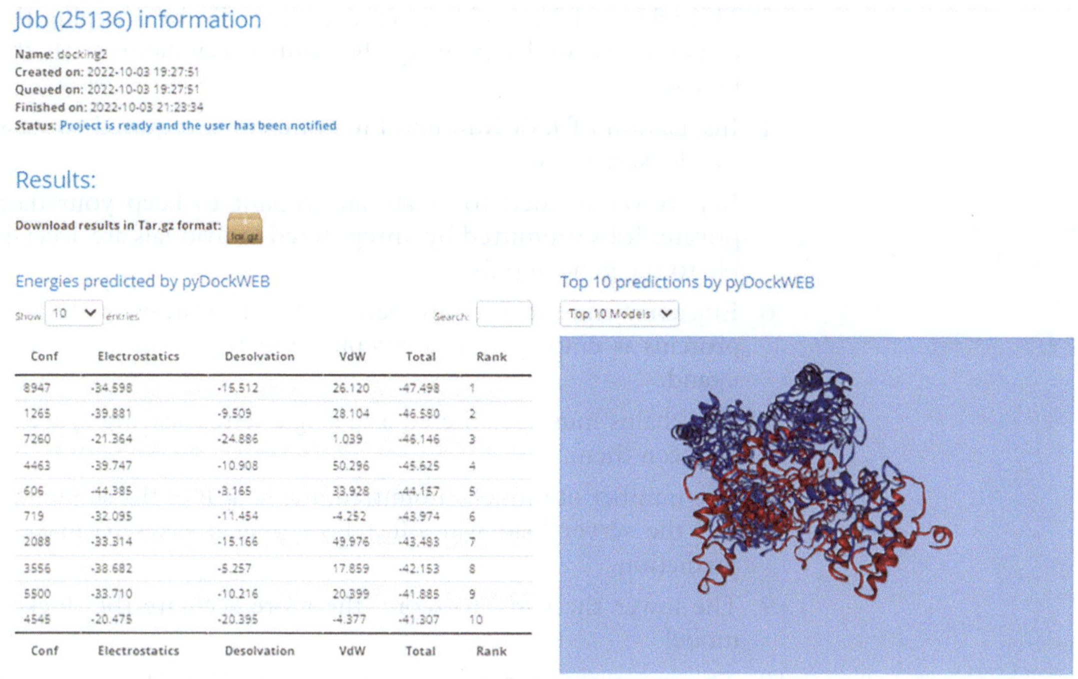

Conf	Electrostatics	Desolvation	VdW	Total	Rank
8947	-34.598	-15.512	26.120	-47.498	1
1265	-39.881	-9.509	28.104	-46.580	2
7260	-21.364	-24.886	1.039	-46.146	3
4463	-39.747	-10.908	50.296	-45.625	4
606	-44.385	-3.165	33.928	-44.157	5
719	-32.095	-11.454	-4.252	-43.974	6
2088	-33.314	-15.166	49.976	-43.483	7
3556	-38.682	-5.257	17.859	-42.153	8
5500	-33.710	-10.216	20.399	-41.885	9
4545	-20.475	-20.395	-4.377	-41.307	10
Conf	Electrostatics	Desolvation	VdW	Total	Rank

Fig. 17 Results generated by pyDockWEB. The result page displays model predictions whose number can be changed from the drop-down menu and a table showing energy predictions of the docked models

13. You will be directed to a new page that shows the status of your job and from where result files can be downloaded once docking is completed.
14. You will also be notified via an email containing a URL to view the results.
15. The result page displays the top 10 model predictions along with a table having information like conformation number, electrostatic energy, desolvation energy, Van der Waals energy, total binding energy, and RANK of the docked pose (Fig. 17).

5 Notes

1. A maximum of six proteins can be used for docking in Multi-LZERD.
2. Protein 1 entered should be the larger of the two proteins as in the docked structure, protein 1 remains stationary while protein 2 is movable (ZDOCK).

3. Multiple residues can be selected from the list by dragging the cursor down or by pressing the control key along with the mouse.
4. Installation of Java Runtime Environment is required to view the docked models.
5. It is recommended to create an account to keep your data private. Jobs submitted by unregistered individuals are accessible to the general public.
6. Efficiency and accuracy are better when the larger of the two proteins is employed as a receptor and the smaller one as a ligand.
7. The chains must be entered as a single letter, leaving space in between them.
8. The number of models submitted, the length of the sequence, and the server load will affect how long it takes to make a prediction.
9. The lower the ClusPro score, the more accurate the docked model.
10. The recommended value of Clustering RMSD is 4 Å for protein–protein docking.
11. The top-ranked model may not always be the best one. The additional knowledge about the system of interest should be taken into consideration when selecting a specific model from the docking result.
12. Upper limit of the number of residues in the protein structure is set to 1000. Before submitting the job, the user may selectively delete extra domains to satisfy the criteria.

References

1. Yan Y, Tao H, Jiahua H, Huang SY (2020) The HDOCK server for integrated protein–protein docking. Nat Protoc 15(5):1829–1852
2. Weng G, Wang E, Wang Z et al (2019) HawkDock: a web server to predict and analyze the protein–protein complex based on computational docking and MM/GBSA. Nucleic Acids Res 47(W1):W322–W330
3. Stanzione F, Giangreco I, Cole JC (2021) Use of molecular docking computational tools in drug discovery. Prog Med Chem 60:273–343
4. Esquivel-Rodríguez J, Yang YD, Kihara D (2012) Multi-LZerD: multiple protein docking for asymmetric complexes. Protein Struct Funct Bioinformatics 80(7):1818–1833
5. Huang SY, Zou X (2007) Ensemble docking of multiple protein structures: considering protein structural variations in molecular docking. Proteins 66(2):399–421
6. Pierce BG, Hourai Y, Weng Z (2011) Accelerating protein docking in ZDOCK using an advanced 3D convolution library. PLoS One 6(9):e24657
7. Kozakov D, Hall DR, Xia B et al (2017) The ClusPro web server for protein–protein docking. Nat Protoc 12(2):255–278
8. Schneidman-Duhovny D, Inbar Y, Nussinov R, Wolfson HJ (2005) PatchDock and SymmDock: servers for rigid and symmetric docking. Nucleic Acids Res 33(2):W363–W367

PyDOCK

9. Jiménez-García B, Pons C, Fernández-Recio J (2013) pyDockWEB: a web server for rigid-body protein–protein docking using electrostatics and desolvation scoring. Bioinformatics 29(13):1698–1699

LZERD

10. Christoffer C, Chen S, Bharadwaj V et al (2021) LZerD web server for pairwise and multiple protein–protein docking. Nucleic Acids Res 49(W1):W359–W365
11. Esquivel-Rodríguez J, Yang YD, Kihara D (2012) Multi-LZerD: multiple protein docking for asymmetric complexes. Proteins 80(7): 1818–1833
12. Esquivel-Rodriguez J, Filos-Gonzalez V, Li B, Kihara D (2014) Pairwise and multimeric protein–protein docking using the LZerD program suite. In: Protein structure prediction. Methods in molecular biology, vol 1137. Humana Press, New York, pp 209–234

ZDOCK

13. Pierce BG, Wiehe K, Hwang H et al (2014) ZDOCK server: interactive docking prediction of protein–protein complexes and symmetric multimers. Bioinformatics 30(12):1771–1773

Cluspro

14. Desta IT, Porter KA, Xia B, Kozakov D, Vajda S (2020) Performance and its limits in rigid body protein-protein docking. Structure 28(9):1071–1081
15. Vajda S, Yueh C, Beglov D et al (2017) New additions to the Clus Pro server motivated by CAPRI. Proteins 85(3):435–444
16. Kozakov D, Beglov D, Bohnuud T et al (2013) How good is automated protein docking? Proteins 81(12):2159–2166

PatchDOCK

17. Duhovny D, Nussinov R, Wolfson HJ (2002) Efficient unbound docking of rigid molecules. In: International workshop on algorithms in bioinformatics. Algorithms in bioinformatics, vol 2452. Springer, Berlin/Heidelberg, pp 185–200
18. Schneidman-Duhovny D, Inbar Y, Nussinov R, Wolfson HJ (2005) Geometry-based flexible and symmetric protein docking. Proteins 60(2):224–231

Chapter 6

Molecular Dynamics Simulations in Protein–Protein Docking

Dominika Cieślak, Ivo Kabelka, and Damian Bartuzi

Abstract

Concerted interactions between all the cell components form the basis of biological processes. Protein–protein interactions (PPIs) constitute a tremendous part of this interaction network. Deeper insight into PPIs can help us better understand numerous diseases and lead to the development of new diagnostic and therapeutic strategies. PPI interfaces, until recently, were considered undruggable. However, it is now believed that the interfaces contain "hot spots," which could be targeted by small molecules. Such a strategy would require high-quality structural data of PPIs, which are difficult to obtain experimentally. Therefore, in silico modeling can complement or be an alternative to in vitro approaches. There are several computational methods for analyzing the structural data of the binding partners and modeling of the protein–protein dimer/oligomer structure. The major problem with in silico structure prediction of protein assemblies is obtaining sufficient sampling of protein dynamics. One of the methods that can take protein flexibility and the effects of the environment into account is Molecular Dynamics (MD). While sampling of the whole protein–protein association process with plain MD would be computationally expensive, there are several strategies to harness the method to PPI studies while maintaining reasonable use of resources. This chapter reviews known applications of MD in the PPI investigation workflows.

Key words Molecular dynamics, Protein–protein interactions, PPI, Protein–protein docking, Molecular docking, Enhanced sampling, Coarse-graining, CG-MD, Metadynamics, Gaussian-accelerated Molecular Dynamics

1 Introduction

Concerted interactions between all the cell components form the basis of biological processes. Among them, protein–protein interactions (PPIs) emerge as one of the crucial elements, as they constitute a tremendous part of this interaction network. Being such an important aspect of the cell's inner workings, PPIs remain in the spotlight of life sciences, raising the need for suitable methods to investigate them. For instance, PPIs may drive the formation of liquid–liquid phase-separated liquid droplets and membraneless organelles, which received a lot of attention in recent years

Agnieszka A. Kaczor (ed.), *Protein-Protein Docking: Methods and Protocols*, Methods in Molecular Biology, vol. 2780,
https://doi.org/10.1007/978-1-0716-3985-6_6,
© The Author(s), under exclusive license to Springer Science+Business Media, LLC, part of Springer Nature 2024

[1]. Such biomolecular condensates have been suggested to be widespread [2] and involved in many processes essential for life. Moreover, the lipid environment can determine the formation PPIs [3], or the lipids themselves can be a part of protein–lipid–protein complexes [4, 5]. Proteins represent roughly 50% of mass of biological membranes [6] and occupy 30–50% of the membrane area [7, 8]. Lipids can drive protein sorting by a curvature preference (and subsequent enrichment of proteins in certain areas), or lipid-mediated protein association via membrane stress, hydrophobic mismatch, or protein preference for a particular lipid phase [3, 9, 10].

PPIs have a crucial role in many life processes, which implies the possibility of their exploitation for medical intervention. Detecting PPIs and identifying protein interaction partners can reveal much about the proteins under investigation [11]. Broadening the knowledge of PPIs can help us better understand numerous diseases and lead to the development of new diagnostic and therapeutic strategies [12, 13]. The medicines could either work as disruptors of protein–protein binding or as scaffolding structures that stabilize PPIs—PROTACs, for example, enable targeted degradation of disease-associated proteins [14, 15]. Unfortunately, the development of such medicines is still challenging.

PPIs may involve (1) domain–domain interactions, (2) domain coupling with less structured stretches of amino acid residues, which can have simple secondary structures, like helices, (3) interactions between domains and disordered regions (peptide–domain interactions), [13] and, surprisingly, (4) interactions between disordered regions of interacting proteins [16]. Many PPI interfaces may seem featureless and "flat," with no obvious pockets that could be easily targeted with small drug-like molecules. Therefore, until recently, PPIs were considered undruggable and their modulation with small molecules was not considered feasible. However, it is now believed that PPI sites contain "hot spots" governing protein association. These spots usually contain bulky amino acids, e.g., tryptophan or arginine, that might have a substantial effect on the whole interaction. The identification of such hot sites and modulation of their interactions is one of the potential strategies for harnessing PPIs in medicine, as small molecules that target the interaction interface could be used [13]. While the strategy seems promising, it requires high-quality structural data of PPIs, which are difficult to obtain experimentally [17]. Therefore, in silico modeling can complement or be an alternative to in vitro approaches [18, 19]. There are several computational methods for analyzing the structural data of the binding partners and modeling of the protein–protein dimer/oligomer structure. Many of these methods, however, suffer from insufficient sampling of protein flexibility.

One of the methods that can take protein flexibility and the effects of the environment into account is Molecular Dynamics (MD). In 1977, the first MD simulation of a protein was performed. It was 3.2 ps long and concerned bovine pancreatic trypsin inhibitor, a small protein, consisting of only 58 residues [20]. Nowadays, thanks to technological development and the advent of supercomputers, in silico methods are becoming more and more powerful. It is now possible to simulate huge biological structures with millions of atoms, like viral particles or ribosomes [21]. MD follows the principles of classical dynamics. The foundation is surprisingly simple, as it requires "only" solving Newton's motion equation over and over again for each atom present in the simulation box. In this way, the motion of a whole system is simulated [22]. As the available computational resources have been increasing, the landscape of possible MD applications has been getting broader. Nowadays, it is possible to perform simulations of processes that occur over relatively long timescales of micro- or even milliseconds and which require enormous sampling, e.g., spontaneous ligand binding to the protein active site [23] or simulations of cytoplasm to investigate effects of protein crowding [24]. Yet, as powerful as the technique is, unbiased all-atom MD simulation is not able to capture the assembly of two proteins randomly oriented in the solvent and the formation of a native complex within reasonable time and resources. Admittedly, conventional MD simulation at elevated temperatures has been recently able to investigate multiple binding and unbinding events of an intrinsically disordered protein fragment to its target and thus shed light on "folding-upon-binding" mechanism. However, the simulation involved only a fragment of the protein and required the use of Anton supercomputer [25].

Sufficient sampling of the whole process of protein–protein aggregation seems to be beyond the reach of all-atom, unbiased MD, as the systems usually tend to stay in suboptimal wells of potentials for most production runs and would require enormous computational time to converge. However, several methods to improve the sampling are available. For example, several simulations can be performed in parallel and utilize a replica exchange scheme [26]. Conventionally, each replica is performed at a different temperature, and system configurations are exchanged between neighboring replicas at set intervals, leading to faster exploration of phase space. The Hamiltonian replica exchange (H-REMD) method extends this scheme by performing exchanges between replicas with modified force field/interaction potentials, e.g., by softening non-bonded or torsional potentials. The efficiency of this method can be further improved by applying the modified potentials only to a part of the system (e.g., proteins) instead of the whole system.

Other strategies that overcome limitations of conventional MD and enable studying long-timescale processes on large systems include end-state methods (such as MM/GBSA), simplified representation of the system by joining atoms into beads (coarse-graining), including additional potentials to the system (metadynamics, accelerated molecular dynamics, potential flooding) and others [27].

2 Methods

2.1 Molecular Dynamics and MM/GB(PB)SA

MD can be used in combination with protein–protein docking algorithms to obtain the initial configurations of protein complexes. The most common docking tools treat the protein backbone as rigid, which minimizes the complexity of the task at hand. Obtained models are subsequently subjected to scoring and refinement. More sophisticated tools consider protein backbone flexible and to manage the computational challenge they use coarse-grain models of proteins. Examples of such algorithms are RosettaDock [28], ATTRACT [29], and CABS [30].

In studies of protein–protein interactions, the choice of force field can have a significant impact on the simulation outcome. The description of proteins with unstructured domains and intrinsically disordered proteins can be especially problematic. Older force fields had a tendency to excessively form helical secondary structures [31, 32]. The choice of water model can also be important for a better description of unfolded states [33]. For Amber ff03w force field, it was suggested to upscale the protein–water interaction to improve the description of unfolded and disordered states. At the same time, increasing the protein–water interaction by a factor of 1.1 mitigated the problem with overestimated non-specific protein–protein attraction [34]. CHARMM family of force fields uses a modified version of TIP3P water with modified LJ parameters on hydrogen atoms. For the description of disordered proteins, the CHARMM36m force field was also developed with further modified protein–water interactions. In this case, the water model was modified by changing the ε on hydrogen atoms by -0.1 kcal/mol. For these reasons, MD simulations should be performed only after an informed choice of a force field is made.

To evaluate the stability of protein–protein complexes or to assess the quality of the docking results, it is also possible to use the so-called end-state methods. Commonly, the Molecular Mechanics/generalized Born Surface Area or Molecular Mechanics/Poisson-Boltzmann Surface Area (MM/GBSA and MM/PBSA, respectively) method is used [35]. With this approach, several independent simulations are performed to create a Gibbs cycle—with individual proteins and the protein complex. These systems are simulated both in solution and in vacuum to include the contribution of solvation energy. As a result, the interaction free energy of the complex under investigation is obtained.

2.2 Coarse-Grain Molecular Dynamics

One of the strategies to decrease the computational cost of simulations is to use coarse-grain molecular dynamics (CG-MD). CG-MD provides an interesting way of improving the sampling and providing access to larger systems and longer time scales. To achieve this, several atoms are coalesced together and considered as a single interacting particle, a so-called bead. This approach can drastically reduce the number of pair interactions and degrees of freedom, simplify the interaction potentials, and smoothen the potential energy surface (due to isotropic interactions of functional groups). The removal of the fastest atomic motions in the system (bond vibrations) also opens the possibility to use a significantly longer simulation time step. CG-MD is particularly useful for studying large structures, such as viruses, or complex sub-cellular systems like nuclear pores or small organelles [21]. Thanks to decreased computational demands, CG-MD enables exploration of molecular events that take place even at millisecond time scale [36]. This feature makes CG-MD a perfect fit for de novo simulating protein–protein complex formation. However, preexisting knowledge on protein binding may be combined with CG to obtain even better results [37].

MARTINI is one of the most popular force fields for CG-MD. In this model, four heavy atoms are mapped to one bead with the exception of ring-like molecules, which are represented with higher resolution. There are several bead types that describe moieties characterized by different physicochemical properties and various degrees of polarity. The latest version of MARTINI has introduced a separate bead type for water molecules, which allows the manipulation of these beads without affecting the whole system. Based on the comparison of diffusion rates, the dynamics in the MARTINI force field is fourfold faster compared to atomistic simulations [38]. The computational efficiency of the MARTINI force field is further improved by neglecting the long-ranged electrostatic interactions (roughly matching the electrostatic screening under physiological conditions). Nonbonded interactions are described by Lennard-Jones potential, as well as Coulomb potentials for charged beads. MARTINI was first used to simulate the assembly of small lipid vesicles, but its applications have become significantly wider since then and include studies of membrane protein–lipid interactions, protein oligomerization, protein and peptide self-assembly, and many more [39, 40].

While reduced computational requirements would make MARTINI seem to be an excellent method for de novo modeling of protein–protein complexes, one should consider its serious limitations. The applied strategy of reducing the resolution of the system results in the requirement of additional restraints mimicking the interactions that otherwise would be difficult to describe in such a setup, for instance, hydrogen bonds that maintain particular types of secondary structure. At the same time, those restrictions limit structure flexibility and thus exclude some uses of the MARTINI method. Processes that include changes in secondary structure, like protein folding, cannot be properly simulated in MARTINI v2 [39], and for this purpose some other models should be considered instead [41, 42]. However, some progress has been made in this area recently and a version of the foldable MARTINI was developed [43].

Another noteworthy issue is that MARTINI v2 can overestimate the association of proteins in solution. In one particular case, to bring the results to a better agreement with experimental data, non-specific attraction had to be reduced by down-scaling the vdW interactions to 30–40% of the original value [44]. On the other hand, the beta version of MARTINI v3 was shown to systematically underestimate the association of proteins in solution [45]. Calculations of interaction free energies related to residue-specific effects were also shown to be problematic in older versions of the force field, which was addressed in the development of MARTINI v3. In a study by Schäfer et al., it was shown that MARTINI v2 can capture both the experimental dimerization free energy and correctly decompose the enthalpic and entropic contributions [46]. However, other studies revealed that MARTINI v2 underestimates the repulsive interactions and irreversible formation of protein clusters can occur [47]. A solution proposed by Javanainen et al., i.e. uniform down-scaling of protein–protein interactions by 10%, successfully brought the dimerization free energies of transmembrane helices close to the experimental values [48]. Other limitations of the MARTINI force field can be also overcome with the use of polarizable MARTINI. Corrections for a better description of cation–π interactions have been published, which, for instance, enabled proper simulation of the adsorption of proteins at the membrane surface [49].

Other force fields for coarse-grained MD are Associative memory, Water mediated, Structure and Energy Model (AWSEM) [50, 51] and its modification, Atomistic Associative memory, Water mediated, Structure and Energy Model (AAWSEM) [41]. AAWSEM can be considered a modification of AWSEM, which does not require bioinformatic input [52]. In contrast to MARTINI, AWSEM and AAWSEM can be used for de novo structure prediction by simulation of protein folding. AWSEM was recently used to explore the interaction of mutated superoxide dismutase 1 (SOD1) with the MATH region of TNF receptor-associated factor 6 (TRAF6) [53]. SOD1 mutations are one of the most important causes of amyotrophic lateral sclerosis. Simulation of SOD1 native homodimer formation from monomers was used to validate the method; obtained structures were clustered and compared with experimentally resolved ones. The authors concluded that even though docking poses were a little bit different in the simulation when compared to the experimental data, the same residues were involved in the interaction. Subsequently, simulations of wild type as well as mutated and/or misfolded variants of SOD1 with TRAF1 MATH domain were performed. Overall, six dimeric systems were examined using CG and metadynamics. The systems were set up in a way that one binding partner (TRAF6) was placed in the center of the simulation box and the other one (SOD1) was placed randomly around it. HDBSCAN [54], a clustering algorithm, was used to examine obtained trajectories in a search for the most populated poses that studied proteins adopted during the dimerization process. Such an approach allowed to identify residues involved in binding between those two proteins. The results were also confirmed by co-immunoprecipitation assay [53].

SIRAH is another recently reported CG model [55]. Using a top-down approach, its developers managed to craft force field parameters independent from the secondary structure, enabling the folding and unfolding of peptides, and simulate protein-peptide binding [42]. This was achieved by applying nonuniform granularity, with atomistic sizes of beads corresponding to the main chain atoms. This property of the model would be especially valuable when changes in secondary structure are expected in the protein–protein binding region.

There are examples of successful use of CG models in studies of membrane protein dimerization/oligomerization. While membrane proteins are confined within the plane of a lipid bilayer, which should facilitate finding a favorable interaction interface, predicting their interactions can be challenging due to various effects of the lipid environment, for instance, membrane curvature [9, 10] or lipid-mediated bridging [5]. CG-MD provides a way to tackle this challenge and it was used, for example, to elucidate the mechanism of GPCR self-assembly in the plasma membrane. CG model of rhodopsin was prepared based on a crystal structure, and special care was taken to maintain the molecule properties like shape, surface polarity, and helix and loop orientation. The use of the CG approach allowed to reach 8 μs of effective simulation time. The results confirmed that hydrophobic mismatch, which is a difference between protein and bilayer hydrophobic thickness, promotes rhodopsin oligomerization [56]. Moreover, protein–protein interactions inside a membrane bilayer show a site preference related to the localized mismatch.

Mesoscopic CG models for proteins have also been developed. For example, each residue can be described by only a single particle [57]. Such a model was recently used to study PPIs in dilute and semi-concentrated solutions [58]. Mesoscopic CG models can also be used to simulate very complex structures, for example, viral capsid. HIV capsid is an assembly of more than a thousand capsid protein monomers, which have to be dimerized in order to enable further assembly. Even though there are many capsid protein structures available, not much is known about the mechanism of their assembly. In one study, each helix in the capsid protein structure was represented by a cylinder and interactions between subunits were considered as either non-specific volume exclusion between all cylinders or pair-wise attractive interactions between specific cylinder pairs in the interface between subunits. The study showed that capsid proteins of HIV assemble into hexamers created by trimers of dimeric subunits [59]. In another study, a model of viral capsid was specifically developed to study the conditions for viral genome release. In this model, each capsid was built from indivisible pentameric subunits [60].

2.3 Gaussian Accelerated Molecular Dynamics

One of the reasons to conduct simulations over very long timescales is to capture biologically relevant rare events. The energy landscape of biological processes can be very rough. It is characteristic for the landscape to have many metastable states separated by high energetic barriers. The probability of overcoming those barriers is low, which extends the simulation time needed to observe these events. Thus, one can add a biasing potential to the system in order to overcome the energy barriers and facilitate the transition between different states. In Gaussian accelerated molecular dynamics (Ga-MD), reduction in energy barriers is achieved by adding harmonic boost potential, which smoothens the energy landscape [61]. In contrast to techniques like umbrella sampling [62] or metadynamics [63], to perform a Ga-MD study one does not have to define reaction coordinates, for example, dihedral angles or RMSD in relation to reference configuration. The use of reaction coordinates can be a disadvantage as it requires some a priori understanding of the process under investigation [64]. Ga-MD is available in the Amber suite. Users of Gromacs might be interested in the Gromacs counterpart of this method, which is called "flooding" [65]. Notably, although similar to the former, the latter is much less frequently used.

2.3.1 Protein–Peptide Interactions

Peptides can have a wide range of biological activities (e.g., antimicrobial, hormonal, or immunomodulatory), which makes them promising drug candidates. However, they are distinct from both proteins and small ligands, when it comes to target binding. Often, peptides do not have a well-defined structure before binding, which makes predicting protein–peptide interaction even more challenging. Protein–peptide interactions are transient and weaker than protein–protein interactions due to the decreased surface of contact between molecules [66]. There are many pieces of software that may be used to predict protein–peptide interactions, for example, GalaxyPepDock [67], Rosetta FlexPepDock [68], HADDOCK

[69], or PeptiDock [70]. Despite the fact that the above-mentioned tools incorporate different approaches to the problem, they are often only moderately successful in predicting the structure of the complex. However, these tools provide a good starting point for further refinement and evaluation of the stability using MD simulations. PepATTRACT is a tool that incorporates a round of refinement using MD simulation [71].

Peptide docking can also be combined with enhanced sampling methods to improve the accuracy of the results. One example of such an approach was adopted by Wang and coworkers [66]. First, the authors obtained an initial model of the complex with a medium-quality resolution using PeptiDock software [70] and subsequently refined it using Ga-MD. The peptide backbone RMSD of the resulting model was in the range of 0.6–2.7 Å. Moreover, such a simulation allowed to obtain more information about the mechanism of this interaction, for example, the conformational changes of the protein. This approach proposed by Wang and coworkers was successfully used by another group to check whether epitopes of myelin proteins can be presented by MHC I/II, and explore the mechanism of molecular mimicry, by which viral proteins presented by the MHC complexes can affect multiple sclerosis (MS) course or even induce the disease [72]. The authors used HPEPDOCK Server [73] for initial site-specific docking of MHC I/II proteins and myelin and viral peptides and refined obtained models using MD simulations. Ga-MD was also successfully used in combination with rigid body global docking of a cyclic peptide to MDM2 and MDMX proteins using ClusPro Dock [70, 74, 75].

The Ga-MD method was also extended for the particular use to study protein–peptide interaction. Peptide Gaussian-accelerated molecular dynamics (Pep-Ga-MD) incorporates the dual-boost potential to boost both: (1) peptide–protein potential to enhance the peptide binding and (2) protein–solvent potential to enhance the protein conformational sampling. The technique allows for extensive thermodynamic and kinetic examination of protein–peptide complexes. In a demonstrative study conducted by the authors of the method, they proved that their approach is able to provide kinetic parameters of the binding that are in line with available experimental data within 1 μs simulation [76]. In another study using Pep-Ga-MD, researchers probed the mechanism of differential agonism of peptide ligands toward the CXCR4 receptor. They have found specific interactions between particular residues responsible for the phenomenon [77].

2.3.2 Protein–Protein Interactions

Similar to Pep-Ga-MD, PPI-Ga-MD was developed to facilitate calculations of binding free energy and kinetics of protein–protein interactions. The method can provide one with a trajectory with multiple binding and unbinding events of two proteins, which is not readily feasible to obtain using conventional MD or Ga-MD [78]. The latter method, however, has been successfully used to predict the binding of two proteins, e.g., G-protein mimetic nanobody and M_2 muscarinic receptor. In the simulation, allosterically bound nanobody caused the closing of the orthosteric binding pocket of the receptor, which was in line with experimental data obtained for nanobody binding to β_2 adrenergic receptor allosteric site. This study also revealed participation of the intracellular loops of the protein in the binding [79].

2.4 Metadynamics

In contrast to Ga-MD, metadynamics involves the use of user-defined reaction coordinates (also called collective variables, abbreviated CVs). This method biases the system toward visiting unexplored regions of the CVs. Metadynamics has been described as filling free energy wells with "computational sand," effectively flattening the surface. Ultimately, the system should be able to freely diffuse along the whole range of the CVs. Then, the free energy can be calculated as the negative sum of the added biasing potential. During the course of the simulation, biasing potential is added, with Gaussian distribution, at the actual position of the CVs [63]. In a more computationally efficient implementation of the method, the Gaussians are deposited at the nearest position on a grid. A good choice of CVs, in a system without major orthogonal degrees of freedom, can significantly improve the sampling. On the other hand, if there are one or more crucial degrees of freedom that are not captured by the CVs, the simulations may poorly sample the CVs or not converge. Therefore, metadynamics requires prior knowledge of the system and careful system setup in order to be successful.

To improve the convergence, many extensions to the metadynamics method have been described. These extensions include (1) the use of multiple walkers (independent simulations that share the biasing potential), (2) adaptive change of the Gaussian heights (so-called well-tempered metadynamics) [80], adaptive change of the Gaussian deposition frequency, transition tempered metadynamics to study state-to-state transitions [81], parallel tempering (PT-MetaD), and others.

In order to increase sampling even further, metadynamics can be combined with CG models. The reduced roughness of a potential energy surface can improve the sampling of orthogonal degrees of freedom. A study on protein aggregation was performed by Ko et al. using the CG model and metadynamics. Recombinant GCSF is prone to aggregation in physiological pH. To elucidate the mechanism of this unfavorable process, eight CG monomers were placed in a simulation box. Three pH values as well as salt influence were studied. This experiment together with biophysical techniques like intrinsic fluorescence spectroscopy, nanoDSF, and DLS, allowed to elucidate the exact mechanism of aggregation.

Intramolecular interaction responsible for the conformational stability of the protein was indicated, and moreover, a shift of one of the interacting residues at a specific pH value was noticed. CG simulation and the number of protein–protein interactions extracted from it indicated that indeed what occurred in the samples was aggregation and not the temporal association of the molecules because this number increased with time during the simulation. Using CG results back-mapped into atomic resolution, together with SAXS (small-angle X-ray scattering) measurements, the authors also proposed a few models of GCSF dimers in various pH and salt conditions, stating that multiple forms of dimers may exist and aggregation is not specific [19]. In another study, metadynamics combined with CG and umbrella sampling was also used to compute the homodimerization free energy landscapes of β_1 and β_2 adrenergic receptors for two putative dimerization interfaces, involving either transmembrane helix (TM) 1 interacting with helix 8 (H8) of the partner, or TM4 interacting with TM3 of the other dimer subunit [82]. The study led to the conclusion that the TM1/H8 dimerization interface is more stable than the TM4/TM3 interface. Very recently, a metadynamics-based protocol for protein–protein and protein-ligand binding free energy calculation was reported by Wang et al. [83]. The technique involves performing a series of metadynamics simulations, focusing on a one-way dissociation process. Unfortunately, the protocol assumes prior knowledge of the bound configuration, so its application to the prediction of protein–protein complexes comes down to scoring of complexes generated using other methods.

2.4.1 Oligomerization

In the study by Barducci and co-workers, metadynamics was used to investigate the oligomerization of fibritin foldon. In the beginning, monomer folding was simulated using a combination of metadynamics with different sampling methods, mainly parallel tempering (PTMetD). Obtained monomer structures in different activation states were almost identical to the experimentally resolved structure of the protein trimer (RMSD <0.2 nm). In a dimerization step, another technique was added to facilitate the simulation—well-tempered ensemble (WTE). The study indicated significant differences in the free-energy landscape between those two seemingly similar processes. The trimerization proceeds through a rather smooth funnel toward a global free-energy minimum, which coincides with the experimental structure. Instead, the dimerization leads to a broad ensemble of almost equally populated structures that include configurations observed in the trimeric experimental structure. The most favored dimeric structure, which has been so far elusive to experimental investigation, corresponds to an almost antiparallel packing of the two monomers. This scenario suggests that the addition of a third monomer operates a conformational selection of the dimer ensemble by stabilizing the conformation observed in the native trimer. The authors managed to identify the N-terminal disordered tail as an important factor in both monomer–monomer and monomer–dimer recognition [84].

2.4.2 Protein–Peptide Interactions

Metadynamics was also used by della Longa et al. to investigate the process of nociceptin binding [85]. Nociceptin peptide structure modeled using Chimera UCSF package [86] was rigidly docked into the nociceptin receptor X-ray structure to obtain the protein–peptide complex. The best complex was chosen on the basis of the literature data, embedded in a phospholipid bilayer, and further subjected to all-atom MD. Such an approach sheds some light on the mechanism in which the transition between the active and inactive state of the nociceptin receptor takes place. Interestingly, the results of the study suggested the existence of a putative molecular microswitch between those two states [85].

3 Conclusions

While all-atom, unbiased MD could hypothetically be used to sample the whole process of protein–protein association, such simulation would require considerable computational resources. Meanwhile, there are several strategies that allow more efficient sampling at much lower computational cost. Coarse-graining seems to be the most broadly used approach, frequently mixed with other techniques, e.g., metadynamics. Keeping in mind their limitations, MD-related methods are a noteworthy tool that, if carefully used, can provide insights into the cell's inner workings.

Acknowledgments

The work was supported by the Polish National Agency for Academic Exchange within the Bekker NAWA Programme, project PANALLOS, grant number BPN/BEK/2021/1/00408/U/00001.

References

1. Ma W, Mayr C (2018) A Membraneless Organelle associated with the endoplasmic reticulum enables 3'UTR-mediated protein-protein interactions. Cell 175:1492–1506.e19. https://doi.org/10.1016/j.cell.2018.10.007
2. Hardenberg M, Horvath A, Ambrus V et al (2020) Widespread occurrence of the droplet state of proteins in the human proteome. Proc Natl Acad Sci 117:33254–33262. https://doi.org/10.1073/pnas.2007670117
3. Javanainen M, Hammaren H, Monticelli L et al (2013) Anomalous and normal diffusion of proteins and lipids in crowded lipid membranes. Faraday Discuss 161:397–417. https://doi.org/10.1039/C2FD20085F
4. Webby MN, Oluwole AO, Pedebos C et al (2022) Lipids mediate supramolecular outer membrane protein assembly in bacteria. Sci Adv 8:eadc9566. https://doi.org/10.1126/sciadv.adc9566

5. Gupta K, Donlan JAC, Hopper JTS et al (2017) The role of interfacial lipids in stabilizing membrane protein oligomers. Nature 541: 421–424. https://doi.org/10.1038/nature20820
6. Guidotti G (1972) Membrane Proteins. Annu Rev Biochem 41:731–752. https://doi.org/10.1146/annurev.bi.41.070172.003503
7. Sowers AE, Hackenbrock CR (1981) Rate of lateral diffusion of intramembrane particles: measurement by electrophoretic displacement and rerandomization. Proc Natl Acad Sci 78: 6246–6250. https://doi.org/10.1073/pnas.78.10.6246
8. Lindén M, Sens P, Phillips R (2012) Entropic tension in crowded membranes. PLoS Comput Biol 8:e1002431. https://doi.org/10.1371/journal.pcbi.1002431
9. Tian A, Baumgart T (2009) Sorting of lipids and proteins in membrane curvature gradients. Biophys J 96:2676–2688. https://doi.org/10.1016/j.bpj.2008.11.067
10. Parton DL, Klingelhoefer JW, Sansom MSP (2011) Aggregation of model membrane proteins, modulated by hydrophobic mismatch, membrane curvature, and protein class. Biophys J 101:691–699. https://doi.org/10.1016/j.bpj.2011.06.048
11. Rao VS, Srinivas K, Sujini GN, Kumar GNS (2014) Protein-protein interaction detection: methods and analysis. Int J Proteomics 2014: 1–12. https://doi.org/10.1155/2014/147648
12. Nevola L, Giralt E (2015) Modulating protein–protein interactions: the potential of peptides. Chem Commun 51:3302–3315. https://doi.org/10.1039/C4CC08565E
13. London N, Raveh B, Schueler-Furman O (2013) Druggable protein–protein interactions – from hot spots to hot segments. Curr Opin Chem Biol 17:952–959. https://doi.org/10.1016/j.cbpa.2013.10.011
14. Neklesa TK, Winkler JD, Crews CM (2017) Targeted protein degradation by PROTACs. Pharmacol Ther 174:138–144. https://doi.org/10.1016/j.pharmthera.2017.02.027
15. Lee J, Lee Y, Jung YM et al (2022) Discovery of E3 ligase ligands for target protein degradation. Molecules 27:6515. https://doi.org/10.3390/molecules27196515
16. Tompa P, Fuxreiter M, Oldfield CJ et al (2009) Close encounters of the third kind: disordered domains and the interactions of proteins. BioEssays 31:328–335. https://doi.org/10.1002/bies.200800151
17. Rakers C, Bermudez M, Keller BG et al (2015) Computational close up on protein–protein interactions: how to unravel the invisible using molecular dynamics simulations? WIREs Comput Mol Sci 5:345–359. https://doi.org/10.1002/wcms.1222
18. Zhao C, Shukla D (2018) SAXS-guided enhanced unbiased sampling for structure determination of proteins and complexes. Sci Rep 8:17748. https://doi.org/10.1038/s41598-018-36090-z
19. Ko SK, Berner C, Kulakova A et al (2022) Investigation of the pH-dependent aggregation mechanisms of GCSF using low resolution protein characterization techniques and advanced molecular dynamics simulations. Comput Struct Biotechnol J 20:1439–1455. https://doi.org/10.1016/j.csbj.2022.03.012
20. McCammon JA, Gelin BR, Karplus M (1977) Dynamics of folded proteins. Nature 267:585–590. https://doi.org/10.1038/267585a0
21. Perilla JR, Goh BC, Cassidy CK et al (2015) Molecular dynamics simulations of large macromolecular complexes. Curr Opin Struct Biol 31:64–74. https://doi.org/10.1016/j.sbi.2015.03.007
22. Hollingsworth SA, Dror RO (2018) Molecular dynamics simulation for all. Neuron 99:1129–1143. https://doi.org/10.1016/j.neuron.2018.08.011
23. Shan Y, Kim ET, Eastwood MP et al (2011) How does a drug molecule find its target binding site? J Am Chem Soc 133:9181–9183. https://doi.org/10.1021/ja202726y
24. Oliveira Bortot L, Bashardanesh Z, van der Spoel D (2020) Making soup: preparing and validating models of the bacterial cytoplasm for molecular simulation. J Chem Inf Model 60: 322–331. https://doi.org/10.1021/acs.jcim.9b00971
25. Robustelli P, Piana S, Shaw DE (2020) Mechanism of coupled folding-upon-binding of an intrinsically disordered protein. J Am Chem Soc 142:11092–11101. https://doi.org/10.1021/jacs.0c03217
26. Sugita Y, Okamoto Y (1999) Replica-exchange molecular dynamics method for protein folding. Chem Phys Lett 314:141–151. https://doi.org/10.1016/S0009-2614(99)01123-9
27. Hénin J, Lelièvre T, Shirts MR et al (2022) Enhanced sampling methods for molecular dynamics simulations. Liv J Comput Mol Sci 4(1):1583
28. Lyskov S, Gray JJ (2008) The RosettaDock server for local protein-protein docking. Nucleic Acids Res 36:W233–W238. https://doi.org/10.1093/nar/gkn216
29. Glashagen G, de Vries S, Uciechowska-Kaczmarzyk U et al (2020) Coarse-grained

and atomic resolution biomolecular docking with the ATTRACT approach. Proteins Struct Funct Bioinforma 88:1018–1028. https://doi.org/10.1002/prot.25860

30. Kurcinski M, Badaczewska-Dawid A, Kolinski M et al (2020) Flexible docking of peptides to proteins using CABS-dock. Protein Sci Publ Protein Soc 29:211–222. https://doi.org/10.1002/pro.3771
31. Best RB, Buchete N-V, Hummer G (2008) Are current molecular dynamics force fields too helical? Biophys J 95:L07–L09. https://doi.org/10.1529/biophysj.108.132696
32. Patapati KK, Glykos NM (2011) Three force fields' views of the 310 Helix. Biophys J 101: 1766–1771. https://doi.org/10.1016/j.bpj.2011.08.044
33. Best RB, Mittal J (2010) Protein simulations with an optimized water model: cooperative Helix formation and temperature-induced unfolded state collapse. J Phys Chem B 114: 14916–14923. https://doi.org/10.1021/jp108618d
34. Best RB, Zheng W, Mittal J (2015) Correction to balanced protein–water interactions improve properties of disordered proteins and non-specific protein association. J Chem Theory Comput 11:1978–1978. https://doi.org/10.1021/acs.jctc.5b00219
35. Chen F, Liu H, Sun H et al (2016) Assessing the performance of the MM/PBSA and MM/GBSA methods. 6. Capability to predict protein–protein binding free energies and re-rank binding poses generated by protein–protein docking. Phys Chem Chem Phys 18: 22129–22139. https://doi.org/10.1039/C6CP03670H
36. Ingólfsson HI, Lopez CA, Uusitalo JJ et al (2014) The power of coarse graining in biomolecular simulations. WIREs Comput Mol Sci 4: 225–248. https://doi.org/10.1002/wcms.1169
37. Baaden M, Marrink SJ (2013) Coarse-grain modelling of protein–protein interactions. Curr Opin Struct Biol 23:878–886. https://doi.org/10.1016/j.sbi.2013.09.004
38. Marrink SJ, Risselada HJ, Yefimov S et al (2007) The MARTINI force field: coarse grained model for biomolecular simulations. J Phys Chem B 111:7812–7824. https://doi.org/10.1021/jp071097f
39. Marrink SJ, Tieleman DP (2013) Perspective on the Martini model. Chem Soc Rev 42:6801. https://doi.org/10.1039/c3cs60093a
40. Souza PCT, Alessandri R, Barnoud J et al (2021) Martini 3: a general purpose force field for coarse-grained molecular dynamics. Nat Methods 18:382–388. https://doi.org/10.1038/s41592-021-01098-3
41. Davtyan A, Schafer NP, Zheng W et al (2012) AWSEM-MD: protein structure prediction using coarse-grained physical potentials and bioinformatically based local structure biasing. J Phys Chem B 116:8494–8503. https://doi.org/10.1021/jp212541y
42. Machado MR, Barrera EE, Klein F et al (2019) The SIRAH 2.0 Force field: Altius, Fortius. Citius J Chem Theory Comput 15:2719–2733. https://doi.org/10.1021/acs.jctc.9b00006
43. Sahoo A, Lee P-Y, Matysiak S (2022) Transferable and polarizable coarse grained model for proteins—ProMPT. J Chem Theory Comput 18:5046–5055. https://doi.org/10.1021/acs.jctc.2c00269
44. Stark AC, Andrews CT, Elcock AH (2013) Toward optimized potential functions for protein–protein interactions in aqueous solutions: osmotic second virial coefficient calculations using the MARTINI Coarse-Grained Force Field. J Chem Theory Comput 9:4176–4185. https://doi.org/10.1021/ct400008p
45. Lamprakis C, Andreadelis I, Manchester J et al (2021) Evaluating the efficiency of the Martini Force Field to study protein dimerization in aqueous and membrane environments. J Chem Theory Comput 17:3088–3102. https://doi.org/10.1021/acs.jctc.0c00507
46. Schäfer LV, de Jong DH, Holt A et al (2011) Lipid packing drives the segregation of transmembrane helices into disordered lipid domains in model membranes. Proc Natl Acad Sci 108:1343–1348. https://doi.org/10.1073/pnas.1009362108
47. Jarin Z, Newhouse J, Voth GA (2021) Coarse-Grained Force Fields from the perspective of statistical mechanics: better understanding of the origins of a MARTINI hangover. J Chem Theory Comput 17:1170–1180. https://doi.org/10.1021/acs.jctc.0c00638
48. Javanainen M, Martinez-Seara H, Vattulainen I (2017) Excessive aggregation of membrane proteins in the Martini model. PLoS One 12: e0187936. https://doi.org/10.1371/journal.pone.0187936
49. Khan HM, Souza PCT, Thallmair S et al (2020) Capturing choline–aromatics cation−π interactions in the MARTINI Force Field. J Chem Theory Comput 16:2550–2560. https://doi.org/10.1021/acs.jctc.9b01194
50. Tsai M-Y, Zheng W, Balamurugan D et al (2016) Electrostatics, structure prediction,

and the energy landscapes for protein folding and binding. Protein Sci 25:255–269. https://doi.org/10.1002/pro.2751

51. Zheng W, Schafer NP, Davtyan A et al (2012) Predictive energy landscapes for protein–protein association. Proc Natl Acad Sci 109:19244–19249. https://doi.org/10.1073/pnas.1216215109
52. Chen M, Lin X, Zheng W et al (2016) Protein folding and structure prediction from the ground up: the atomistic associative memory, water mediated, structure and energy model. J Phys Chem B 120:8557–8565. https://doi.org/10.1021/acs.jpcb.6b02451
53. Garg P, Semmler S, Baudouin C et al (2022) Misfolding-associated exposure of natively buried residues in mutant SOD1 facilitates binding to TRAF6. J Mol Biol 434:167697. https://doi.org/10.1016/j.jmb.2022.167697
54. McInnes L, Healy J, Astels S (2017) hdbscan: hierarchical density based clustering. J Open Source Softw 2:205. https://doi.org/10.21105/joss.00205
55. Darré L, Machado MR, Brandner AF et al (2015) SIRAH: a structurally unbiased Coarse-Grained Force Field for proteins with aqueous solvation and long-range electrostatics. J Chem Theory Comput 11:723–739. https://doi.org/10.1021/ct5007746
56. Periole X, Huber T, Marrink S-J, Sakmar TP (2007) G protein-coupled receptors self-assemble in dynamics simulations of model bilayers. J Am Chem Soc 129:10126–10132. https://doi.org/10.1021/ja0706246
57. Lund M, Jönsson B (2003) A Mesoscopic Model for protein-protein interactions in solution. Biophys J 85:2940–2947. https://doi.org/10.1016/S0006-3495(03)74714-6
58. Pasquier C, Midtgaard SR, Polimeni M et al (2023) Anisotropic protein-protein interactions in dilute and concentrated solutions. J Colloid Interface Sci 629:794–804. https://doi.org/10.1016/j.jcis.2022.08.054
59. Qiao X, Jeon J, Weber J et al (2015) Mechanism of polymorphism and curvature of HIV capsid assemblies probed by 3D simulations with a novel coarse grain model. Biochim Biophys Acta BBA - Gen Subj 1850:2353–2367. https://doi.org/10.1016/j.bbagen.2015.08.017
60. Sukeník L, Mukhamedova L, Procházková M et al (2021) Cargo release from nonenveloped viruses and virus-like nanoparticles: capsid rupture or pore formation. ACS Nano 15:19233–19243. https://doi.org/10.1021/acsnano.1c04814
61. Wang J, Arantes PR, Bhattarai A et al (2021) Gaussian accelerated molecular dynamics: principles and applications. WIREs Comput Mol Sci 11. https://doi.org/10.1002/wcms.1521
62. Torrie GM, Valleau JP (1977) Nonphysical sampling distributions in Monte Carlo free-energy estimation: umbrella sampling. J Comput Phys 23:187–199. https://doi.org/10.1016/0021-9991(77)90121-8
63. Laio A, Gervasio FL (2008) Metadynamics: a method to simulate rare events and reconstruct the free energy in biophysics, chemistry and material science. Rep Prog Phys 71:126601. https://doi.org/10.1088/0034-4885/71/12/126601
64. Miao Y, Feher VA, McCammon JA (2015) Gaussian accelerated molecular dynamics: unconstrained enhanced sampling and free energy calculation. J Chem Theory Comput 11:3584–3595. https://doi.org/10.1021/acs.jctc.5b00436
65. Lange OF, Schäfer LV, Grubmüller H (2006) Flooding in GROMACS: accelerated barrier crossings in molecular dynamics. J Comput Chem 27:1693–1702. https://doi.org/10.1002/jcc.20473
66. Wang J, Alekseenko A, Kozakov D, Miao Y (2019) Improved modeling of peptide-protein binding through global docking and accelerated molecular dynamics simulations. Front Mol Biosci 6:112. https://doi.org/10.3389/fmolb.2019.00112
67. Lee H, Heo L, Lee MS, Seok C (2015) GalaxyPepDock: a protein–peptide docking tool based on interaction similarity and energy optimization. Nucleic Acids Res 43:W431–W435. https://doi.org/10.1093/nar/gkv495
68. Raveh B, London N, Zimmerman L, Schueler-Furman O (2011) Rosetta FlexPepDock ab-initio: simultaneous folding, docking and refinement of peptides onto their receptors. PLoS One 6:e18934. https://doi.org/10.1371/journal.pone.0018934
69. van Zundert GCP, Rodrigues JPGLM, Trellet M et al (2016) The HADDOCK2.2 web server: user-friendly integrative modeling of biomolecular complexes. J Mol Biol 428:720–725. https://doi.org/10.1016/j.jmb.2015.09.014
70. Porter KA, Xia B, Beglov D et al (2017) ClusPro PeptiDock: efficient global docking of peptide recognition motifs using FFT. Bioinformatics 33:3299–3301. https://doi.org/10.1093/bioinformatics/btx216
71. de Vries SJ, Rey J, Schindler CEM et al (2017) The pepATTRACT web server for blind, large-scale peptide–protein docking. Nucleic Acids Res 45:W361–W364. https://doi.org/10.1093/nar/gkx335

72. Jakhmola S, Sk MF, Chatterjee A et al (2022) A plausible contributor to multiple sclerosis; presentation of antigenic myelin protein epitopes by major histocompatibility complexes. Comput Biol Med 148:105856. https://doi.org/10.1016/j.compbiomed.2022.105856
73. Zhou P, Jin B, Li H, Huang S-Y (2018) HPEP-DOCK: a web server for blind peptide–protein docking based on a hierarchical algorithm. Nucleic Acids Res 46:W443–W450. https://doi.org/10.1093/nar/gky357
74. Vajda S, Yueh C, Beglov D et al (2017) New additions to the ClusPro server motivated by CAPRI. Proteins 85:435–444. https://doi.org/10.1002/prot.25219
75. Wang Y-T, Cheng T-L (2021) Computational modeling of cyclic peptide inhibitor–MDM2/MDMX binding through global docking and Gaussian accelerated molecular dynamics simulations. J Biomol Struct Dyn 39:4005–4014. https://doi.org/10.1080/07391102.2020.1773317
76. Wang J, Miao Y (2020) Peptide Gaussian accelerated molecular dynamics (Pep-GaMD): enhanced sampling and free energy and kinetics calculations of peptide binding. J Chem Phys 153:154109. https://doi.org/10.1063/5.0021399
77. Pawnikar S, Miao Y (2022) Mechanism of peptide agonist binding in CXCR4 Chemokine receptor. Front Mol Biosci 9:821055. https://doi.org/10.3389/fmolb.2022.821055
78. Wang J, Miao Y (2022) Protein–protein interaction-Gaussian Accelerated Molecular Dynamics (PPI-GaMD): characterization of protein binding thermodynamics and kinetics. J Chem Theory Comput 18:1275–1285. https://doi.org/10.1021/acs.jctc.1c00974
79. Miao Y, McCammon JA (2018) Mechanism of the G-protein mimetic nanobody binding to a muscarinic G-protein-coupled receptor. Proc Natl Acad Sci 115:3036–3041. https://doi.org/10.1073/pnas.1800756115
80. Barducci A, Bussi G, Parrinello M (2008) Well-tempered metadynamics: a smoothly converging and tunable free-energy method. Phys Rev Lett 100:020603. https://doi.org/10.1103/PhysRevLett.100.020603
81. Dama JF, Rotskoff G, Parrinello M, Voth GA (2014) Transition-tempered metadynamics: robust, convergent metadynamics via on-the-fly transition barrier estimation. J Chem Theory Comput 10:3626–3633. https://doi.org/10.1021/ct500441q
82. Johnston JM, Wang H, Provasi D, Filizola M (2012) Assessing the relative stability of dimer interfaces in g protein-coupled receptors. PLoS Comput Biol 8:e1002649. https://doi.org/10.1371/journal.pcbi.1002649
83. Wang J, Ishchenko A, Zhang W et al (2022) A highly accurate metadynamics-based Dissociation Free Energy method to calculate protein–protein and protein–ligand binding potencies. Sci Rep 12:2024. https://doi.org/10.1038/s41598-022-05875-8
84. Barducci A, Bonomi M, Prakash MK, Parrinello M (2013) Free-energy landscape of protein oligomerization from atomistic simulations. Proc Natl Acad Sci 110. https://doi.org/10.1073/pnas.1320077110
85. della Longa S, Arcovito A (2016) A dynamic picture of the early events in Nociceptin binding to the NOP receptor by metadynamics. Biophys J 111:1203–1213. https://doi.org/10.1016/j.bpj.2016.07.004
86. Pettersen EF, Goddard TD, Huang CC et al (2004) UCSF Chimera?A visualization system for exploratory research and analysis. J Comput Chem 25:1605–1612. https://doi.org/10.1002/jcc.20084

Chapter 7

Machine Learning Methods in Protein–Protein Docking

Ilona Michalik and Kamil J. Kuder

Abstract

An exponential increase in the number of publications that address artificial intelligence (AI) usage in life sciences has been noticed in recent years, while new modeling techniques are constantly being reported. The potential of these methods is vast—from understanding fundamental cellular processes to discovering new drugs and breakthrough therapies. Computational studies of protein–protein interactions, crucial for understanding the operation of biological systems, are no exception in this field. However, despite the rapid development of technology and the progress in developing new approaches, many aspects remain challenging to solve, such as predicting conformational changes in proteins, or more "trivial" issues as high-quality data in huge quantities.

Therefore, this chapter focuses on a short introduction to various AI approaches to study protein–protein interactions, followed by a description of the most up-to-date algorithms and programs used for this purpose. Yet, given the considerable pace of development in this hot area of computational science, at the time you read this chapter, the development of the algorithms described, or the emergence of new (and better) ones should come as no surprise.

Key words Protein, Protein–protein docking, Protein–protein interaction, Machine learning, Deep learning, Artificial intelligence, Structural biology

1 Introduction

Intelligence has always aroused the admiration and curiosity of scientists due to the extraordinary range and diversity of intellectual abilities in individuals and the neurobiological and cognitive mechanisms that underlie them [1]. There is no single universally accepted definition. The same is true for the term artificial intelligence (AI): depending on the individual and the field in which it is used at any given time, it can mean something quite different. Nevertheless, intelligence is primarily associated with a human trait, which is why AI, on the one hand, raises concerns among the community when spoken of in the context of machines, while on the other hand, it is undoubtedly an exciting prospect for new

Agnieszka A. Kaczor (ed.), *Protein-Protein Docking: Methods and Protocols*, Methods in Molecular Biology, vol. 2780, https://doi.org/10.1007/978-1-0716-3985-6_7,
© The Author(s), under exclusive license to Springer Science+Business Media, LLC, part of Springer Nature 2024

research, better understanding of the world, and even the very workings of the human mind [2, 3].

The origins of artificial intelligence date back to the mid-twentieth century [4–8], but it is only in recent years that it has led to its widespread adoption. This is due to the rapid development of technology and the vast amount of data being collected. In the field of biological research, there has been an exponential increase in the number of publications that address AI, while new modeling techniques are constantly being reported. The potential of these methods is vast—from understanding basic cellular processes to discovering new drugs and breakthrough therapies [9]. This chapter will describe the most important algorithms used in protein–protein docking, based on machine learning (ML) methods, with examples of their implementation in various bioinformatics tools.

In the most general terms, ML is distinguished as one of the subsets of AI, and, in turn, deep learning (DL) belongs to ML. These terms are not interchangeable, although they can all refer to the ability of computers to learn from existing data. Often, it is not the choice of the algorithm itself, but the collection of data and its proper preparation and understanding, which is key to achieving high accuracy and reliability of prediction [10, 11]. For this purpose, a basic knowledge of statistics, mathematics, and knowledge of the field to which the data relates is important. The results of a 2017 study showed that as many as about 75% of life sciences training programs did not meet the expected requirements of acquiring the right skills for data management [12]. These issues significantly exceed the ability to accurately present the entire theoretical and practical basis in this book. However, in recent years, a number of publications have been published in this area primarily dedicated to biologists in recent years [9, 13–17]. It is also particularly relevant because of the numerous problems that often arise both in the training of the model and in the analysis of the output data. First of all, it is worth mentioning the problem of insufficient, nonrepresentative, or poor quality of learning data, as well as overfitting or underfitting of models. Counter-intuitively, underfitting is a more favorable situation than overfitting, because an overfitting model will have trouble making predictions on new data that it has not been exposed to in the learning process.

Based on the degree and type of supervision of model learning, ML systems are divided into four categories: unsupervised learning, semisupervised learning, supervised learning, and reinforcement learning (Fig. 1). In supervised learning, the input data are labeled with the correct solution to the problem—labels that can contain the result in the form of a discrete or continuous variable. Depending on the type of variable, the prediction becomes a classification or regression task, for which different algorithms have been developed. In semisupervised and unsupervised learning, respectively,

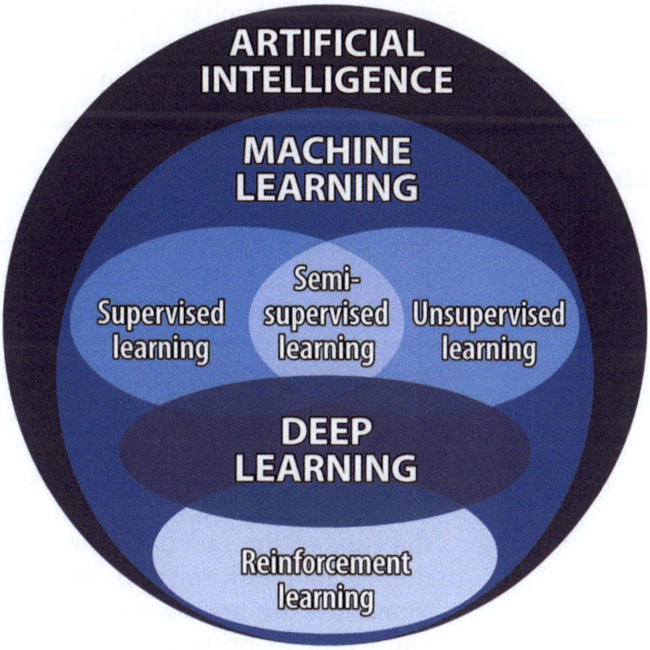

Fig. 1 Basic division of machine learning methods including the most important connections between them

the data passed to the algorithm contain only partial labels or no labels at all. Reinforcement learning is an approach inspired by behavioral psychology, in which the model is forced to learn on its own, based on observations of the environment and receiving rewards or punishments. It uses feedback from the environment to make a subsequent decision [18].

Biological neurons, meanwhile, inspired the development of even more sophisticated ML architectures—artificial neural networks (ANNs), the basis of DL. They are ideally suited for complicated ML tasks because of their versatility and scalability. In the field of structural biology, DL has gained particular recognition and a powerful wave of interest after the success of the AlphaFold2 model to predict the tertiary structure of proteins based on amino acid sequences, developed by London-based DeepMind [19]. As of July 2021, the life sciences world had even been gripped by so-called AlphaFold mania [20]. Although the current AlphaFold neural network has many limitations, the prospects for further evolution are promising, including improvements in the accuracy of protein–protein interaction (PPI) predictions [21], which remain a considerable challenge for computational methods. However, the study of PPI networks and the global physical organization of cells is necessary for a better understanding of basic cellular biochemistry and physiology [22], given that more than 80% of proteins do not perform their functions in vivo independently, but in complexes. Such interactions can be homo- or heterooligomeric, obligatory or nonobligatory, and transient, forming signaling pathways, or persistent, forming stable protein complexes [23].

2 Protein–Protein Docking

The classical approach to the protein–protein docking process begins with identifying the proteins that interact with each other and determining their 3D structures. Appropriate preprocessing of the proteins (protonation states, possible tautomerism, etc.) is also necessary. The next step is to use docking algorithms and energy criteria to extract the most stable binding modes (known as sampling and scoring), followed by refinement of selected protein complexes [24, 25]. With templates available for the components of a given complex, the task is partially facilitated, but there is still the issue of the variety of conformational changes that can occur during complex formation. Therefore, the 3D structures of isolated monomers are not always sufficient for accurate prediction of the entire protein–protein complex [26]. An additional complication is also the problem of distinguishing functional PPI from nonspecific interactions.

2.1 Representation of Input Protein Structures and Feature Descriptors

ML-based computational models are generally applied only at a single specific step, supporting the entire docking process, such as the mere prediction of the probability of interactions between different proteins, determination of binding sites on the surface of proteins, scoring, or refinement of formed protein complexes. However, regardless of the task of the model, choosing the right representation of protein structures and the features that describe their properties are key issues for high performance. Extracting relevant features is particularly important when using traditional ML algorithms. DL methods are more effective at defining relevant features and their relationships independently, thus requiring less human intervention and allowing the implementation of end-to-end models. The downside of DL algorithms, however, is the problem of so-called "black boxes," that is, simply put, the difficulty of interpreting their operation.

In general, there are two ways to represent input data using ML methods: structure-based and sequence-based approaches. In both methods, it is possible to take into account evolutionary information (e.g., from multiple sequence alignments—MSA or conserved features), physicochemical properties (e.g., hydrophobicity of residues, polarity, volume, charge), and composition of amino acid residues. Structure-based approaches, however, usually gain an advantage due to the availability of valuable structural (e.g., secondary structure motif, measures of protein flexibility derived from crystallographic B-factors, disorder, relative solvent accessibility) and geometric (e.g., average depth and protrusion indexes for each surface residue, indices describing local surface shape or curvature, 2D/3D Zernike descriptors of voxelized protein surface representations, and invariant geometric fingerprint descriptors) [26]. An additional problem is the representation of variable-length protein sequences, which require a procedure for transforming them into a fixed-size encoded vector. Many solutions have already been designed for this purpose, such as one-hot encoding, natural language processing (NLP), or autocovariance techniques.

2.2 Traditional Machine Learning Methods

Of the traditional ML algorithms for PPI prediction, Support Vector Machines (SVM), Random Forests (RF), and Shallow Neural Networks (SNN) are the most commonly used.

2.2.1 Support Vector Machines

SVM is one of the most popular and effective algorithms for developing supervised learning models, also widely used in the field of drug design. It is primarily useful in classifying complex, small, or medium-sized datasets by its relatively low tendency to overfit, although it can also be used in a regression or outlier identification task. Its versatility comes from the ability to specify different kernel functions, such as a linear kernel, a polynomial kernel, Gaussian RBF, a sigmoidal kernel, or the less common string kernel, which is sometimes used to classify DNA sequences [27–29]. Critical to the performance of SVM-based models is the selection of appropriate hyperparameters, which can be achieved, for example, by grid search.

One of the SVM models (with the polynomial kernel of degree 1) for predicting protein–protein and protein–peptide interactions is the model developed by Romero-Molina et al. [30] and implemented in the PPI-Detect tool [31]. It uses a numerical procedure to encode polypeptides, which is an extension of the ProtDCal program [32, 33] with the ability to code protein pairs. Pairs of amino acid sequences are transformed into a vector whose elements represent numerical descriptors of the groups of residues in proteins. Training data were collected from three extensive databases containing information on interacting protein domain pairs (3did [34], iPfam [35]), as well as noninteracting (Negatome 2.0 [36]) protein domain pairs, which increased the generalization of the predictor. PPI-Detect achieved higher performance compared to other sequential PPI predictors (PIPE [37], Pred-PPI [38], and SPPS [39]). In addition, because of its speed and use of primary structures, it can be helpful in the screening of protein and peptide libraries to detect potential PPI.

Another tool also based on ProtDCal [32, 33] and the SVM regression algorithm is the PPI-Affinity Web service [40, 41]. Moreover, an ensemble learning approach was used here: the training set was divided into four partially overlapping subsets and each model was trained separately, after which the hyperparameters of each model were optimized using grid search and their independent predictions were combined using the Vote method implemented in Weka 3.8.4 [42], according to ensemble rules based on mean, maximum, and minimum predicted values among the selected models. The final ensemble model was selected by evaluating all possible combinations of models, resulting in a predictor with better results than the standard individual model. PPI-Affinity is used to predict protein–protein and protein–peptide binding affinity, as well as to generate using MODELLER v9.23 [43–46]) and score mutants of a given structure to optimize binding affinity or spot critical mutations that may destabilize PPI. Therefore, it can be used at early stages of the drug design process and its performance is confirmed by results obtained in various tests on external datasets. Deficiencies in the accuracy of predicting the absolute value of the binding free energy of the generated structures are unavoidable because of the biases present in the experimental data, which are used to train the models [40]. Experimental conditions, as well as measurement procedures, can vary from study to study, so despite the best efforts used in selecting training and test datasets, no model is perfect and this should be kept in mind when interpreting the results.

Also noteworthy is the SVM model with Radial Basis Function (RBF) kernel, designed to effectively distinguish between native and non-native homocomplexes and heterocomplexes obtained by protein–protein docking. The classification is based on calculated structural and chemical features of the interfaces, which showed statistically significant differences ($p \leq 0.01$). The model has achieved very good performance in benchmarking and comparative analyses [47], therefore a web server called Protein Complex Prediction by Interface Properties (PCPIP) has been made available [48]. PCPIP is particularly useful for confirming the correctness of the poses of previously docked proteins by the various standard programs available.

2.2.2 Forests of Randomized Trees and Boosting Methods

Decision trees (DTs) are another versatile algorithm for building models on complex datasets, while being extremely intuitive and easy to interpret. They are also components of random forests (RF), part of the aforementioned ensemble learning methods, in which the predictions of a group of predictors are combined, resulting in often better predictions than from a single best model. Ideally, individual predictors should be as independent of each other as possible to increase the likelihood that they will make completely different errors, and thus improve the accuracy of the final ensemble model. Furthermore, randomness in RF contributes to reducing the variance of the estimator, as opposed to individual DT, which typically exhibits high variance and tends to overfit.

Randomness in RF has two main sources: each individual DT is created from a drawn sample from the training set, and during partitioning of each node of a given DT, the best partition is defined either from all input features or just from a random subset. The undoubted advantage of random forests is also the ease of determining the relative importance of each feature of the model.

Even more random DTs are generated by using an ensemble called Extremely Random Trees (Extra-Trees), in which the search for the best distribution for each feature at each node is replaced by random distribution thresholds for each feature. This contributes to an even greater reduction in variance and model learning time relative to RT, but at the expense of an increase in the systematic error.

Examples of the application of the RF algorithm in the field of PPI research are primarily models for the prediction of their interaction sites. One of these is the IntPred method [49], which uses sequence and structure information to create 11 features within the input data. Predictive labels were introduced at the surface patch level, created using a program from the BiopTools toolkit [50], or amino acid residues. In addition to the 100-tree RF algorithm, neural networks (NN) with five hidden layers were also tested, but the performance of the RF-based model significantly outperformed it.

In the model developed by You et al. [51] for PPI prediction using RF, a novel method of representing features extracted from protein sequences using Multi-scale Local Descriptor (MLD) was used. The MLD technique divides sequences into multiple segments of varying lengths to identify multiple overlapping continuous binding patterns. Finally, the descriptors are encoded in the binary system, achieving an 1134-dimensional vector to represent each pair as input for the RF predictor. The authors also further confirmed the significant advantage of the ensemble classifier over the individual classifier that creates it.

A very interesting option for increasing the performance of models is also the use of reinforcement methods for ensemble learning. The concept of operation is based on sequential learning of predictors, instead of training all of them in parallel, using feedback from the predecessors. The most popular reinforcement methods are adaptive boosting (AdaBoost) [52] and gradient boosting [53]. In the AdaBoost method, the classifiers focus on the most difficult examples from the training set, for which the algorithm has been previously untrained, and by modifying the weights gradually aimed to minimize or completely eliminate erroneous predictions. Once all predictors have been trained, the ensemble calculates the final predictions, taking into account the weight of each predictor, which depends on the accuracy it has achieved. Over the past few years, various improvements to the

AdaBoost method have also been developed, such as the PSOPD-AdaBoost-A ensemble algorithm [54] to increase the efficiency of processing unbalanced data. In the gradient boosting method, the weights of examples from the training set are not modified after each successive learned predictor. It is the predictor that is just learned using the residual error made by the predictor preceding it. The prediction of the entire ensemble is the sum of the predictions of the individual predictors. Depending on the library and algorithm you choose, there are different hyperparameters to choose for your model. An optimized version of this method called Extreme Gradient Boosting, available in the XGBoost library [55], has become particularly popular. The idea behind the development of the XGBoost library was to achieve the best possible results with minimal use of computational resources in a short period of time. Although its alternative libraries have been developed, such as CatBoost [56] and LightGBM [57], the XGBoost library is widely spread and is constantly being developed by researchers.

The BIPSPI (xgBoost Interface Prediction of Specific-Partner Interactions) predictor [58], available as a web server [59], and a newer version called BIPSPI+ [60, 61], uses XGBoost models to predict partner-specific PPI sites. The BIPSPI predictor, unlike many other methods, was trained on the sequence and structural features of pairs of protein complex residues, rather than individual protein structures. In addition, the authors designed a scoring function, taking into account scores for all pairs of residues. The improved version, BIPSPI+, not only achieved even higher binding site prediction performance through careful selection of specific datasets, but also new functionalities were introduced, such as the guided protein–protein docking tool (PP-docking), using the PatchDock rigid-body docking algorithm [62] with BIPSPI+ predictions in the form of a binding site search space constraint. It would be relatively easy to obtain better 3D models using more complex docking algorithms. However, the goal of this project was to allow users to obtain very simple and fast initial 3D models of quaternary structures for a better understanding of binding site predictions.

2.2.3 Shallow Neural Networks

As mentioned in this chapter's introduction, artificial neural networks are the result of inspiration from biological neural networks. Although ANN contains many analogous elements in their structure, they are not their precise representation. The basic units of ANNs are perceptrons, which can be considered a reflection of biological neurons, which receive signals and pass them on, serving in this case as the input (training) data. As in any ML model, care must be taken to properly prepare the input data, especially when scaling to a consistent format to compare them freely and to avoid biases associated with high values of certain features that could be

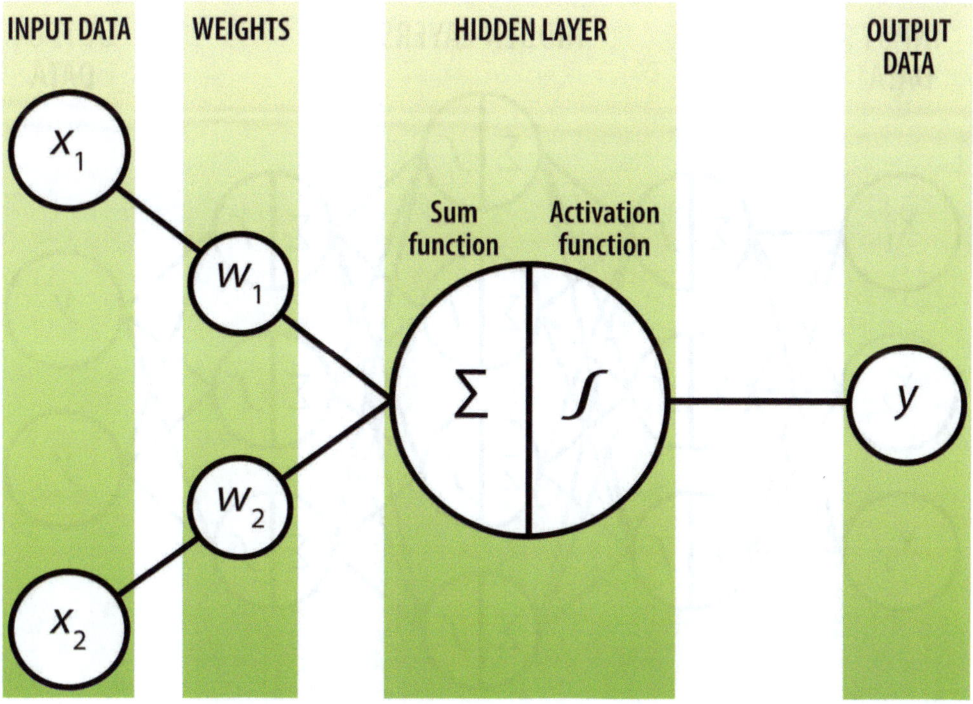

Fig. 2 Basic structure of a perceptron

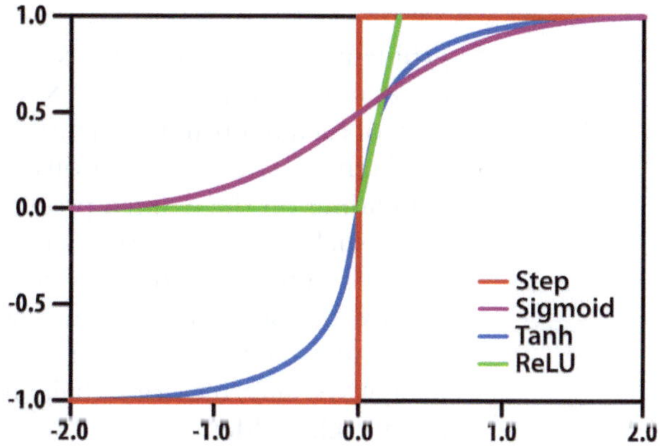

Fig. 3 Examples of the most commonly used activation functions

misinterpreted. Furthermore, an additional neuron is often inserted at the input (the bias input) to send a value of 1 to the output. The data are then modified based on the weights and processed further by a hidden node, responsible for summing the weighted values and applying the activation function to the resulting sum. The final result (expected prediction) is returned as output (Fig. 2). A single perceptron is limited to simple linear problems. However, for nonlinear problems, more perceptron-like neurons are required, which interact with each other to eventually solve the

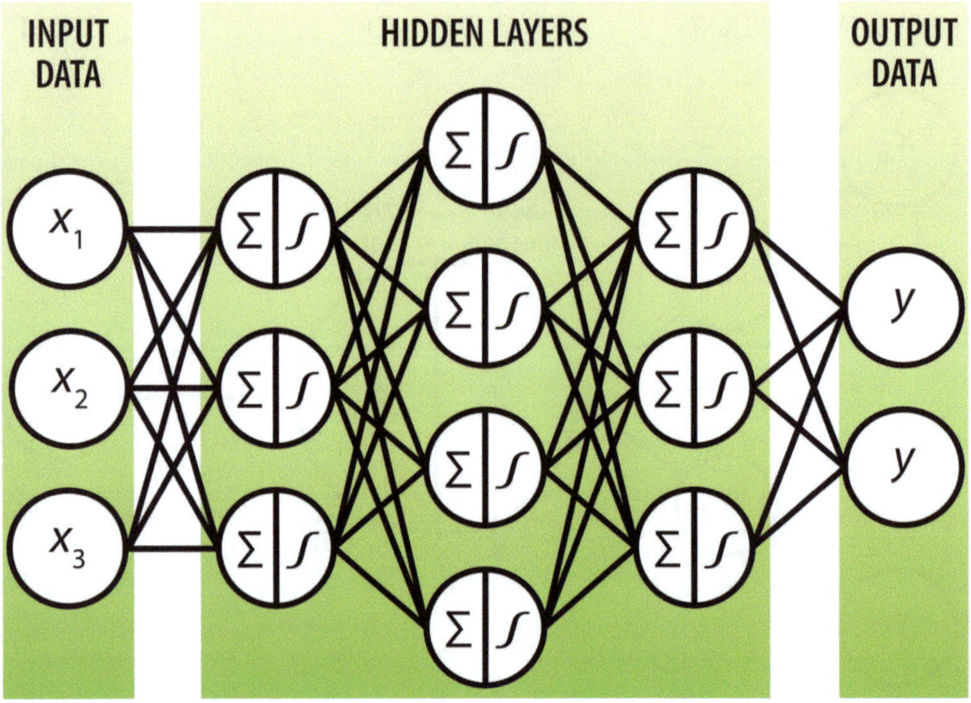

Fig. 4 Basic structure of a shallow neural network with three hidden layers. For clarity, the weights are not labeled in the scheme

problem in multiple dimensions. The activation function introduces nonlinearity into the ANN and its value reflects the activation level of each neuron. Frequently used functions are shown in Fig. 3. Different activation functions can be used in different layers, depending on the defined problem to be solved [63].

A shallow neural network is built with several hidden layers, but each with multiple hidden nodes (Fig. 4). For a simple feedforward SNN with a single hidden layer, after multiplying the input data by different weights (corresponding to different relationships in the data, ways, and rules for initializing them have also been developed) from each hidden node, summing and calculating the activation function, the weighted results from each hidden node are summed, and the final activation function for the output node is determined. This process is called forward propagation, and it also corresponds to the principle of using the already trained ANN. In the model learning phase, additional backpropagation is used [64], involving the calculation of the cost of a given observation (e.g., the difference between the predictive value and the true value in the case of supervised learning, or the mean squared error) and the adjustment of the weights for each hidden node to gradually minimize the cost, e.g., by the simple gradient method with the determination of the learning rate, which determines the rate of modification of the weights. The training phase ends when a predefined termination condition is reached, which can be the number of iterations [63].

Fariselli et al. [65] designed a tool based on a trained feedforward SNN to detect the surface of interacting residues in proteins based on their 3D structure. The SNN was trained with a standard backpropagation algorithm, and the final architecture was selected after initial tests consisting of four nodes in the hidden layer and one neuron in the output layer, representing contact or lack thereof. In turn, it was powered by a 11 amino acid residues wide window, centered on the residue to be predicted at any given time. In turn, each residue was encoded as a 20-element vector. A cross-validation procedure on 226 protein heterodimers from the selected dataset showed a model accuracy of 73%.

Ofran and Rost [66] also proposed a method using feedforward SNN with back-propagation to predict the site of protein–protein interactions, but on a sequence basis. Furthermore, they focused on transient interactions, and the windows contained nine residues. The neural network architecture had 189 input units, one hidden layer with 300 nodes, and two output units in the output layer. Despite the unsatisfactory raw prediction results, after filtering the method, it already performed much better than random, and after calibrating the system to the point of its strongest predictions, 94% of the predicted interaction sites were already correct.

An example of a more advanced and novel sequence-based system is ProNA2020 [67]. It was designed to comprehensively identify protein structures that bind to other proteins, DNA or RNA, and predict the residues involved in binding. Furthermore, the ProtVec [68] method from NLP was incorporated, and homology-based inference was combined with ML algorithms (SVM, RF, and SNN were tested). The system was subjected to tests and comparisons with other contemporary methods, which showed that ProNA2020 generally outperformed, was on par, or at least was not significantly inferior to the other models. For this reason, it can be used for high-throughput analyses, such as whole-proteome level comparisons between organisms [67].

2.3 Deep Learning Methods in Protein–Protein Docking

Deep learning algorithms enable computational models to learn representations of multidimensional data through abstraction, primarily using the power of ANN [69]. Higher-level (more abstract) features are defined in terms of lower-level (less abstract) features [70]. The emergence of deep neural networks (DNNs) has revolutionized the analysis of large datasets. Considerable value was brought about by using ReLu activation functions to avoid the vanishing gradient problems found in SNN [71]. DNNs are usually built with a much larger number of layers (especially hidden layers) than SNNs, which contributes to their versatility, but at the same time reinforces the problem of so-called black boxes. However, as the number of hidden layers increases, their ability to identify more features increases at the same time. Therefore, it is always worth considering the balance of benefits and losses before using a particular method, and the design of ANN itself is therefore an experimental process, requiring a lot of trial and error in terms of changes to the overall architecture and optimization of hyperparameters. The larger hardware requirements, the collection and preparation of large datasets, and the longer time required to learn the model are the main limitations. It is worth starting work by trying simpler ML algorithms, which may prove sufficient. However, for complex problems containing unstructured data, where relationships between data cannot be easily detected, DL techniques are undoubtedly an ideal solution.

In the field of DL techniques, there are many variants of ANN that have also found their way into the PPI field. The most common are the following: Convolutional Neural Networks (CNNs or ConvNets), Recurrent Neural Networks (RNNs, involving Long Short-Term Memory networks—LSTM), Generative networks (e.g., Generative Adversarial Networks—GAN), and Geometric Deep Learning (GDL). DL algorithms can additionally incorporate other ML methods to train a given model. Such hybrid approaches have already been successfully introduced, for example, for de novo protein design [71–73]. However, it is worth keeping track of the latest reports and innovative approaches, as this chapter has been limited to listing the best-known algorithms with single examples. AI, on the other hand, is currently the area with the most dynamic development, and interesting examples of its use in the field of PPI prediction can be found much more, such as DHL-PPI (Deep hash learning protein to protein interaction) [74], or the generative neural network trained on protein structures produced by the molecular simulation to generate new probable protein conformations, useful in protein–protein docking [75].

2.3.1 Convolutional Neural Networks

The Convolutional Neural Network (ConvNet, CNN; not to be confused with the broadcasting station) was inspired by neuroscience, more specifically the organization of the visual cortex, and was originally developed for image and/or object detection. Today, the approach is not limited exclusively to images but is also the basis of speech recognition, video analysis, text mining, and other fields. CNNs demonstrate the ability to learn contextual features and thus overcome the problems of pixel-based classification. Also due to the reduction of the number of required parameters, a greater generalization and reduced requirements for the computing power to process all the data are needed.

The structure of ConvNets is essentially based on four main components: convolution layer, pooling layer, activation function, and fully connected layer. The convolution layer is responsible for identifying features (both low- and high-level) using filters, the pooling layer reduces the spatial dimension of feature maps, and the fully connected layer integrates information from each layer to ultimately generate the output.

CNNs are a very promising prospect in the medical imaging field, including computer-aided diagnosis (CAD) [76, 77], prediction of the toxicity of chemical compounds, virtual screening, and prediction of protein–ligand interactions in the design of new drugs, among others [78–81]. ConvNets has also been used, for example, in an end-to-end model called DeepPPISP [82] to predict PPI sites. The premise of DeepPPISP was to extract both local contextual features and global features from protein sequences and integrate them using deep learning techniques to perform the classification task. Instead of a standard CNN, text-based CNNs (TextCNN) were implemented, which are able to capture global features of the amino acid sequence. The results showed that DeepPPISP outperformed other popular PPI site prediction methods, but also had its limitations, such as a relatively long running time and being not optimal for proteins with long sequences. However, another predictor, DeepPPISP-XGB [83], was developed on it, in which DeepPPISP was used as an amino acid sequence encoder only for feature extraction alone, while the XGBoost method was also implemented to predict binding sites. The DeepPPISP-XGB model also achieved competitive results against other methods. Moreover, the authors confirmed the beneficial effect of adding global features on model performance.

Another model, DeepTrio [84], is used to predict PPI using a masked multi-scale CNN architecture based on protein sequences. It takes a pair of sequences as input and generates a three-dimensional output vector indicating the probability of interaction, lack of interaction, and a single protein for further analysis. In addition to identifying potential interactions, it is also possible to study the effect of each amino acid residue on PPI by creating "importance maps" for each. Experimental evaluations have shown that the model achieves better performance than several other state-of-the-art methods in terms of various quality metrics. DeepTrio is available in a user-friendly form as an online tool [85].

CNN-FSRF [86] was also developed to predict potential PPIs based on protein sequences, but it is based on a hybrid method: it integrates the CNN with the Feature-Selective Rotation Forest (FSRF) classifier. In the CNN-FSRF model, amino acid sequences are transformed into a Position-Specific Scoring Matrix (PSSM) with evolutionary information included, after which the CNN extracts deeply hidden protein features. Finally, the features are passed to the FSRF algorithm, which removes redundant noise information and produces the final prediction results. Like the other models described earlier, CNN-FSRF has been evaluated, the results of which suggest that it can be successfully used for PPI research.

2.3.2 Recurrent Neural Networks

RNNs are capable of accepting input data without a predetermined length and modeling information sequentially, which translates into their usefulness especially in everyday life when converting spoken language into written text or automatic translations between different spoken languages.

RNN uses algorithmic processing to transform the so-called seed entered into the system (which can be, for example, a SMILES string) into a reference vector, which is then used by the hidden layer to generate an output vector. The output vector, in turn, is further optimized through the input training datasets, forming the output data. These data are accumulated in the time-sensitive memory of the hidden layer, allowing it to store information about the relationships between sequences of input data. In addition, the weights of the hidden layers are modified over time using backpropagation [63, 87]. Long Short-Term Memory [88] units are primarily used to build RNN.

DLPred [89] is one example of a tool for predicting PPI binding sites based on sets of protein sequences that use a simplified long-term memory network (SLSTM). In addition, the predicted solvent-accessible surface is used to correct the prediction model's preference for non-interaction sites.

Another application of this type of network can be the efficient extraction of information from databases containing biomedical texts. The Shortest Dependency Path B-LSTM (sdpLSTM) [90] model relies on the technique of deep bidirectional long-term memory (B-LSTM) to identify PPI information from word sequences and data associated with dependency paths. Experiments show that such a model can also be successfully applied in practice due to its relatively high performance.

2.3.3 Generative Adversarial Networks

In the case of GANs, a key role is played by the competition between two networks—the generator and discriminator networks. The generator is responsible for generating a potential solution to the problem, while the discriminator is tasked with evaluating its correctness against the actual solution based on information from the training set. The calculated cost is fed back to the network to optimize the prediction and achieve higher performance. An important advantage of the GAN network is its ability to create complete information from incomplete data, such as generating a 3D image from 2D.

Balogh et al. [91] developed a tool consisting of two algorithms: a first-order search to find and extract induced subgraphs serving as image input and a conditional generative adversarial network (cGAN) to train the final model for the prediction of PPI linkage. cGAN takes as input a combined representation from preprocessed data. The undoubted advantage of this approach is that there is no requirement for molecular attributes from external databases—only raw topological features were used for prediction.

The authors created a representation of a set of proteins in the form of network nodes, while the interactions between them are presented as edges, which may provide an interesting perspective for future research.

2.3.4 Geometric Deep Learning

Geometric deep learning is an umbrella term for emerging techniques that generalize deep ANN models to Euclidean and non-Euclidean models, such as graphs, manifolds, grids, or string representations [92]. In general, GDL includes approaches that incorporate geometric priorities, i.e., structural data and symmetry properties of input variables, to improve the quality of information captured by the model. In addition, it provides new modeling capabilities, such as augmenting the dataset with insufficient data for DL models [93]. Thus, GDL can be considered a successful extension of image-based deep neural network architectures to geometric data [94]. The use of known symmetries of large systems is a classic remedy to the dimensionality problem and is the basis of most physical theories. DL techniques are no exception; researchers have been trying to adapt ANN to take advantage of low-dimensional geometry derived from physical measurements, such as position and momentum in molecules and their associated symmetries, such as translation or rotation [95].

In chemoinformatics, GDL makes it possible to experiment with different molecular representations of the same molecule and use their intrinsic geometric features to enhance model quality. It is now an important alternative to classical methods, which rely primarily on molecular descriptors. Molecular graphs have proven to be particularly useful in predicting properties, due to their human interpretability and the ease of incorporating desired edge and node features [93].

The concept of the Graph Neural Network (GNN) algorithm was based on the mechanism of information diffusion. The input data are processed by a collection of interconnected units, each of which corresponds to a graph node. The units update their states, based on the information received from their "neighbors," until an equilibrium state is reached [96]. GNNs have already been applied, for example, to the conformational de novo design of realistic 3D molecules [97].

The GNN-DOVE [98] model, the successor to DOVE [99], used GNN to increase the accuracy of the evaluations of generated models of protein complexes. DOVE was developed based on 3D CNN and although it performed better than other existing methods, it also had its critical limitations, which often resulted in erroneous predictions. The use of GNN made it possible to overcome them, and as a result, in the tests conducted, the GNN-DOVE model outperformed the DOVE method in terms of performance.

Jha et al. [100] trained a model for PPI prediction using a graph convolutional network (GCN) and graph attention network (GAT) using structural information and sequence features. Protein graphs representing the amino acid network were constructed from their PDB format files, in which each node was a single residue. Two nodes were connected if they had a pair of atoms at a threshold distance. To extract features of the residues from the sequences, language models (SeqVec and ProtBert) were additionally used. The results obtained by the authors of this work confirmed the effectiveness of the proposed approach.

The MaSIF (molecular surface interaction fingerprinting) [94] model provides a general geometric method for deep learning of interaction fingerprints on molecular surfaces of proteins. Networks are trained end-to-end, so intermediate descriptors are not universal, but optimized for specific tasks. The authors have demonstrated MaSIF in three proof-of-concept applications: [1] ligand pocket similarity (MaSIF-ligand); [2] prediction of PPI sites on protein surfaces (MaSIF-site); and [3] ultrafast surface scanning, in which surface fingerprints are used to predict the structural configuration of protein–protein complexes (MaSIF-search). Applications of this approach range from finding interactions between nonhomologous proteins to identifying structural features important for protein function and design [94].

3 Conclusions

The study of protein–protein interactions is crucial to understanding the operation of biological systems, and AI techniques are a very promising method in this regard. However, despite the rapid development of technology and advances in the development of new approaches, many aspects remain challenging to solve, such as predicting conformational changes in proteins. Above all, high-quality data are needed, and in the case of deep learning techniques, additionally in huge amounts, which at this point in time, in the PPI field, is quite a limitation. There are also few integrated tools for directly generating quaternary structures based on the prediction of binding sites and potential interactions generated by trained models [101]. Machine learning methods, let alone deep learning, are a relatively new concept in this topic, so more time is needed to find an optimal solution that would further accelerate the development of biological sciences. However, the most promising method is certainly geometric deep learning, whose full potential has not yet been realized [93] due to the aforementioned limitations.

References

1. Barbey AK (2018) Network neuroscience theory of human intelligence. Trends Cogn Sci 22:8–20
2. Griffiths TL (2020) Understanding human intelligence through human limitations. Trends Cogn Sci 24:873–883
3. Zhaoping L (2020) Artificial and natural intelligence: from invention to discovery. Neuron 105:413–415
4. McCulloch WS, Pitts W (1943) A logical calculus of the ideas immanent in nervous activity. Bull Math Biophys 5(4):115–133
5. Turing AM (2009) Computing machinery and intelligence. In: Epstein R, Roberts G, Beber G (eds) Parsing the turing test. Springer, Dordrecht, pp 23–65
6. Rosenblatt F (1957) The perceptron—a perceiving and recognizing automaton. Cornell Aeronautical Laboratory, Ithaca
7. Samuel AL (1959) Some studies in machine learning using the game of checkers. IBM J Res Dev 3:210–229
8. Kaul V, Enslin S, Gross SA (2020) History of artificial intelligence in medicine. Gastrointest Endosc 92:807–812
9. Walsh I, Fishman D, Garcia-Gasulla D et al (2021) DOME: recommendations for supervised machine learning validation in biology. Nat Methods 18(10):1122–1127
10. Banko M, Brill E (2001) Scaling to very very large corpora for natural language disambiguation. In: ACL '01: proceedings of the 39th annual meeting on association for computational linguistics, pp 26–33
11. Halevy A, Norvig P, Pereira F (2009) The unreasonable effectiveness of data. IEEE Intell Syst 24:8–12
12. Brazas MD, Blackford S, Attwood TK (2017) Plug gap in essential bioinformatics skills. Nature 544(7649):161–161
13. Jones DT (2019) Setting the standards for machine learning in biology. Nat Rev Mol Cell Biol 2011(20):659–660
14. Choi RY, Coyner AS, Kalpathy-Cramer J et al (2020) Introduction to machine learning, neural networks, and deep learning. Transl Vis Sci Technol 9:14
15. Greener JG, Kandathil SM, Moffat L et al (2021) A guide to machine learning for biologists. Nat Rev Mol Cell Biol 23(1):40–55
16. Magnano CS, Mu F, Russ RS et al (2022) An approachable, flexible and practical machine learning workshop for biologists. Bioinformatics 38:i10
17. Adnan N, Umer F (2022) Understanding deep learning—challenges and prospects. J Pak Med Assoc 72:S59–S63
18. Géron A (2019) Hands-on machine learning with Scikit-Learn, Keras, and TensorFlow. O'Reilly Media, Inc, Sebastopol
19. Jumper J, Evans R, Pritzel A et al (2021) Highly accurate protein structure prediction with AlphaFold. Nature 2021:1–11
20. Callaway E (2022) What's next for AlphaFold and the AI protein-folding revolution. Nature 604:234–238
21. Bryant P, Pozzati G, Elofsson A (2022) Improved prediction of protein-protein interactions using AlphaFold2. Nat Commun 13(1):1–11
22. Kuzmanov U, Emili A (2013) Protein-protein interaction networks: probing disease mechanisms using model systems. Genome Med 5:1–12
23. Rao VS, Srinivas K, Sujini GN et al (2014) Protein-protein interaction detection: methods and analysis. Int J Proteomics 2014:1–12
24. Lensink MF, Brysbaert G, Mauri T et al (2021) Prediction of protein assemblies, the next frontier: the CASP14-CAPRI experiment. Proteins Struct Funct Bioinform 89: 1800–1823
25. Sunny S, Jayaraj PB (2021) Protein–protein docking: past, present, and future. Protein J 41(1):1–26
26. Casadio R, Martelli PL, Savojardo C (2022) Machine learning solutions for predicting protein–protein interactions. Wiley Interdiscip Rev Comput Mol Sci 2022:e1618
27. Kuksa P, Pavlovic V (2007) Fast kernel methods for SVM sequence classifiers. Lect Notes Comput Sci (including Subser Lect Notes Artif Intell Lect Notes Bioinformatics) 4645 (LNBI):228–239
28. Vert JP (2006) Kernel methods in genomics and computational biology. Kernel Methods Bioeng Signal Image Process 2006:42–63
29. Leslie C, Eskin E, Noble WS (2002) The spectrum kernel: a string kernel for SVM protein classification. Pac Symp Biocomput 2002:564–575
30. Romero-Molina S, Ruiz-Blanco YB, Harms M et al (2019) PPI-Detect: a support vector machine model for sequence-based prediction of protein-protein interactions. J Comput Chem 40:1233–1242
31. PPI-Detect. https://ppi-detect.zmb.uni-due.de/. Accessed 20 Jul 2022

32. Ruiz-Blanco YB, Paz W, Green J et al (2015) ProtDCal: a program to compute general-purpose-numerical descriptors for sequences and 3D-structures of proteins. BMC Bioinform 16:1–15
33. Romero-Molina S, Ruiz-Blanco YB, Green JR et al (2019) ProtDCal-Suite: a web server for the numerical codification and functional analysis of proteins. Protein Sci 28:1734
34. Mosca R, Céol A, Stein A et al (2014) 3did: a catalog of domain-based interactions of known three-dimensional structure. Nucleic Acids Res 42:D374
35. Finn RD, Miller BL, Clements J et al (2014) iPfam: a database of protein family and domain interactions found in the Protein Data Bank. Nucleic Acids Res 42:D364
36. Blohm P, Frishman G, Smialowski P et al (2014) Negatome 2.0: a database of non-interacting proteins derived by literature mining, manual annotation and protein structure analysis. Nucleic Acids Res 42:D396
37. Pitre S, Dehne F, Chan A et al (2006) PIPE: a protein-protein interaction prediction engine based on the re-occurring short polypeptide sequences between known interacting protein pairs. BMC Bioinform 7:1–15
38. Guo Y, Yu L, Wen Z et al (2008) Using support vector machine combined with auto covariance to predict protein–protein interactions from protein sequences. Nucleic Acids Res 36:3025–3030
39. Liu X, Liu B, Huang Z et al (2012) SPPS: a sequence-based method for predicting probability of protein-protein interaction partners. PLoS One 7:e30938
40. Romero-Molina S, Ruiz-Blanco YB, Mieres-Perez J et al (2022) PPI-affinity: a web tool for the prediction and optimization of protein-peptide and protein-protein binding affinity. J Proteome Res 21:1829
41. PPI-Affinity. https://protdcal.zmb.uni-due.de/PPIAffinity. Accessed 20 July 2022
42. Frank E, Hall MA, Witten IH (2016) The WEKA workbench data mining: practical machine learning tools and techniques. Morgan Kaufmann, 4th
43. Webb B, Sali A (2016) Comparative protein structure modeling using MODELLER. Curr Protoc Bioinformatics 2016:5.6.1–5.6.37
44. Martí-Renom MA, Stuart AC, Fiser A et al (2000) Comparative protein structure modeling of genes and genomes. Annu Rev Biophys Biomol Struct 29:291–325
45. Šali A, Blundell TL (1993) Comparative protein modelling by satisfaction of spatial restraints. J Mol Biol 234:779–815
46. Fiser A, Do RKG, Šali A (2000) Modeling of loops in protein structures. Protein Sci 9: 1753–1773
47. Das S, Chakrabarti S (2021) Classification and prediction of protein–protein interaction interface using machine learning algorithm. Sci Rep 11(11):1–12
48. PCPIP | Protein Complex Prediction by Interface Properties. http://www.hpppi.iicb.res.in/pcpip/. Accessed 25 July 2022
49. Northey TC, Bareši A, Martin ACR (2018) IntPred: a structure-based predictor of protein–protein interaction sites. Bioinformatics 34:223
50. Porter CT, Martin ACR (2015) BiopLib and BiopTools—a C programming library and toolset for manipulating protein structure. Bioinformatics 31:4017–4019
51. You ZH, Chan KCC, Hu P (2015) Predicting protein-protein interactions from primary protein sequences using a novel multi-scale local feature representation scheme and the random Forest. PLoS One 10:e0125811
52. Freund Y, Schapire RE (1997) A decision-theoretic generalization of on-line learning and an application to boosting. J Comput Syst Sci 55:119–139
53. Friedman JH (2001) Greedy function approximation: a gradient boosting machine. Ann Stat 29:1189–1232
54. Li K, Zhou G, Zhai J et al (2019) Improved PSO_AdaBoost ensemble algorithm for imbalanced data. Sensors (Basel) 19:1476
55. XGBoost Documentation—xgboost 1.6.2 documentation. https://xgboost.readthedocs.io/en/stable/. Accessed 16 Aug 2022
56. CatBoost—open-source gradient boosting library. https://catboost.ai/. Accessed 28 Jul 2022
57. Welcome to LightGBM's documentation!—LightGBM 3.3.2 documentation. https://lightgbm.readthedocs.io/en/v3.3.2/. Accessed 16 Aug 2022
58. Sanchez-Garcia R, Sorzano COS, Carazo JM et al (2019) BIPSPI: a method for the prediction of partner-specific protein–protein interfaces. Bioinformatics 35:470
59. xgbPredApp. https://biocomp.cnb.csic.es/bipspi1/. Accessed 16 Aug 2022
60. Sanchez-Garcia R, Macias JR, Sorzano COS et al (2022) BIPSPI+: mining type-specific datasets of protein complexes to improve protein binding site prediction. J Mol Biol 434: 167556

61. xgbPredApp. https://bipspi.cnb.csic.es/. Accessed 16 Aug 2022
62. Schneidman-Duhovny D, Inbar Y, Nussinov R et al (2005) PatchDock and SymmDock: servers for rigid and symmetric docking. Nucleic Acids Res 33:W363
63. Hurbans R (2020) Grokking artificial intelligence algorithms. Manning Publications Co., Shelter Island
64. Rumelhart DE, Hinton GE, Williams RJ (1986) Learning representations by back-propagating errors. Nature 323(6088): 533–536
65. Fariselli P, Pazos F, Valencia A et al (2002) Prediction of protein–protein interaction sites in heterocomplexes with neural networks. Eur J Biochem 269:1356–1361
66. Ofran Y, Rost B (2003) Predicted protein–protein interaction sites from local sequence information. FEBS Lett 544:236–239
67. Qiu J, Bernhofer M, Heinzinger M et al (2020) ProNA2020 predicts protein-DNA, protein-RNA, and protein-protein binding proteins and residues from sequence. J Mol Biol 432:2428–2443
68. Asgari E, Mofrad MRK (2015) Continuous distributed representation of biological sequences for deep proteomics and genomics. PLoS One 10:e0141287
69. Lecun Y, Bengio Y, Hinton G (2015) Deep learning. Nature 521(7553):436–444
70. Najafabadi MM, Villanustre F, Khoshgoftaar TM et al (2015) Deep learning applications and challenges in big data analytics. J Big Data 2:1–21
71. Tripathi MK, Nath A, Singh TP et al (2021) Evolving scenario of big data and Artificial Intelligence (AI) in drug discovery. Mol Divers 25:1439–1460
72. Sabban S, Markovsky M, Valdenegro-Toro M et al (2020) RamaNet: computational *de novo* helical protein backbone design using a long short-term memory generative neural network. F1000Research 9:298
73. Karimi M, Zhu S, Cao Y et al (2020) De novo protein design for novel folds using guided conditional wasserstein generative adversarial networks. J Chem Inf Model 60:5667–5681
74. Jiang Y, Wang Y, Shen L et al (2022) Identification of all-against-all protein–protein interactions based on deep hash learning. BMC Bioinform 23:266
75. Degiacomi MT (2019) Coupling molecular dynamics and deep learning to mine protein conformational space. Structure 27:1034–1040.e3
76. Tajbakhsh N, Shin JY, Gurudu SR et al (2016) Convolutional neural networks for medical image analysis: full training or fine tuning? IEEE Trans Med Imaging 35:1299–1312
77. Chougrad H, Zouaki H, Alheyane O (2018) Deep convolutional neural networks for breast cancer screening. Comput Methods Prog Biomed 157:19–30
78. Pereira JC, Caffarena ER, Dos SCN (2016) Boosting docking-based virtual screening with deep learning. J Chem Inf Model 56: 2495–2506
79. Ragoza M, Hochuli J, Idrobo E et al (2017) Protein-ligand scoring with convolutional neural networks. J Chem Inf Model 57:942
80. Stepniewska-Dziubinska MM, Zielenkiewicz P, Siedlecki P (2018) Development and evaluation of a deep learning model for protein–ligand binding affinity prediction. Bioinformatics 34:3666
81. Fernandez M, Ban F, Woo G et al (2018) Toxic colors: the use of deep learning for predicting toxicity of compounds merely from their graphic images. J Chem Inf Model 58: 1533–1543
82. Zeng M, Zhang F, Wu FX et al (2020) Protein–protein interaction site prediction through combining local and global features with deep neural networks. Bioinformatics 36:1114–1120
83. Wang P, Zhang G, Yu ZG et al (2021) A deep learning and XGBoost-based method for predicting protein-protein interaction sites. Front Genet 12:752732
84. Hu X, Feng C, Zhou Y et al (2022) DeepTrio: a ternary prediction system for protein–protein interaction using mask multiple parallel convolutional neural networks. Bioinformatics 38:694
85. No Title. http://bis.zju.edu.cn/deeptrio/. Accessed 17 Aug 2022
86. Wang L, Wang HF, Liu SR et al (2019) Predicting protein-protein interactions from matrix-based protein sequence using convolution neural network and feature-selective rotation forest. Sci Rep 9:9848
87. Patel L, Shukla T, Huang X et al (2020) Machine learning methods in drug discovery. Molecules 25(22):5277
88. Hochreiter S, Schmidhuber J (1997) Long short-term memory. Neural Comput 9: 1735–1780
89. Zhang B, Li J, Quan L et al (2019) Sequence-based prediction of protein-protein interaction sites by simplified long short-term memory network. Neurocomputing 357:86–100

90. Yadav S, Kumar A, Ekbal A et al (2018) Feature Assisted bi-directional LSTM model for protein-protein interaction identification from biomedical texts. arXiv 2018, arXiv: abs/1807.02162
91. Balogh OM, Benczik B, Horváth A et al (2022) Efficient link prediction in the protein–protein interaction network using topological information in a generative adversarial network machine learning model. BMC Bioinform 23:1–19
92. Bronstein MM, Bruna J, Lecun Y et al (2016) Geometric deep learning: going beyond Euclidean data. IEEE Signal Process Mag 34:18–42
93. Atz K, Grisoni F, Schneider G (2021) Geometric deep learning on molecular representations. Nat Mach Intell 3(12):1023–1032
94. Gainza P, Sverrisson F, Monti F et al (2020) Deciphering interaction fingerprints from protein molecular surfaces using geometric deep learning. Nat Methods 17:184–192
95. Bronstein MM, Bruna J, Cohen T et al (2021) Geometric deep learning: grids, groups, graphs, geodesics, and gauges
96. Scarselli F, Gori M, Tsoi AC et al (2009) The graph neural network model. IEEE Trans Neural Netw 20:61–80
97. Satorras VG, Hoogeboom E, Fuchs FB et al (2021) E(n) equivariant normalizing flows. Adv Neural Inf Proces Syst 6:4181–4192
98. Wang X, Flannery ST, Kihara D (2021) Protein docking model evaluation by graph neural networks. Front Mol Biosci 8:647915
99. Wang X, Terashi G, Christoffer CW et al (2020) Protein docking model evaluation by 3D deep convolutional neural networks. Bioinformatics 36:2113
100. Jha K, Saha S, Singh H (2022) Prediction of protein–protein interaction using graph neural networks. Sci Rep 12:8360
101. Quadir F, Roy RS, Soltanikazemi E et al (2021) DeepComplex: a web server of predicting protein complex structures by deep learning inter-chain contact prediction and distance-based modelling. Front Mol Biosci 8:716973

Part III

Protein–Protein Docking Tools

Chapter 8

Software and Databases for Protein–Protein Docking

Małgorzata Jarończyk, Ruben Abagyan, and Maxim Totrov

Abstract

Protein–protein interactions (PPIs) provide valuable insights for understanding the principles of biological systems and for elucidating causes of incurable diseases. One of the techniques used for computational prediction of PPIs is protein–protein docking calculations, and a variety of software has been developed. This chapter is a summary of software and databases used for protein–protein docking.

Key words Protein–protein docking, Protein–protein interactions, PPI, Protein–protein interface, Fast Fourier transform, Protein bioinformatics, Protein databases, Monte Carlo

1 Introduction

Protein–protein interactions (PPIs) play a crucial role in cellular functions and biological processes, and their study is critical to drug discovery and bioengineering efforts [1–3]. Historically, X-ray crystallography, isothermal titration calorimetry, and other biophysical methods have been used to study PPIs, but these techniques were expensive and time consuming and covered only a small portion of the complete PPI networks. Therefore, computational tools, such as protein–protein docking, which predict the binding mode and estimate binding free energy between individual protein structures, are needed to complement the experimental methods for the identification of protein–protein interactions and determination of their complex structures. The complete analysis of PPIs can enable a better understanding of cellular organization, processes, and functional states and even help to design modulators that target some of those PPIs. Despite so many efforts made to solve the problem completely, a large number of difficult cases, including those with

The original version of the chapter has been revised. A correction to this chapter can be found at https://doi.org/10.1007/978-1-0716-3985-6_19

Agnieszka A. Kaczor (ed.), *Protein-Protein Docking: Methods and Protocols*, Methods in Molecular Biology, vol. 2780, https://doi.org/10.1007/978-1-0716-3985-6_8,
© The Author(s), under exclusive license to Springer Science+Business Media, LLC, part of Springer Nature 2024, Corrected Publication 2025

large size or complex composition, dependence on large-scale conformational rearrangements, initially flexible regions, ligands, metals, or post-translational modifications, remain challenging or unsolved [4].

Protein–protein interactions (PPIs) occur between identical or non-identical chains (i.e., homo- or hetero-oligomers) [5]. In addition, based on the interaction stability, PPIs are classified into obligate and non-obligate ones. Obligate interactions are usually considered to be permanent, while non-obligate interactions can be either permanent or transient [5, 6].

In 2001, Critical Assessment of Predicted Interactions (CAPRI) (http://pdbe.org/capri/; http://www.capri-docking.org/) community-wide initiative was established with the mission of offering computational biologists the opportunity to test their algorithms in blind predictions of experimentally determined 3D structures of protein complexes [7]. Among different protein–protein docking programs, the ICM-Pro (Molsoft) has been used successfully to predict complex structures in various rounds of CAPRI [8, 9].

Accumulating data on protein structures and interactions, improved energy functions, and powerful techniques to accelerate the sampling process provide the opportunity to elucidate protein–protein structural complexes via computational docking. There are two types of docking methodologies used for studying protein–protein interactions: (1) template-based docking, where similar complexes (templates) guide the docking pose prediction; (2) template-free, where sampling of the binding modes is performed with no prior knowledge of similar experimentally determined structures of the complex, and template-based or comparative.

Protein–protein docking process is divided into two steps: sampling and scoring, and refinement of the top-scoring solutions. The essential part of the docking protocol is to sample potential protein–protein binding orientations using global or local searches. In a global search, one protein acts as the receptor and the other as the ligand moves around the receptor, exploring all potential orientations between protein partners in 3D space. Several docking algorithms have been developed; e.g., the fast Fast-Fourier transform (FFT)-based algorithms [10] are implemented to reduce docking complexity. There are two main protein–protein docking problems: the efficient sampling of the conformational and orientation space in search of near-native structures (sampling), and the identification of such near-native structures among the many models generated (scoring). Rigid-body docking is generally faster than flexible docking [11]. Still, flexible docking (that allows intra-protein conformational degrees of freedom) better reflects the dynamic nature of the proteins. Limitations of these methods are implicit in the necessity to generate large amounts of the docking models (usually on the order of hundreds of thousands) to have a significant chance of generating at least one near-native docking model. Another common strategy is reducing the number of considered docking poses by performing clustering and only applying a scoring function to the cluster representatives. In most of the cases, the applicability of a scoring function is strongly dependent on the sampling approaches used [12].

In the biomedical and translational medicine fields, the study of PPI networks and protein complexes may yield practical advances in the understanding of the molecular basis of disease processes, which in turn can lead to improvements in diagnostics and therapeutics.

2 Materials

Proteins and their complex structures can be extracted from available resources collecting structural information obtained by X-ray crystallography, NMR, and electron microscopy of proteins such as Protein Data Bank (PDB) [13]. If no experimentally solved structures are available, homology modeling methods are applied. PPI databases collect and store data on known interactions between proteins (Table 1) [14]. The data are often gathered either through direct user submissions or by surveys in publications or in the literature that describe PPIs [15]. Several protein–ligand docking software applications and online web services for protein–protein docking are available (Table 2).

3 Methods

3.1 Protein Preparation

As a first step in the docking protein–protein procedure, the structures of interacting entities are required. As most docking algorithms rely on atom-level representations of the structures, X-ray crystal structures of the proteins are most useful, as they can provide single structures (versus multiple possible models) at high resolution (<2.5 Å). A protein structure with co-factors, metals, and other co-crystal molecules (excluding crystallographic waters) is imported from the PDB database into ICM and converted to a model ready for energy calculations. Amino acid protonation states are assigned, and partial charges of protein atoms are taken from a library of Empirical Conformational Energy Program for Peptides (ECEPP/3) residue templates [16]. When the crystal structure of a protein is not available, the ICM homology modeling approach is applied [17]. In the first steps, a starting model is created based on the homologous structure with the conserved portion fixed and the non-conserved portion having standard covalent geometry and free torsion angles. In the second step, the Biased Probability Monte Carlo (BPMC) sampling and global energy optimization procedure is applied to search the subspaces of either all the non-conservative side-chain torsion angles or torsion angles in a loop backbone and surrounding side chains.

Table 1
List of protein–protein interaction databases

	PPI database	Website	References
BioGRID	The Biological General Repository for Interaction Datasets	http://www.thebiogrid.org	[28]
BIND	Biomolecular Interaction Network Database	http://bond.unleashedinformatics.com	[29]
DIP	Database of Interacting Proteins	http://dip.doe-mbi.ucla.edu	[30]
HPRD	Human Protein Reference Database	http://www.hprd.org	[31]
IntAct	IntAct molecular interaction database	http://www.ebi.ac.uk/intact	[32]
InterProSurf	InterProSurf	http://curie.utmb.edu	[33]
MINT	Molecular Interaction Database	http://mint.bio.uniroma2.it/mint	[34]
MIPS	Mammalian Protein–Protein Interaction Database	https://mips.helmholt	[35]
MPPI	A Database Extension to Visualize Structural Interactome in One-to-many Manner	http://bis.zju.edu.cn/mppi	[36]
PPI-Hotspot	Database of Protein–Protein Interaction Hot Spots	https://ppihotspot.limlab.dnsalias.org	[37]
SpotOn	Hot SPOTs ON protein complexes	http://milou.science.uu.nl/services/SPOTON	[38]
STRING	Functional Protein Association Networks	https://string-db.org	[39]

3.2 Protein–Protein Site Prediction

Proteins interact with one another via certain interfaces; hence, correct identification of PPI binding sites may be critical for accurate docking. Most PPI site predictors fall into two major categories: residue- and patch-based methods. Residue-based techniques assign a score to each residue in the target protein based on its likelihood of being part of the interface [18]. The results of such methods frequently include raw interface/non-interface scores calculated for all residues in the target protein, as well as a separate list of predicted interface residues with scores greater than a predefined threshold. Patch-based methods divide a target protein surface into discrete patches/clusters [19]. These surface patches are then analyzed and ranked using a combined score calculated from individual features, with the top-ranked group selected as the predicted interface. Patch-based methods frequently produce a confidence score for the entire cluster of residues in addition to interface/non-interface scores assigned to

Table 2
List of protein–protein docking software or servers

Software or server	Description	References
AutoDock	https://autodock.scripps.edu	[40]
ATTRACT	http://www.attract.ph.tum.de/services/ATTRACT	[41]
ClusPro	https://cluspro.org	[42]
DockRosetta	https://rosie.rosettacommons.org/docking	[43]
DOT	https://www.sdsc.edu/CCMS/DOT	[44]
FroDock	https://frodock.iqfr.csic.es	[45]
FTDock	http://www.sbg.bio.ic.ac.uk/docking/ftdock.html	[46]
GRAMM	http://vakser.bioinformatics.ku.edu/resources/gramm/grammx	[47]
HADDOCK	http://hdock.phys.hust.edu.cn	[48]
HEX	http://hex.loria.fr/	[49]
ICM-Pro	https://www.molsoft.com	[8]
LightDock	https://life.bsc.es/pid/lightdock	[50]
LZerD	https://lzerd.kiharalab.org	[51]
MEGADOCK	http://www.bi.cs.titech.ac.jp/megadock	[52]
MolFit	http://www.weizmann.ac.il/Chemical_Research_Support//molfit/home.html	[53]
PatchDock	http://bioinfo3d.cs.tau.ac.il	[54]
PIPER	https://www.schrodinger.com/products/piper	[55]
pyDock	https://life.bsc.es/pid/pydock	[56]
UNRES	http://unres-server.chem.ug.edu.pl	[57]
UDOCK	http://udock.fr	[58]
ZDOCK	https://zdock.umassmed.edu	[59]

individual residues. Many patch-based strategies have a flaw in that the predicted patches are generally circular, whereas biological interfaces are more irregular in shape. Furthermore, these methods require estimating the size of a putative interaction site.

The Optimal Docking Area (ODA) method [20] (http://www.molsoft.com/oda) implemented in the ICM is a useful way of predicting likely protein–protein interaction interfaces. The method generates protein surface patches of different sizes and analyzes energy required for "de-solvation" (a part of the protein binding event) of those patches, based on atomic solvation parameters previously derived from octanol/water transfer experiments and adjusted for protein–protein binding. These areas of low docking surface energy would correspond to regions likely to be buried in the interaction with other proteins and the patches correlate well with known protein interaction sites in transient hetero-complexes.

3.3 Protein–Protein Docking

Protein–protein docking is the method that in the initial stage treats the proteins as rigid bodies, allowing for fast and efficient search of the six-dimensional (6D) space (3D of translational freedom and 3D of rotational freedom). The 6D space is searched for regions of high shape and biochemical complementarity, using a "soft" scoring function that allows for some clashes between atoms. One of the most popular search techniques for low-energy conformations is Fast Fourier transform [10]; other methods include geometric hashing [21] or Monte Carlo sampling [22]. The protein–protein docking methodology used a combination of multiple starting conformations covering the interaction space of interest, followed by a BPMC procedure with pseudo-Brownian random as implemented in the ICM [7, 8, 23]. An alternative FFT-based search procedure was also implemented in ICM. Minima of the potential that evaluates simplified steric fit and solvation terms are identified via FFT-accelerated grid convolutions [24].

The next step is the ranking of docked poses using a scoring function, for which many methods have been developed [25]. Scoring functions can be classified into three different types: force-field-based, knowledge-based, and empirical-based scoring functions [26]. Force-field-based scoring functions use non-bonded terms of classical mechanics force fields. Knowledge-based scoring functions are based on statistical observations of intermolecular contacts identified from structure databases. Empirical scoring functions make use of several intermolecular interaction terms that are calibrated through a linear regression procedure, where theoretical values are fitted to be closest as possible to experimental data.

The ICM Score is calculated as the weighted sum of scores describing the energy terms evaluated during docking simulations [27]. Grid score is the score calculated as the sum of internal torsion and van der Waals interaction energy for the obtained ligand conformation and the interaction energy with the five grid potential maps subtracted with the energy for the free ligand in aqueous solution. The ligand is subsequently combined with a full atom receptor model for electrostatic calculations. As a result, the hydrogen bond score is calculated. The electrostatic score is calculated as the difference between the electrostatic interaction energy evaluated by applying the boundary surface method and the electrostatic interaction energy of the free ligand in the solution.

3.4 Protein–Protein Docking Refinement

The conformations from the first rigid body step are further optimized by an ICM global optimization algorithm [23], with flexible interface ligand side-chains and a grid map representation of the protein [8]. In this refinement step, the internal energy for the ligand interface side-chains is also considered, including the van der Waals, hydrogen bonding, and torsion energy calculated with ECEPP/3 parameters and the Coulomb electrostatic energy.

4 Summary

The study of protein–protein interactions is essential in understanding biological processes and human diseases as well as developing new therapeutic agents. A number of computational tools and protein interaction databases have been developed. In the near future, the challenging task will be closer integration of docking algorithms with protein interface prediction software, structural databases, and sequence analysis techniques to produce better predictions of PPI networks.

Acknowledgments

This work was supported by Polish Ministry of Science and Higher Education (financial support for statutory activity of National Medicines Institute).

References

1. Bonetta L (2010) Interactome under construction. Nature 468:851–852. https://doi.org/10.1038/468851a
2. Kuzmanov U, Emili A (2013) Protein-protein interaction networks: probing disease mechanisms using model systems. Genome Med 5:37. https://doi.org/10.1186/gm441
3. Kaczor AA, Bartuzi D, Stępniewski TM et al (2018) Protein–protein docking in drug design and discovery. Methods Mol Biol 1762:285–305
4. Wodak SJ, Vlasblom J, Turinsky AL, Pu S (2013) Protein–protein interaction networks: the puzzling riches. Curr Opin Struct Biol 23:941–953. https://doi.org/10.1016/j.sbi.2013.08.002
5. Nooren IMA (2003) NEW EMBO MEMBER'S REVIEW: diversity of protein-protein interactions. EMBO J 22:3486–3492. https://doi.org/10.1093/emboj/cdg359
6. Janin J, Bahadur RP, Chakrabarti P (2008) Protein–protein interaction and quaternary structure. Q Rev Biophys 41:133–180. https://doi.org/10.1017/S0033583508004708
7. Janin J, Henrick K, Moult J et al (2003) CAPRI: a critical assessment of PRedicted interactions. Proteins Struct Funct Genet 52:2–9. https://doi.org/10.1002/prot.10381
8. Fernández-Recio J, Totrov M, Abagyan R (2003) ICM-DISCO docking by global energy optimization with fully flexible side-chains. Proteins Struct Funct Bioinform 52:113–117. https://doi.org/10.1002/prot.10383
9. Fernández-Recio J, Abagyan R, Totrov M (2005) Improving CAPRI predictions: optimized desolvation for rigid-body docking. Proteins Struct Funct Bioinform 60:308–313. https://doi.org/10.1002/prot.20575
10. Katchalski-Katzir E, Shariv I, Eisenstein M et al (1992) Molecular surface recognition: determination of geometric fit between proteins and their ligands by correlation techniques.

Proc Natl Acad Sci 89:2195–2199. https://doi.org/10.1073/pnas.89.6.2195

11. Desta IT, Porter KA, Xia B et al (2020) Performance and its limits in rigid body protein-protein docking. Structure 28:1071–1081.e3. https://doi.org/10.1016/j.str.2020.06.006
12. Pozzati G, Kundrotas P, Elofsson A (2022) Scoring of protein–protein docking models utilizing predicted interface residues. Proteins Struct Funct Bioinform 90:1493–1505. https://doi.org/10.1002/prot.26330
13. Bernstein FC, Koetzle TF, Williams GJB et al (1977) The protein data bank: a computer-based archival file for macromolecular structures. J Mol Biol 112:535–542. https://doi.org/10.1016/S0022-2836(77)80200-3
14. Gonzalez O (2013) Protein–protein interaction databases. In: Encyclopedia of systems biology. Springer, New York, pp 1786–1790
15. Stacey RG, Skinnider MA, Chik JHL, Foster LJ (2018) Context-specific interactions in literature-curated protein interaction databases. BMC Genomics 19:758. https://doi.org/10.1186/s12864-018-5139-2
16. Nemethy G, Gibson KD, Palmer KA et al (1992) Energy parameters in polypeptides. 10. Improved geometrical parameters and non-bonded interactions for use in the ECEPP/3 algorithm, with application to proline-containing peptides. J Phys Chem 96:6472–6484. https://doi.org/10.1021/j100194a068
17. Cardozo T, Totrov M, Abagyan R (1995) Homology modeling by the ICM method. Proteins Struct Funct Genet 23:403–414. https://doi.org/10.1002/prot.340230314
18. Ma B, Elkayam T, Wolfson H, Nussinov R (2003) Protein–protein interactions: structurally conserved residues distinguish between binding sites and exposed protein surfaces. Proc Natl Acad Sci 100:5772–5777. https://doi.org/10.1073/pnas.1030237100
19. Jones S, Thornton JM (1997) Prediction of protein-protein interaction sites using patch analysis 1 1Edited by G. von Heijne. J Mol Biol 272:133–143. https://doi.org/10.1006/jmbi.1997.1233
20. Fernandez-Recio J, Totrov M, Skorodumov C, Abagyan R (2004) Optimal docking area: a new method for predicting protein-protein interaction sites. Proteins Struct Funct Bioinform 58:134–143. https://doi.org/10.1002/prot.20285
21. Fischer D, Lin SL, Wolfson HL, Nussinov R (1995) A geometry-based suite of molecular-docking processes. J Mol Biol 248:459–477. https://doi.org/10.1016/S0022-2836(95)80063-8
22. Abagyan R, Totrov M, Kuznetsov D (1994) ICM?A new method for protein modeling and design: applications to docking and structure prediction from the distorted native conformation. J Comput Chem 15:488–506. https://doi.org/10.1002/jcc.540150503
23. Totrov M, Abagyan R (1994) Detailed ab initio prediction of lysozyme–antibody complex with 1.6 Å accuracy. Nat Struct Mol Biol 1:259–263. https://doi.org/10.1038/nsb0494-259
24. Arnautova YA, Abagyan R, Totrov M (2018) Protein-RNA docking using ICM. J Chem Theory Comput 14:4971–4984. https://doi.org/10.1021/acs.jctc.8b00293
25. Moal IH, Torchala M, Bates PA, Fernández-Recio J (2013) The scoring of poses in protein-protein docking: current capabilities and future directions. BMC Bioinform 14:286. https://doi.org/10.1186/1471-2105-14-286
26. Liu J, Wang R (2015) Classification of current scoring functions. J Chem Inf Model 55:475–482. https://doi.org/10.1021/ci500731a
27. Neves MAC, Totrov M, Abagyan R (2012) Docking and scoring with ICM: the benchmarking results and strategies for improvement. J Comput Aided Mol Des 26:675–686. https://doi.org/10.1007/s10822-012-9547-0
28. Oughtred R, Rust J, Chang C et al (2021) The BioGRID database: a comprehensive biomedical resource of curated protein, genetic, and chemical interactions. Protein Sci 30:187–200. https://doi.org/10.1002/pro.3978
29. Bader GD (2001) BIND—the biomolecular interaction network database. Nucleic Acids Res 29:242–245. https://doi.org/10.1093/nar/29.1.242
30. Xenarios I (2000) DIP: the database of interacting proteins. Nucleic Acids Res 28:289–291. https://doi.org/10.1093/nar/28.1.289
31. Hermjakob H (2004) IntAct: an open source molecular interaction database. Nucleic Acids Res 32:452D–455D. https://doi.org/10.1093/nar/gkh052
32. Orchard S, Ammari M, Aranda B et al (2014) The MIntAct project—IntAct as a common curation platform for 11 molecular interaction databases. Nucleic Acids Res 42:D358–D363. https://doi.org/10.1093/nar/gkt1115
33. Negi SS, Schein CH, Oezguen N et al (2007) InterProSurf: a web server for predicting interacting sites on protein surfaces. Bioinformatics 23:3397–3399. https://doi.org/10.1093/bioinformatics/btm474

34. Chatr-aryamontri A, Ceol A, Palazzi LM et al (2007) MINT: the molecular INTeraction database. Nucleic Acids Res 35:D572–D574. https://doi.org/10.1093/nar/gkl950
35. Pagel P, Kovac S, Oesterheld M et al (2005) The MIPS mammalian protein-protein interaction database. Bioinformatics 21:832–834. https://doi.org/10.1093/bioinformatics/bti115
36. Zhou Y, Chen H, Li S, Chen M (2021) mPPI: a database extension to visualize structural interactome in a one-to-many manner. Database 2021. https://doi.org/10.1093/database/baab036
37. Chen YC, Chen Y-H, Wright JD, Lim C (2022) PPI-hotspot DB: database of protein–protein interaction hot spots. J Chem Inf Model 62:1052–1060. https://doi.org/10.1021/acs.jcim.2c00025
38. Moreira IS, Koukos PI, Melo R et al (2017) SpotOn: high accuracy identification of protein-protein Interface hot-spots. Sci Rep 7: 8007. https://doi.org/10.1038/s41598-017-08321-2
39. Szklarczyk D, Gable AL, Nastou KC et al (2021) The STRING database in 2021: customizable protein–protein networks, and functional characterization of user-uploaded gene/measurement sets. Nucleic Acids Res 49: D605–D612. https://doi.org/10.1093/nar/gkaa1074
40. Morris GM, Huey R, Lindstrom W et al (2009) AutoDock4 and AutoDockTools4: automated docking with selective receptor flexibility. J Comput Chem 30:2785–2791. https://doi.org/10.1002/jcc.21256
41. Zacharias M (2003) Protein-protein docking with a reduced protein model accounting for side-chain flexibility. Protein Sci 12:1271–1282. https://doi.org/10.1110/ps.0239303
42. Kozakov D, Hall DR, Xia B et al (2017) The ClusPro web server for protein–protein docking. Nat Protoc 12:255–278. https://doi.org/10.1038/nprot.2016.169
43. Chaudhury S, Berrondo M, Weitzner BD et al (2011) Benchmarking and analysis of protein docking performance in Rosetta v3.2. PLoS One 6:e22477. https://doi.org/10.1371/journal.pone.0022477
44. Roberts VA, Thompson EE, Pique ME et al (2013) DOT2: macromolecular docking with improved biophysical models. J Comput Chem 34:1743–1758. https://doi.org/10.1002/jcc.23304
45. Ramírez-Aportela E, López-Blanco JR, Chacón P (2016) FRODOCK 2.0: fast protein–protein docking server. Bioinformatics 32: 2386–2388. https://doi.org/10.1093/bioinformatics/btw141
46. Gabb HA, Jackson RM, Sternberg MJE (1997) Modelling protein docking using shape complementarity, electrostatics and biochemical information 1 1Edited by J. Thornton. J Mol Biol 272:106–120. https://doi.org/10.1006/jmbi.1997.1203
47. Tovchigrechko A, Vakser IA (2006) GRAMM-X public web server for protein-protein docking. Nucleic Acids Res 34:W310–W314. https://doi.org/10.1093/nar/gkl206
48. Yan Y, Zhang D, Zhou P et al (2017) HDOCK: a web server for protein–protein and protein–DNA/RNA docking based on a hybrid strategy. Nucleic Acids Res 45:W365–W373. https://doi.org/10.1093/nar/gkx407
49. Macindoe G, Mavridis L, Venkatraman V et al (2010) HexServer: an FFT-based protein docking server powered by graphics processors. Nucleic Acids Res 38:W445–W449. https://doi.org/10.1093/nar/gkq311
50. Jiménez-García B, Roel-Touris J, Romero-Durana M et al (2018) LightDock: a new multi-scale approach to protein–protein docking. Bioinformatics 34:49–55. https://doi.org/10.1093/bioinformatics/btx555
51. Christoffer C, Bharadwaj V, Luu R, Kihara D (2021) LZerD protein-protein docking webserver enhanced with de novo structure prediction. Front Mol Biosci 8. https://doi.org/10.3389/fmolb.2021.724947
52. Ohue M, Shimoda T, Suzuki S et al (2014) MEGADOCK 4.0: an ultra–high-performance protein–protein docking software for heterogeneous supercomputers. Bioinformatics 30: 3281–3283. https://doi.org/10.1093/bioinformatics/btu532
53. Berchanski A, Eisenstein M (2003) Construction of molecular assemblies via docking: modeling of tetramers with D2 symmetry. Proteins Struct Funct Genet 53:817–829. https://doi.org/10.1002/prot.10480
54. Schneidman-Duhovny D, Inbar Y, Nussinov R, Wolfson HJ (2005) PatchDock and SymmDock: servers for rigid and symmetric docking. Nucleic Acids Res 33:W363–W367. https://doi.org/10.1093/nar/gki481
55. Kozakov D, Brenke R, Comeau SR, Vajda S (2006) PIPER: an FFT-based protein docking program with pairwise potentials. Proteins Struct Funct Bioinform 65:392–406. https://doi.org/10.1002/prot.21117
56. Cheng TM-K, Blundell TL, Fernandez-Recio J (2007) pyDock: electrostatics and desolvation for effective scoring of rigid-body protein-

protein docking. Proteins Struct Funct Bioinform 68:503–515. https://doi.org/10.1002/prot.21419

57. Krupa P, Karczyńska AS, Mozolewska MA et al (2021) UNRES-Dock—protein–protein and peptide–protein docking by coarse-grained replica-exchange MD simulations. Bioinformatics 37:1613–1615. https://doi.org/10.1093/bioinformatics/btaa897

58. Levieux G, Tiger G, Mader S et al (2014) Udock, the interactive docking entertainment system. Faraday Discuss 169:425–441. https://doi.org/10.1039/C3FD00147D

59. Pierce BG, Wiehe K, Hwang H et al (2014) ZDOCK server: interactive docking prediction of protein-protein complexes and symmetric multimers. Bioinformatics 30:1771–1773. https://doi.org/10.1093/bioinformatics/btu097

Chapter 9

Software for Predicting Binding Free Energy of Protein–Protein Complexes and Their Mutants

Małgorzata Jarończyk

Abstract

Protein–protein binding affinity prediction is important for understanding complex biochemical pathways and to uncover protein interaction networks. Quantitative estimation of the binding affinity changes caused by mutations can provide critical information for protein function annotation and genetic disease diagnoses. The binding free energies of protein–protein complexes can be predicted using several computational tools. This chapter is a summary of software developed for the prediction of binding free energies for protein–protein complexes and their mutants.

Key words Protein–protein binding affinity, Free energy calculations, Protein–protein interaction, PPI

1 Introduction

Protein–protein interactions are crucial for cell biology and play an important role in the regulation of various biological processes such as signal transduction, transport, immunity, and metabolism [1]. Accurate prediction of binding free energies enables us to understand how these affinities can be altered, leading to a more complete knowledge of protein interactions in living organisms. Many protein–protein interactions are sensitive to mutations that might change the binding affinity. In fact, it has been found that mutations at protein–protein interfaces are more common in disease-causing variants [2]. Since new synthetic protein–protein complexes with the required function and novel medications are being developed, it is crucial to measure the change in binding affinity upon mutation in order to evaluate the mutational impacts. Many experimental biophysical and biochemical techniques have

The original version of the chapter has been revised. A correction to this chapter can be found at https://doi.org/10.1007/978-1-0716-3985-6_19

Agnieszka A. Kaczor (ed.), *Protein-Protein Docking: Methods and Protocols*, Methods in Molecular Biology, vol. 2780, https://doi.org/10.1007/978-1-0716-3985-6_9,
© The Author(s), under exclusive license to Springer Science+Business Media, LLC, part of Springer Nature 2024, Corrected Publication 2025

been employed to determine the protein–protein binding free energy. Since these methods are laborious, expensive, and time-consuming, computational techniques have been developed and tested for the prediction of protein–protein binding affinity and its change upon mutation [3–5]. One of the proposed methods is the prediction of binding affinity with docking scoring functions. However, docking scoring functions cannot predict experimentally measured binding affinity with sufficient precision [6]. The development of databases with binding affinity data has allowed the development of alternative methods [7, 8]. The first proposed model for protein–protein binding affinity prediction takes into account only two features of the complex: the size of its interface and the amplitude of the conformation change between the free and bound subunit [9]. In this review, tools for affinity prediction of protein–protein complexes and for mutants were described.

2 Materials and Methods

2.1 Databases for Binding Affinity

Databases of experimentally established thermodynamic data for protein–protein complexes are important for studying the factors that regulate binding affinity and allow for developing effective computational tools. Table 1 lists some of the available databases for the binding affinity of protein–protein complexes and their mutants. The Protein–protein Interactions Thermodynamic Database (PINT) includes thermodynamic data such as dissociation constant, binding free energy, enthalpy, and heat capacity change, as well as experimental settings and protein and peptide information [10]. The Alanine Scanning Energetics Database (ASEdb) contains single alanine mutations in protein–protein, protein–nucleic acid, and protein–small molecule interactions with experimentally confirmed binding affinities [11]. Structural database of

Table 1
Databases for binding affinity of protein–protein complexes and mutants

	PPI database	Website	References
ASEdb	Alanine Scanning Energetics database	http://www.asedb.org	[11]
SKEMPI2	Structural database of Kinetics and Energetics of Mutant Protein Interactions	https://life.bsc.es/pid/skempi2	[12]
PINT	Protein–protein Interactions Thermodynamic Database	http://www.bioinfodatabase.com/pint/	[10]
dbMPIKT	Database of kinetic and thermodynamic mutant protein interactions	https://www.dlearningapp.com/web/dbMPIKT/index.php	[15]
PROXiMATE	PROtein–protein compleX MutAtion ThErmodynamics	http://www.iitm.ac.in/bioinfo/PROXiMATE/	[14]

Kinetics and Energetics of Mutant Protein Interactions (SKEMPI 2) includes thermodynamic data such as experimentally measured changes in binding free energy, entropy, enthalpy, and rate constants for mutant protein–protein complexes [12, 13]. SKEMPI 2 is one of the most often used databases for training and/or testing novel methods for the prediction of protein–protein binding affinity and its change upon mutation. PROtein–protein compleX MutAtion ThErmodynamics (PROXiMATE) is a thermodynamic database of missense mutations in heterodimeric protein–protein complexes that include structural and functional information as well as links to important databases for protein sequence, structure, and interaction networks [14]. A kinetic and thermodynamic database of mutant protein interaction (dbMPIKTs) provides many mutants that have thermodynamics and dynamics data [15]. It contains 5291 mutants in protein interactions collected from previous databases and the literature.

2.2 Software for Prediction of Binding Affinity

Many computational methods have been designed to help predict the binding affinity of protein–protein complexes and change upon mutation in the complex. The available software and web servers used for binding affinity prediction for protein–protein complexes are listed in Table 2.

2.2.1 Structure-Based Prediction of Binding Affinity

Structure-based methods use the 3D structure of a protein–protein complex as input to predict the binding affinity. FoldX predicts the binding affinity of the complex through the linear combination of multiple energy terms weighted to experimental determinations [16]. For the energy calculation, FoldX accounts for van der Waals contributions, solvation of polar and apolar groups, hydrogen bonds, electrostatic contribution of charged groups, and entropic cost penalties. Zlab Affinity for Protein–Protein interaction (ZAPP) uses the three-dimensional structures of the complex and unbound proteins as input and predicts protein–protein binding free energies using a linear combination of nine energy terms [17]. More recently, EnCPdock server was developed that predicts binding energy by computing the overall shape and electrostatic complementarities (Sc, EC) at the protein–protein interface and plots them as an ordered pair {Sc, EC} in the two-dimensional plot [18]. In addition, the PPI-Affinity tool predicts binding energy by using the crystal structure of protein–protein complex or sequence [19] and by using 3D-structure descriptors implemented in the protein codification package ProtDCal [20].

2.2.2 Sequence-Based Prediction of Binding Affinity

Most computer prediction approaches need 3D protein structures, limiting their application to only known protein complexes. Therefore, methods based on protein sequence information have been developed for binding affinity prediction. Yugandhar and Gromiha [21] developed the first sequence-based method (PPA-Pred2) for predicting the binding affinity of protein–protein complexes. They

Table 2
List of software or web servers for prediction of protein–protein binding affinity and change in binding affinity upon mutation

Software/server	Website or link to software	References
ISLAND	https://sites.google.com/view/wajidarshad/software	[22]
AREA-Affinity	https://affinity.cuhk.edu.cn/	[28]
ICM-Pro	https://www.molsoft.com	[29]
LISA	https://www.lcqb.upmc.fr/LISA/	[27]
PPI-Affinity	https://protdcal.zmb.uni-due.de/PPIAffinity	[19]
PANDA	https://github.com/wajidarshad/panda	[36]
PRODIGY	http://milou.science.uu.nl/services/PRODIGY	[26]
SAAMBE-SEQ	http://compbio.clemson.edu/saambe_webserver/	[38]
SAAMBE-3D	http://compbio.clemson.edu/saambe_webserver/	[37]
BindProfX	https://zhanggroup.org/BindProfX/	[39]
Zapp	–	[17]
SSIPe	https://zhanglab.ccmb.med.umich.edu/SSIPe	[40]
PPA-Pred2	https://www.iitm.ac.in/bioinfo/PPA_Pred/	[21]
mCSM-PPI2	http://biosig.unimelb.edu.au/mcsm_ppi2/	[34]
MutaBind2	https://mutabind.org/	[32]
BeAtMuSiC	http://babylone.ulb.ac.be/beatmusic/index.php	[33]
ProAffiMuSeq	https://web.iitm.ac.in/bioinfo2/proaffimuseq/	[35]
ProBAPred	http://lightning.med.monash.edu/probapred/	[23]
EnCPdock	https://www.scinetmol.in/EnCPdock/	[18]
FOLDX	http://foldx.embl.de/	[16]
DeepPPAPred	https://web.iitm.ac.in/bioinfo2/DeepPPAPred/	[24]
Flex ddG	https://www.rosettacommons.org/	[30]

have classified the protein–protein complexes based on their function, molecular weight, and percentage of binding site residues and analyzed the relationship between sequence and structural properties of interacting free proteins and binding affinity. The function-based classification has been made with seven groups: (i) antigen–antibody complexes, (ii) enzyme–inhibitor complexes, (iii) other enzymes; complexes of enzymes with proteins other than inhibitors, (iv) G-protein-containing complexes, (v) receptor-containing complexes, (vi) non-cognate complexes, and (vii) miscellaneous complexes. The sequence-only regression model called In SiLico protein AffiNity preDictor (ISLAND) for prediction of the protein–protein binding affinity was developed [22]. This approach uses a local alignment kernel machine learning method for comparing sequences of different lengths that share common parts.

Protein–protein Binding Affinity Predictor (ProBAPred) is a useful ensemble regression-based computational framework that integrates a variety of protein sequence-derived and structural features to accurately estimate the binding affinity [23]. More recently, the deep learning-based model DeepPPAPred was developed to predict binding affinity using sequence and predicted structure-based features [24].

2.2.3 Contact-Based Prediction of Binding Affinity

The network of contacts at the interface of a protein–protein complex can help in describing the binding affinity of the interaction. Based on the number of contacts at the interface, an innovative and very simple method to predict binding affinity was developed by Vangone and Bonvin [25]. The model uses inter-residue contacts (ICs) contact types: polar/polar, polar/charged, polar/apolar, charged/charged, charged/apolar, and apolar/apolar between two interacting proteins and the effect of the non-interacting surface. This approach was implemented in the PRODIGY web server [26], an online tool to predict the binding affinity of a protein–protein complex given its three-dimensional structure.

Local interaction signal analysis (LISA) localizes the forces at the interface that influence the affinity between two proteins and suggests general principles of the distribution of these forces on interaction surfaces [27]. In LISA, the effect of atom–atom contacts on binding affinity is studied through an analysis of the geometry of the atomic interactions realized at three different levels (local, medium, and global) of granularity. First, LISA explores local interferences among co-localized atom–atom contacts to distinguish favorable from non-favorable contacts. Second, it investigates hot-sites of interactions by identifying regions in the protein interfaces with high or medium interaction strength, and high or medium number of contacts, and third, it explores the global distribution of contacts over the entire interaction surface.

2.2.4 Area-Based Prediction of Binding Affinity

The important role in determining protein–protein interactions and their binding affinity play geometric characteristics such as area (both interface and surface areas) in the structure of a protein–protein complex. The area-based models for the prediction of protein–protein or antibody–protein antigen binding affinity were implemented in the AREA-AFFINITY web server [28]. According to the nature of different amino acid types, the interface and surface areas are divided into 8 and 10 different types, respectively. Then, 60 representative models for protein–protein binding affinity prediction or 37 representative models specific for antibody–protein antigen-binding affinity prediction were constructed.

2.3 Protein–Protein Binding Affinity Change Upon Mutations

Mutation of amino acid residues at protein–protein interfaces alters the binding affinity of protein–protein complexes and may lead to diseases. Estimation of the changes of two interacting proteins upon mutations is commonly carried out by computational approaches.

MERSi is a new algorithm available in ICM-Pro (Molsoft) that allows the prediction of the effect of a mutation on protein stability and protein–protein binding free energy [29]. Monte Carlo simulations are performed to account for structural relaxation caused by mutation. Following the mutation of a specific residue, Monte Carlo simulations with flexible side chains for the mutated residue and its adjacent residues are run. Backbone flexibility is also taken into account in rare circumstances such as proline and disulfide-forming/breaking cysteine alterations. The change of binding free energy upon mutation is determined as the weighted total of all physically important free-energy contributions, which include van der Waals attractions, repulsions, electrostatics, hydrogen bonding, and the solvation component.

The flex ddG method within the Rosetta macromolecular modeling suite samples local side chains and includes backbone conformational changes. Then, side chain conformations for the mutant sequence are optimized, models are minimized, and finally interface score $\Delta\Delta G$ is calculated [30, 31]. MutaBind2 is a method that estimates the impact of single as well as multiple mutations on protein–protein interactions using crystal structure information of protein–protein complexes [32]. BeAtMuSiC and mCSM use coarse-grained statistical potentials generated from known 3D protein structures and machine learning [33, 34]. These statistical potentials describe correlations between amino acids, pair-wise inter-residue distances, torsion angles of the backbone, and the accessibility of the solvent.

ProAffiMuSeq is a tool that predicts the change in binding free energy using sequence-based features and functional classification of proteins [35]. There are six functional classes, namely (i) antigen–antibody complexes, (ii) enzyme–inhibitor complexes, (iii) G-protein complexes, (iv) receptor complexes, (v) other enzyme complexes (i.e., enzymes in complex with any protein other than an inhibitor), and (vi) miscellaneous complexes. One of the other tools used for the prediction of a change in protein binding affinity upon mutations that use protein sequence is a predictor called Predict chAnge iN binDing Affinity (PANDA) [36].

Single Amino Acid Mutation change of Binding Energy (SAAMBE-3D) is a machine learning algorithm to predict the effects of single amino acid mutation on protein–protein binding affinity that incorporates only knowledge-based features [37]. Overall, the features related to the X-ray structure of protein–protein complex as well as related to the mutation site such as the change in molar volume of mutated residue, hydrophobicity, flexibility, hydrogen bond donor/acceptor, polarity, mutation type, chemical nature of the amino acid residue, and size of the amino acid due to mutation. In addition, sequence-based features related to inter-chain interaction around the mutation site are used. SAAMBE-SEQ can be applied to protein complexes without known structure [38]. This method uses features that include the average Position-Specific Scoring Matrix (PSSM) for mutant and interaction chain, conservation score at mutation site, change in molar volume, hydrophobicity, flexibility, hydrogen bonds, polarity, mutation type, chemical nature, and size of the mutated amino acid.

BindProfX calculates the binding affinity for a given complex structure by extracting the protein–protein interaction interface structure and then searching the interface against a protein complex structure database to find homologous interfaces [39]. Fix-number, gap penalty, and evolutionary composition are three pseudo-counts that were introduced in the approach to counteract the present structure databases' poor ability to count interface analogies.

SSIPe combines protein interface profiles, collected from structural and sequence homology searches, with a physics-based energy function for accurate binding affinity estimation [40]. Detailed data analyses revealed that the major advantage of SSIPe over other traditional approaches lies in the novel combination of the physical energy function with the new knowledge-based interface profile. SSIPe also considerably outperformed a former profile-based method (BindProfX) due to the newly introduced sequence profiles and optimized pseudocount technique that allows for consideration of amino acid-specific prior mutation probabilities.

3 Summary

Accurate determination of protein–protein binding affinity and its change due to mutation can greatly aid in the characterization of novel interacting protein partners and the generation of new biological hypotheses. Despite the fact that several computational methods for predicting binding affinity are presently available, they still need to be improved.

Acknowledgments

This work was supported by Polish Ministry of Science and Higher Education (financial support for statutory activity of National Medicines Institute).

References

1. Gonzalez MW, Kann MG (2012) Chapter 4: protein interactions and disease. PLoS Comput Biol 8:e1002819. https://doi.org/10.1371/journal.pcbi.1002819
2. David A, Sternberg MJE (2015) The contribution of missense mutations in core and rim residues of protein–protein interfaces to human disease. J Mol Biol 427:2886–2898. https://doi.org/10.1016/j.jmb.2015.07.004
3. Gonzalez TR, Martin KP, Barnes JE et al (2020) Assessment of software methods for estimating protein-protein relative binding affinities. PLoS One 15:e0240573. https://doi.org/10.1371/journal.pone.0240573
4. Guo Z, Yamaguchi R (2022) Machine learning methods for protein-protein binding affinity prediction in protein design. Front Bioinform 2:1065703. https://doi.org/10.3389/fbinf.2022.1065703
5. Siebenmorgen T, Zacharias M (2020) Computational prediction of protein–protein binding affinities. WIREs Comput Mol Sci 10:e1448. https://doi.org/10.1002/wcms.1448
6. Kastritis PL, Bonvin AMJJ (2010) Are scoring functions in protein−protein docking ready to predict Interactomes? Clues from a novel binding affinity benchmark. J Proteome Res 9: 2216–2225. https://doi.org/10.1021/pr9009854
7. Kastritis PL, Moal IH, Hwang H et al (2011) A structure-based benchmark for protein-protein binding affinity. Protein Sci 20:482–491. https://doi.org/10.1002/pro.580
8. Vreven T, Moal IH, Vangone A et al (2015) Updates to the integrated protein–protein interaction benchmarks: docking benchmark version 5 and affinity benchmark version 2. J Mol Biol 427:3031–3041. https://doi.org/10.1016/j.jmb.2015.07.016
9. Janin J (2014) A minimal model of protein-protein binding affinities. Protein Sci 23: 1813–1817. https://doi.org/10.1002/pro.2560
10. Kumar MDS (2006) PINT: protein-protein interactions thermodynamic database. Nucleic Acids Res 34:D195–D198. https://doi.org/10.1093/nar/gkj017
11. Thorn KS, Bogan AA (2001) ASEdb: a database of alanine mutations and their effects on the free energy of binding in protein interactions. Bioinformatics 17:284–285. https://doi.org/10.1093/bioinformatics/17.3.284
12. Jankauskaite J, Jiménez-García B, Dapkunas J et al (2019) SKEMPI 2.0: an updated benchmark of changes in protein-protein binding energy, kinetics and thermodynamics upon mutation. Bioinformatics 35:462–469. https://doi.org/10.1093/bioinformatics/bty635
13. Moal IH, Fernández-Recio J (2012) SKEMPI: a structural kinetic and energetic database of mutant protein interactions and its use in empirical models. Bioinformatics 28:2600–2607. https://doi.org/10.1093/bioinformatics/bts489
14. Jemimah S, Yugandhar K, Michael Gromiha M (2017) PROXiMATE: a database of mutant protein–protein complex thermodynamics and kinetics. Bioinformatics 33:2787–2788. https://doi.org/10.1093/bioinformatics/btx312
15. Liu Q, Chen P, Wang B et al (2018) dbMPIKT: a database of kinetic and thermodynamic mutant protein interactions. BMC Bioinform 19:455. https://doi.org/10.1186/s12859-018-2493-7
16. Schymkowitz J, Borg J, Stricher F et al (2005) The FoldX web server: an online force field. Nucleic Acids Res 33:W382–W388. https://doi.org/10.1093/nar/gki387
17. Vreven T, Hwang H, Pierce BG, Weng Z (2012) Prediction of protein-protein binding free energies. Protein Sci 21:396–404. https://doi.org/10.1002/pro.2027
18. Biswas G, Mukherjee D, Dutta N et al (2023) EnCPdock: a web-interface for direct conjoint comparative analyses of complementarity and binding energetics in inter-protein associations. J Mol Model 29:239. https://doi.org/10.1007/s00894-023-05626-0
19. Romero-Molina S, Ruiz-Blanco YB, Mieres-Perez J et al (2022) PPI-affinity: a web tool for the prediction and optimization of protein–peptide and protein–protein binding affinity. J Proteome Res 21:1829–1841. https://doi.org/10.1021/acs.jproteome.2c00020
20. Ruiz-Blanco YB, Paz W, Green J, Marrero-Ponce Y (2015) ProtDCal: a program to compute general-purpose-numerical descriptors for sequences and 3D-structures of proteins. BMC Bioinform 16:162. https://doi.org/10.1186/s12859-015-0586-0
21. Yugandhar K, Gromiha MM (2014) Protein–protein binding affinity prediction from amino acid sequence. Bioinformatics 30:3583–3589. https://doi.org/10.1093/bioinformatics/btu580
22. Abbasi WA, Yaseen A, Hassan FU et al (2020) Island: in-silico proteins binding affinity prediction using sequence information. BioData

Min 13:20. https://doi.org/10.1186/s13040-020-00231-w
23. Lu B, Li C, Chen Q, Song J (2018) ProBAPred: inferring protein–protein binding affinity by incorporating protein sequence and structural features. J Bioinforma Comput Biol 16:1850011. https://doi.org/10.1142/S0219720018500117
24. Nikam R, Yugandhar K, Gromiha MM (2023) Deep learning-based method for predicting and classifying the binding affinity of protein-protein complexes. Biochim Biophys Acta, Proteins Proteomics 1871:140948. https://doi.org/10.1016/j.bbapap.2023.140948
25. Vangone A, Bonvin AM (2015) Contacts-based prediction of binding affinity in protein–protein complexes. elife 4:e07454. https://doi.org/10.7554/eLife.07454
26. Xue LC, Rodrigues JP, Kastritis PL et al (2016) PRODIGY: a web server for predicting the binding affinity of protein–protein complexes. Bioinformatics 32:3676–3678. https://doi.org/10.1093/bioinformatics/btw514
27. Raucci R, Laine E, Carbone A (2018) Local interaction signal analysis predicts protein-protein binding affinity. Structure 26:905–915.e4. https://doi.org/10.1016/j.str.2018.04.006
28. Yang YX, Huang JY, Wang P, Zhu BT (2022) AREA-AFFINITY: a web server for machine learning-based prediction of protein-protein and antibody-protein antigen binding affinities. J Chem Inf Model. https://doi.org/10.1021/acs.jcim.2c01499
29. Schapira M, Totrov M, Abagyan R (1999) Prediction of the binding energy for small molecules, peptides and proteins. J Mol Recognit 12:177–190. https://doi.org/10.1002/(SICI)1099-1352(199905/06)12:3<177::AID-JMR451>3.0.CO;2-Z
30. Barlow KA, Conchúir Ó, Thompson S et al (2018) Flex ddG: Rosetta ensemble-based estimation of changes in protein–protein binding affinity upon mutation. J Phys Chem B 122:5389–5399. https://doi.org/10.1021/acs.jpcb.7b11367
31. Sora V, Laspiur AO, Degn K et al (2023) RosettaDDGPrediction for high-throughput mutational scans: from stability to binding. Protein Sci 32:e4527. https://doi.org/10.1002/pro.4527
32. Zhang N, Chen Y, Lu H et al (2020) MutaBind2: predicting the impacts of single and multiple mutations on protein-protein interactions. iScience 23:100939. https://doi.org/10.1016/j.isci.2020.100939
33. Dehouck Y, Kwasigroch JM, Rooman M, Gilis D (2013) BeAtMuSiC: prediction of changes in protein–protein binding affinity on mutations. Nucleic Acids Res 41:W333–W339. https://doi.org/10.1093/nar/gkt450
34. Rodrigues CHM, Myung Y, Pires DE V, Ascher DB (2019) mCSM-PPI2: predicting the effects of mutations on protein–protein interactions. Nucleic Acids Res 47:W338–W344. https://doi.org/10.1093/nar/gkz383
35. Jemimah S, Sekijima M, Gromiha MM (2020) ProAffiMuSeq: sequence-based method to predict the binding free energy change of protein–protein complexes upon mutation using functional classification. Bioinformatics 36:1725–1730. https://doi.org/10.1093/bioinformatics/btz829
36. Abbasi WA, Abbas SA, Andleeb S (2021) PANDA: predicting the change in proteins binding affinity upon mutations by finding a signal in primary structures. J Bioinforma Comput Biol 19:2150015. https://doi.org/10.1142/S0219720021500153
37. Pahari S, Li G, Murthy AK et al (2020) SAAMBE-3D: predicting effect of mutations on protein–protein interactions. Int J Mol Sci 21:2563. https://doi.org/10.3390/ijms21072563
38. Li G, Pahari S, Murthy AK et al (2021) SAAMBE-SEQ: a sequence-based method for predicting mutation effect on protein–protein binding affinity. Bioinformatics 37:992–999. https://doi.org/10.1093/bioinformatics/btaa761
39. Xiong P, Zhang C, Zheng W, Zhang Y (2017) BindProfX: assessing mutation-induced binding affinity change by protein interface profiles with pseudo-counts. J Mol Biol 429:426–434. https://doi.org/10.1016/j.jmb.2016.11.022
40. Huang X, Zheng W, Pearce R, Zhang Y (2020) SSIPe: accurately estimating protein–protein binding affinity change upon mutations using evolutionary profiles in combination with an optimized physical energy function. Bioinformatics 36:2429–2437. https://doi.org/10.1093/bioinformatics/btz926

Chapter 10

Assessment of Protein–Protein Docking Models Using Deep Learning

Yuanyuan Zhang, Xiao Wang, Zicong Zhang, Yunhan Huang, and Daisuke Kihara

Abstract

Protein–protein interactions are involved in almost all processes in a living cell and determine the biological functions of proteins. To obtain mechanistic understandings of protein–protein interactions, the tertiary structures of protein complexes have been determined by biophysical experimental methods, such as X-ray crystallography and cryogenic electron microscopy. However, as experimental methods are costly in resources, many computational methods have been developed that model protein complex structures. One of the difficulties in computational protein complex modeling (protein docking) is to select the most accurate models among many models that are usually generated by a docking method. This article reviews advances in protein docking model assessment methods, focusing on recent developments that apply deep learning to several network architectures.

Key words Protein–protein interaction, Protein–protein docking, Docking evaluation, Protein structure prediction, Deep learning

1 Introduction

Protein–protein interactions (PPIs) play vital roles in many biological processes and determine the molecular functions of proteins. In order to understand the functional mechanisms of protein complexes, it is crucial to obtain their three-dimensional (3D) structures. Although protein complex structures would need to be ultimately determined by means of experimental techniques, such as cryo-electron microscopy, nuclear magnetic resonance (NMR) spectroscopy, and X-ray crystallography, computational approaches are also widely used for modeling protein complex structures. Protein complex structure modeling is often quoted as protein docking and has been actively developed in protein bioinformatics over the past two decades [1, 2].

Agnieszka A. Kaczor (ed.), *Protein-Protein Docking: Methods and Protocols*, Methods in Molecular Biology, vol. 2780,
https://doi.org/10.1007/978-1-0716-3985-6_10,
© The Author(s), under exclusive license to Springer Science+Business Media, LLC, part of Springer Nature 2024

Protein docking methods can be roughly divided into two main categories: template-based docking and ab initio docking. A template-based approach uses a known protein complex structure as a scaffold for building a docking model. It can be further classified into global structure-based [3] and local structure-based [4] approaches. An ab initio docking method takes individual protein structures to dock as input and makes many assemblies of the input protein structures without referring to known complex structures. Numerous ab initio docking methods were developed with various algorithms. Algorithms used for generating docking conformations include Fast Fourier Transform (FFT) [5, 6], geometric hashing [7–11], and particle swarm optimization [12]. Protein representations used include atomic coordinates [13], molecular surface [7], and voxels [14]. Also, sometimes the flexibility of individual proteins [11, 15, 16] is taken into consideration for more accurate docking.

Our lab has developed a series of protein docking methods, called LZerD, which include pairwise protein docking [17], multiple protein docking [9, 18], docking with disordered proteins [19–21], and predictions of assembly order of protein complexes [22]. Some of these methods are available as a web server [23, 24].

One of the challenges in ab initio docking is a selection of near-native (i.e., correct) models out of a large number of generated docking models (referred to as decoys). It is a nontrivial task because it is not uncommon that out of tens of thousands of decoys generated there are only around 100 or fewer correct models included (Fig. 1). In recognition of the significance of protein docking model evaluation (scoring), the Critical Assessment of Prediction of Interactions (CAPRI) [25], which is the community-based protein docking prediction experiment, has set a specific category for evaluating scoring methods.

Traditional methods for docking model selection estimate the binding affinity of proteins using physics-based potentials or scoring functions that capture preferred atom or residue interactions. These are briefly reviewed in the subsequent section. More recently, deep learning has been introduced for protein docking model evaluation. In the protein bioinformatics field, deep learning has been successfully applied in various tasks that deal with 3D data and structures, including capturing structure features in cryo-EM density data [26–30], identifying protein complexes in cryo-ET maps [31], protein structure prediction [32–34], and quality evaluation of single protein structure models [35, 36].

The application of deep learning to protein docking model selection has started quite recently. The focus of this article is to review deep learning-based methods developed for protein docking model selection. After briefly reviewing conventional methods for docking model evaluation, we will discuss deep neural network-

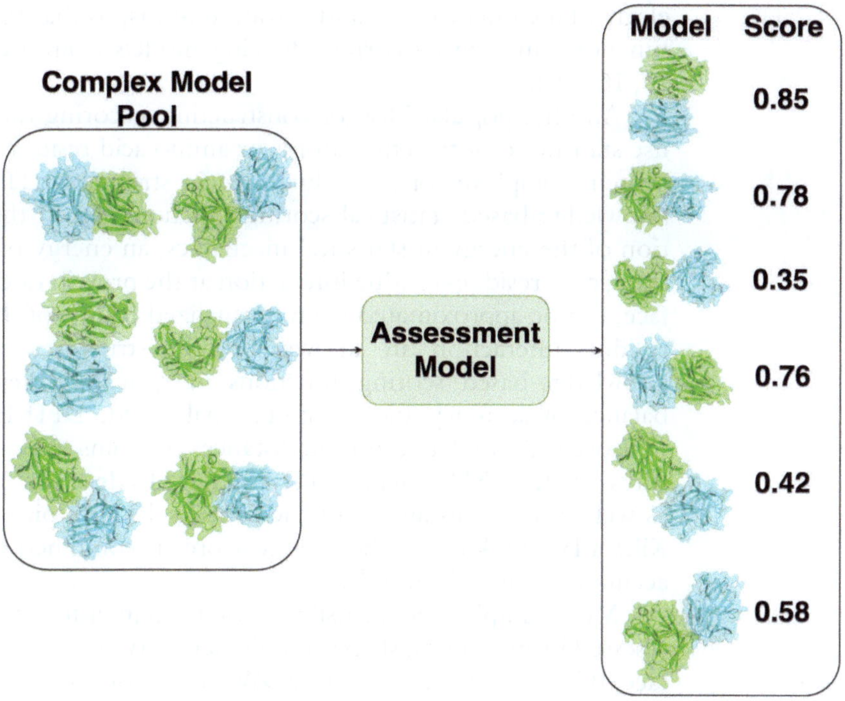

Fig. 1 Problem of docking model selection. Out of many docking models, the aim is to select the most accurate models, which are close to the native structure. A docking model assessment method will rank models and provide a score that predicts the quality (accuracy) of the models

based methods. We also mention docking methods that use AlphaFold2 [33], a recent successful protein structure prediction method, as the base.

2 Conventional Methods for Scoring Docking Models

In principle, the correct binding pose of proteins should be identified by evaluating the binding free energy of the proteins. With that in mind, Kingsley et al. performed a steered molecular dynamics (MD) simulation to separate interacting proteins apart and computed the potential of mean force along the pulling path [37]. In the method evERdock, the binding free energy was computed in an explicit water environment using short MD [38, 39]. Although using MD makes it possible to approximate the binding free energy of protein interaction, it is too time-consuming to process thousands of decoys, thus not practical in real-life ab initio docking situations. Alternatively, it is common to use a potential function of MD to evaluate docking decoys. To tune the potential function for docking decoy selection, it is typical to calibrate weights for potential terms, including van der Waals (VDW) interactions, electrostatic interactions, and solvation terms, so that the resulting function can identify correct docking models from a decoy pool [9, 16, 40].

Another popular idea for constructing a scoring function is to use statistics of interacting atoms or amino acid residues at known protein complexes or individual protein structures. This is called knowledge-based statistical scoring functions. From the formulation of the energy in statistical mechanics, an energy of a physical state, e.g., residue–residue interaction at the protein docking interface, can be approximated from normalized counts of the residue–residue interaction in known protein structures [41]. The knowledge-based scoring functions in general achieve a good balance of accuracy and computational speed. DFIRE [42] and ITScore [43] are based on the distances of atoms of protein structures, while GOAP [44] considers angles of side-chain interactions as well as atom distances. In the protein docking pipeline by the Kihara lab, ranking by these three scores is combined to produce accurate model selection [45].

More simply, the plausibility of protein interaction can be assessed by examining shape complementarity at the docking interface. It is essentially evaluating VDW interactions between proteins. In the LZerD protein docking method [7, 9, 23, 46], surface complementarity was quantified by a mathematical representation of 3D shape, the 3D Zernike descriptors (3DZD) [47]. A voxel representation is used for ZDOCK [48] and ClusPro [49] protein docking programs.

As proteins in a protein complex mutate under the constraint to maintain the interaction during evolution, mutation patterns in the multiple sequence alignments (MSAs) of interacting can give a useful hint to identify correct docking models. InterEvDock [50] computes a docking score from co-evolutionary information based on MSAs.

3 Deep Learning-Based Methods

In recent years, we have observed many applications of deep learning in protein bioinformatics, with protein docking being one such area. Below, we discuss the applications of deep learning in assessing docking models. The discussion is divided into two subsections based on the architecture of deep neural networks used in the methods: those that use Convolutional Neural Networks (CNNs) and methods that use Graph Neural Networks (GNNs). At last, we discuss methods that use AlphaFold2, which is a deep learning-based method for predicting single protein structures and protein complex structures.

The way deep learning handles protein complex models is conceptually different from the above-mentioned conventional methods. In conventional methods, researchers identify features, e.g., atom/residue/surface interaction patterns, sequence mutation patterns, in protein complex models that can be used for distinguishing correct models from wrong models, and compute a score (e.g. energy) based on the features. Such features are called hand-crafted features. The advantage is that the meaning of scores is understandable and usually relatively straightforward to compute. On the other hand, since such features are selected "subjectively" by researchers, a method does not fully capture the relationships of atoms potentially missing other important features. In contrast, deep learning-based methods represent and take the input data, protein structure complex model, as a whole with the hope that any even unobvious features to human researchers can be captured. An input data is processed through layers of networks with many parameters to output the conclusion (e.g., the likelihood that the input protein complex model is correct). As a result, a disadvantage is that the outcome may not be interpretable; i.e., it is not easy for users to understand the reasons why the neural network produces the result.

3.1 CNN-Based Methods

3D Convolutional Neural Network (3D-CNN) is a type of deep neural network architecture suitable to handle 3D data, including protein 3D structures. Besides protein docking model assessment, it has been applied to other biological 3D structures, such as drug–protein interaction scoring [51], protein structure quality assessment [35, 36], and protein/DNA/RNA structure detection in cryo-EM maps [26, 28]. A general workflow of CNN for assessing protein docking is shown in Fig. 2.

Docking decoy selection with a Voxel-based deep neural nEtwork (DOVE) [52] is the first 3D CNN-based method for protein docking model evaluation. From a queried 3D protein complex structure, DOVE first extracts the protein–protein interface, comprising atoms within 10 Å from other proteins, and superimposes them on a 3D cube of a 40^3 Å^3 size. The cube consists of voxels

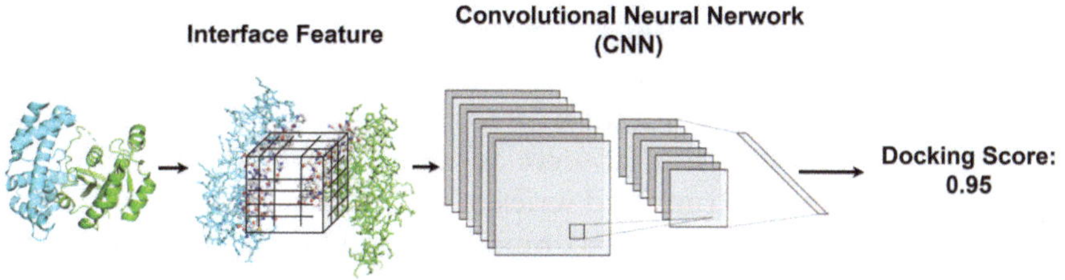

Fig. 2 Overview of CNN-based methods. Typically, atoms or residues at the protein interaction interface in a model are captured by voxels in an input box and processed through layers of a convolutional neural network. A method usually outputs a likelihood (a value between 0 and 1) that the model is correct

measuring 2^3 Å^3 each, and within each voxel, the number of heavy atoms—carbon, oxygen, nitrogen, and other atoms—is counted. Also, atom-level potentials computed with statistical potential functions [43, 44] are computed and they are considered as features of the input protein complex model. The architecture of DOVE consists of two convolutional layers, two max pooling layers, and three fully connected layers.

DOVE was trained on a protein docking benchmark dataset [53], which includes 178 protein complexes with around 54,000 models for each target. During the training process, input protein complex data were rotated in 24 different orientations so that output is less affected by the orientation of the input structure model (i.e., to achieve rotation invariance). DOVE has benchmarked on two datasets [53, 54] showing a clear improvement over conventional scoring approaches.

DeepRank [55] uses a similar 3D-CNN architecture for assessing the accuracy of docking models. It has two convolutional layers, two max pooling layers, three batch normalization layers, and two fully connected layers. With a box of a size of 30^3 Å^3, it captures features at the protein binding interface, which is defined by atoms within 5.5 Å of the other protein. It takes two types of features to a voxel of 1 Å^3: atom-level features include atom density, atomic charges, intermolecular electrostatic energy, and intermolecular van der Waals energy. Residue-level features include the number of residue–residue contacts, buried surface area, and position-specific scoring matrix. Features are smoothed by a Gaussian filter before mapping onto each voxel. In addition to predicting the likelihood of an input model being correct, it can also predict if an interface is biological or an artifact from crystallography.

PAUL uses a hierarchical, rotation-equivariant neural network to distinguish accurate docking models [56]. Mathematically, equivariance is the function that can yield the same outputs given different conformations of the input. Such property has been found to not only improve the performance of neural networks but also increase the interpretability of the model [57, 58]. More importantly, equivariance in rotation and translation is crucial in determining reliable 3D structures [59, 60]. Therefore, it is intuitive to introduce equivariance into deep learning frameworks for protein structure prediction tasks as well as structure assessment. Convolutional filters in the PAUL network architecture provide the equivariance to rotation, translation, and permutation of the atomic coordinates that serve as inputs to the network [56]. PAUL introduces a truncated series of spherical harmonics with learnable radial functions by decomposing traditional convolution kernels, to achieve rotation-translation equivalence instead of defining any fixed size of voxels to process input features. The input features are the 3D coordinates of atom, atom types, and a Boolean flag to indicate which chain the atom belongs to. PAUL can perform two types of different tasks. It classifies a model being correct or incorrect that is distinguished by a cutoff value of a ligand root mean square deviation (LRMSD) of 10 Å. It also predicts LRMSD value of a model. The classification and regression are trained separately with the same network architecture.

3.2 GNN-Based Methods

Although CNN showed good performance in protein complex model classification, a major theoretical drawback of conventional CNN is their incapability to be rotation-equivariant, which poses challenges to learning data features that have 3D rotations. Another drawback of CNN is that the input feature size and shape are fixed by the input box, which is placed at the protein docking interface. Since the size is fixed, often not all the interface residues and atoms fit into the box. The GNN architecture overcomes these drawbacks of CNN in an intuitive architecture of using graphs. In GNN, a protein complex structure is represented by graphs, where atoms are represented as nodes and interactions are represented by edges. Figure 3 shows a general overview of GNN-based methods. By the graph representation, the structure captured is rotation-invariant, and also interfaces of various shapes can be properly handled.

GNN-DOVE [61] is a method developed by our group that uses GNN to capture atom interactions at the protein docking interface. Atom interactions in a protein complex model are represented by two graphs, one representing the entire complex and the other representing only the interface. Atoms are nodes in the graph

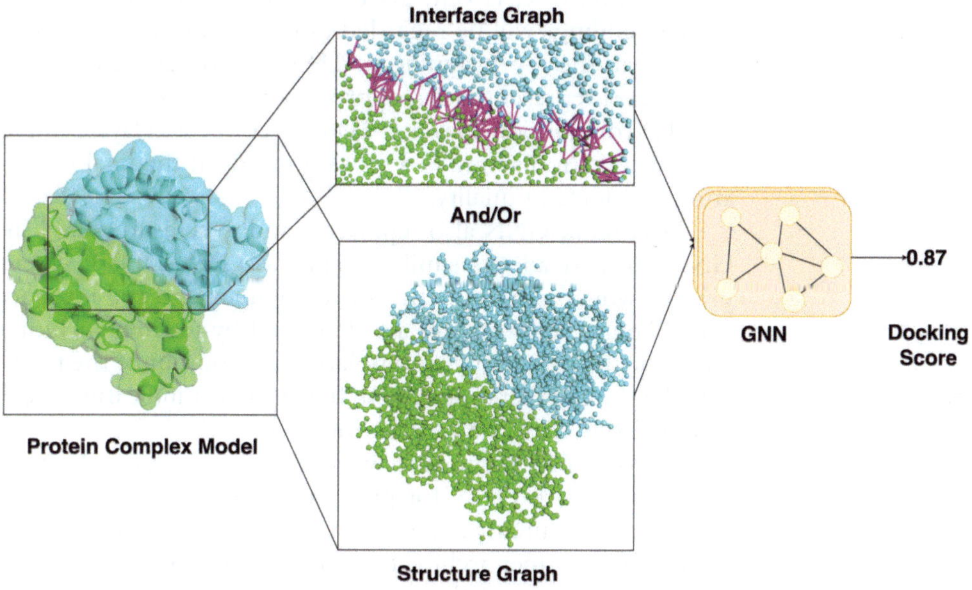

Fig. 3 Overview of GNN-based methods. A protein docking model is represented as a graph of connected and interacting atoms or residues. Typically, a method considers the entire structure model and/or the interactions at the interface

with atom types, atom degrees, the number of connected hydrogen atoms, the number of implicit valence electrons, and aromatic properties used as features. Edges represent covalent bonds of atoms or interface atoms that are close to other atoms from the other protein.

Both graphs are processed by a graph attention neural network (GAT) to output an interaction score, which indicates the possibility that the docking decoy is correct (acceptable or better quality based on the CAPRI protein docking contest criteria [25]). In each graph, atom features are updated by GAT layers through accumulating information from connected neighboring nodes. To focus only on the intermolecular interactions, the final node (atom) embedding of each layer is obtained by subtracting the node embedding of the interface graph from the full structure graph (Fig. 3).

Another method, DGANN [62], constructs graphs by measuring the Euclidean distance between amino acids instead of atoms. An edge is formed between a pair of amino acids if they are closer than 5 Å from each other. An amino acid node has 26-dimensional features, which include five physio-chemical properties and 21 Position-Specific Scoring Matrix (i.e., amino acid frequencies in an MSA) features. DGANN is adapted from DGCNN [63] with a replacement of Graph Convolutional layers (GCN) with Graph Attention layers (GAT). GAT aggregates the information from neighbors with different weights by self-attention. DGANN includes two layers of GAT to encode structural information and the final layer of GAT outputs 1D embedding of each node. Those 1D embeddings will be regarded as the importance of each amino acid. Next, the top-k significant 1D embeddings are chosen as the graph-level representations of protein–protein interfaces, followed by a 1D convolutional layer, a 1D fully connected layer, and lastly a sigmoid function acting as a scoring metric at the final layer evaluating the docking quality.

DeepRank-GNN also represents amino acids in graphs in their GNN [64]. It adopts similar graphs as GNN-DOVE where two graphs, which represent the entire complex and the interface, are used. Amino acid residues are described by features including the amino acid type, charge, and polarity. The output of the network is a docking score that ranges from 0 to 1, with 1 indicating the model is most likely correct and 0 for incorrect.

DProQ [65] is the first work that incorporates the idea of a transformer with GNN for protein docking model assessment. It has an architecture of Graph Transformer [66] with a gated mechanism. Transformer architectures are a type of neural network that has achieved state-of-the-art results in multiple natural language processing (NLP) tasks. Transformer adopts a technique called multi-head self-attention to capture global and local information in an efficient manner. Thus, a graph transformer combines both

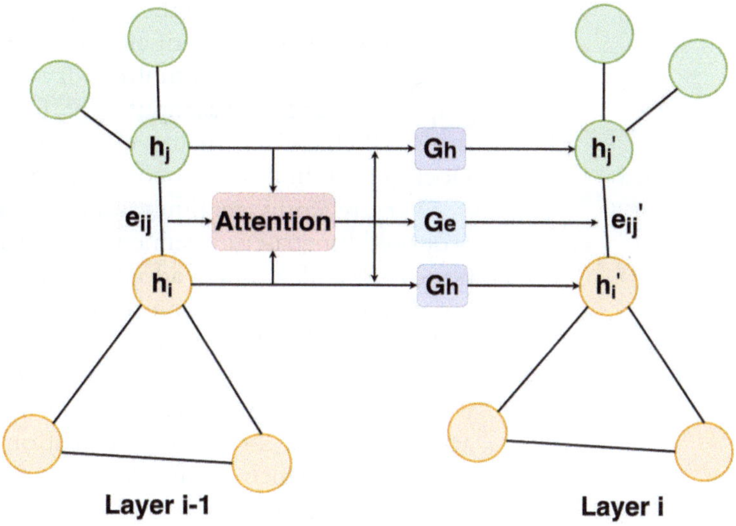

Fig. 4 Explanation of Gated Graph Transformer (GGT). Nodes (residues) of two proteins are shown in green and orange nodes. Embeddings (encoded features) of nodes and edges are updated by using the attention mechanism, which is designed to give different weights to different features according to their importance. Gated mechanism, Gh and Ge, consider the previous values of embeddings and mix them with the current updated embedding values. Layer denotes a time stamp of embedding updates

the strength of GNN and transformer to gain a stronger power to represent protein structures, where the relationships between graph nodes can be established by projecting them into a comparable space to compute attention. DProQ applies a k-NN method to construct graphs by connecting each node (amino acid residue) to its 10 closest neighbors in the 3D space. Therefore, the number of connections for each node in DProQ is consistent. Each node has a vector of 35-dimensional features, which include the residue type, the secondary structure type, relatively accessible surface area, and phi and psi dihedral angles, which are computed by DSSP [67] and positional encoding. The distance between $C_\alpha - C_\alpha$, $C_\beta - C_\beta$, and $N - O$ are included as edge features.

Gated Graph Transformer (GGT) is shown in Fig. 4. For a pair of nodes (residues) i, j connected by an edge at layer l, it first computes the attention score at edge i, j dependent on node embeddings (encoded features) h_i, h_j and edge embedding (encoded feature) $e_{i,\, j}$. Then, $e_{i,\, j}$ is updated by the attention score obtained from the previous step with the application of an element-wise edge gate G_e, which determines how much edge information and weighs from the edge of each node neighbor. A similar node gate G_h controls how to modulate information from node neighbors while updating node embeddings. By using multiple layers, more distant connections of atoms in the network can be considered to update embeddings of nodes and edges. After two layers of GGT, the sum of all node embeddings is regarded as the graph representation. A multiple-layer perceptron (MLP) takes the graph representation as input to perform multi-task learning, one is the CAPRI model quality class prediction and the other is the DockQ protein docking score [68] prediction.

Very recently, an Equivariant Graph Refiner (EGR) has been proposed for structure refinement and assessment for protein complexes [69]. The model adopts an equivariant graph message passing mechanism in addition to the conventional graph neural work architecture. EGR leverages E(3)-equivariance as inductive prior, a common technique of incorporating equivariance into neural networks [70–72]. By doing so, it allows the network to have more accurate spatial structure information in terms of rotation and translations, which would contribute to more accurate structure refinement and model quality assessment.

The input to EGR is a graph representing a protein complex model, where nodes represent atoms in the complex and each atom is connected to its 20 closest neighbors. Node features include atom type and atom-to-surface distances. Edge features are atom local distance, orientations, and positional encoding of the two nodes for a given edge as features. In addition, two binary features are added as edge features to indicate if a pair of connected atoms forms a covalent bond and if they belong to the same chain. There are a total of 39 node features and 15 edge features. During training, the model is supervised under a multi-task fashion: the main task is structure refinement, where the predicted refinement structure is compared against the ground truth structure. Another task is the structure assessment, where the node features are used to predict the Cα atom local Distance Difference Test (LDDT) value [73] (LDDT evaluates the correctness of the local structure around each residue in a protein model). Compared to previous deep learning methods on quality assessment for protein complexes, EGR is able to handle structure refinement and assessment at the same time in an end-to-end fashion.

3.3 AlphaFold-Multimer

At last, we mention AlphaFold2 [33] and AlphaFold-Multimer [34]. Since its successful debut in the CASP protein structure prediction contest in 2020 [74], AlphaFold2 has substantially changed the protein structure prediction field, including protein docking, as the accuracy it achieves in general is substantially higher than existing methods. Precomputed structure models for virtually all the known proteins are made available in the AlphaFold Database [75] and searchable by structural similarity at 3D-AF-Surfer [76]. Soon after the paper on AlphaFold2 was published, the authors at DeepMind released AlphaFold-Multimer, a protein complex structure prediction method from their amino acid sequence information.

AlphaFold2 outputs a confidence score for each residue in a model, pLDDT scores, and a predicted LDTT value. AlphaFold-Multimer outputs another confidence score, named predicted interface TM score, ipTM, which assesses accuracy at interface regions of a protein complex model, along with pLDDT for each protein. pLDDT and ipTM scores seem to be reasonably accurate, and practically users can somewhat trust them when assessing models generated by AlphaFold2 works [77].

4 Summary

In protein docking, selecting correct models from a large number of alternative models is an important and difficult task. Assessment of models is needed in ab initio docking, where models from the same method are subject to comparison, as well as cases where models by various methods need to be selected. Here, we reviewed methods that use deep learning, which emerged as a powerful approach to machine learning. We hope this article provides a snapshot of this rapidly evolving field, which would be useful for researchers who are interested in coming into this field as well as biologists who are in need of using a docking model assessment method.

Acknowledgments

This work was partly supported by the National Institutes of Health (R01GM133840, 3R01 GM133840-02S1) and the National Science Foundation (DMS2151678, DBI2003635, CMMI1825941, MCB2146026, and MCB1925643). XW is a recipient of the MolSSI graduate fellowship.

References

1. Aderinwale T, Christoffer CW, Sarkar D et al (2020) Computational structure modeling for diverse categories of macromolecular interactions. Curr Opin Struct Biol 64:1–8
2. Sunny S, Jayaraj P (2022) Protein–protein docking: past, present, and future. Protein J 41(1):1–26
3. Anishchenko I, Kundrotas PJ, Tuzikov AV et al (2015) Structural templates for comparative protein docking. Proteins: Struct, Funct, Bioinf 83(9):1563–1570
4. Tuncbag N, Gursoy A, Nussinov R et al (2011) Predicting protein-protein interactions on a proteome scale by matching evolutionary and structural similarities at interfaces using PRISM. Nat Protoc 6(9):1341
5. Katchalski-Katzir E, Shariv I, Eisenstein M et al (1992) Molecular surface recognition: determination of geometric fit between proteins and their ligands by correlation techniques. Proc Natl Acad Sci USA 89(6):2195–2199
6. Padhorny D, Kozakov D, Kazennov A et al (2016) Protein-protein docking by fast generalized Fourier transforms on 5D rotational manifolds. Proc Natl Acad Sci USA 113(30):E4286–E4293
7. Venkatraman V, Yang YD, Sael L et al (2009) Protein-protein docking using region-based

3D Zernike descriptors. BMC Bioinform 10(1):407
8. Fischer D, Lin SL, Wolfson HL et al (1995) A geometry-based suite of moleculardocking processes. J Mol Biol 248(2):459–477
9. Esquivel-Rodríguez J, Yang YD, Kihara D (2012) Multi-LZerD: multiple protein docking for asymmetric complexes. Proteins: Struct, Funct, Bioinf 80(7):1818–1833
10. Peterson LX, Shin WH, Kim H et al (2018) Improved performance in CAPRI round 37 using LZerD docking and template-based modeling with combined scoring functions. Proteins: Struct, Funct, Bioinf 86:311–320
11. Christoffer C, Kihara D (2022) Domain-based protein docking with extremely large conformational changes. J Mol Biol 434(21):167820
12. Moal IH, Bates PA (2010) SwarmDock and the use of normal modes in protein-protein docking. Int J Mol Sci 11(10):3623–3648
13. Lyskov S, Gray JJ (2008) The RosettaDock server for local protein–protein docking. Nucleic Acids Res 36(suppl_2):W233–W238
14. Pierce BG, Hourai Y, Weng Z (2011) Accelerating protein docking in ZDOCK using an advanced 3D convolution library. PLoS One 6(9):e24657
15. Oliwa T, Shen Y (2015) cNMA: a framework of encounter complex-based normal mode analysis to model conformational changes in protein interactions. Bioinformatics 31(12):i151–i160
16. Gray JJ, Moughon S, Wang C et al (2003) Protein–protein docking with simultaneous optimization of rigid-body displacement and side-chain conformations. J Mol Biol 331(1): 281–299
17. Venkatraman V, Yang YD, Sael L et al (2009) Protein-protein docking using region-based 3D Zernike descriptors. BMC Bioinform 10(1):1–21
18. Aderinwale T, Christoffer C, Kihara D (2022) RL-MLZerD: multimeric protein docking using reinforcement learning. Front Mol Biosci 9:969394
19. Peterson LX, Roy A, Christoffer C et al (2017) Modeling disordered protein interactions from biophysical principles. PLoS Comput Biol 13(4):e1005485
20. Verburgt J, Zhang Z, Kihara D (2022) Multi-level analysis of intrinsically disordered protein docking methods. Methods (San Diego, Calif) 204:55–63
21. Christoffer C, Kihara D (2020) IDP-LZerD: software for modeling disordered protein interactions. In: Protein structure prediction. Springer, pp 231–244
22. Peterson LX, Togawa Y, Esquivel-Rodriguez J et al (2018) Modeling the assembly order of multimeric heteroprotein complexes. PLoS Comput Biol 14(1):e1005937
23. Christoffer C, Chen S, Bharadwaj V et al (2021) LZerD webserver for pairwise and multiple protein–protein docking. Nucleic Acids Res 49(W1):W359–W365
24. Christoffer C, Bharadwaj V, Luu R et al (2021) LZerD protein-protein docking webserver enhanced with de novo structure prediction. Front Mol Biosci 8:750
25. Lensink MF, Velankar S, Baek M et al (2018) The challenge of modeling protein assemblies: the CASP12-CAPRI experiment. Proteins: Struct, Funct, Bioinf 86:257–273
26. Wang X, Alnabati E, Aderinwale T et al (2021) Detecting protein and DNA/RNA structures in cryo-EM maps of intermediate resolution using deep learning. Nat Commun 12(1): 2302–2302
27. Subramaniya SRMV, Terashi G, Kihara D (2021) Super resolution Cryo-EM maps with 3D deep generative networks. Biophys J 120(3):283a
28. Subramaniya SRMV, Terashi G, Kihara D (2019) Protein secondary structure detection in intermediate-resolution cryo-EM maps using deep learning. Nat Methods 16(9): 911–917
29. Terashi G, Wang X, Maddhuri Venkata Subramaniya SR et al (2022) Residue-wise local quality estimation for protein models from cryo-EM maps. Nat Methods 19(9):1116–1125
30. Alnabati E, Terashi G, Kihara D (2022) Protein structural modeling for electron microscopy maps using VESPER and MAINMAST. Curr Protocols 2(7):e494
31. Gubins I, Chaillet ML, van Der Schot G et al (2020) SHREC 2020: classification in cryo-electron tomograms. Comput Graph 91:279–289
32. Jain A, Terashi G, Kagaya Y et al (2021) Analyzing effect of quadruple multiple sequence alignments on deep learning based protein inter-residue distance prediction. Sci Rep 11(1):1–13
33. Jumper J, Evans R, Pritzel A et al (2021) Highly accurate protein structure prediction with AlphaFold. Nature 596(7873):583–589
34. Evans R, O'Neill M, Pritzel A et al (2022) Protein complex prediction with AlphaFold-Multimer. BioRxiv:2021.2010. 2004.463034
35. Derevyanko G, Grudinin S, Bengio Y et al (2018) Deep convolutional networks for quality assessment of protein folds. Bioinformatics 34(23):4046–4053

36. Pagès G, Charmettant B, Grudinin S (2019) Protein model quality assessment using 3D oriented convolutional neural networks. Bioinformatics 35(18):3313–3319
37. Kingsley LJ, Esquivel-Rodríguez J, Yang Y et al (2016) Ranking protein–protein docking results using steered molecular dynamics and potential of mean force calculations. J Comput Chem 37(20):1861–1865
38. Takemura K, Guo H, Sakuraba S et al (2012) Evaluation of protein-protein docking model structures using all-atom molecular dynamics simulations combined with the solution theory in the energy representation. J Chem Phys 137(21):215105
39. Takemura K, Matubayasi N, Kitao A (2018) Binding free energy analysis of protein-protein docking model structures by evERdock. J Chem Phys 148(10):105101
40. Chéron J-B, Zacharias M, Antonczak S et al (2017) Update of the ATTRACT force field for the prediction of protein-protein binding affinity. J Comput Chem 38(21):1887–1890
41. Lu H, Lu L, Skolnick J (2003) Development of unified statistical potentials describing protein-protein interactions. Biophys J 84(3): 1895–1901
42. Zhang C, Liu S, Zhu Q et al (2005) A knowledge-based energy function for protein–ligand, protein–protein, and protein–DNA complexes. J Med Chem 48(7): 2325–2335
43. Huang SY, Zou X (2008) An iterative knowledge-based scoring function for protein–protein recognition. Proteins: Struct, Funct, Bioinf 72(2):557–579
44. Zhou H, Skolnick J (2011) GOAP: a generalized orientation-dependent, all-atom statistical potential for protein structure prediction. Biophys J 101(8):2043–2052
45. Christoffer C, Terashi G, Shin WH et al (2020) Performance and enhancement of the LZerD protein assembly pipeline in CAPRI 38-46. Proteins: Struct, Funct, Bioinf 88(8):948–961
46. Li B, Kihara D (2012) Protein docking prediction using predicted protein-protein interface. BMC Bioinform 13(1):1–17
47. Kihara D, Sael L, Chikhi R et al (2011) Molecular surface representation using 3D Zernike descriptors for protein shape comparison and docking. Curr Protein Pept Sci 12(6):520–530
48. Pierce BG, Wiehe K, Hwang H et al (2014) ZDOCK server: interactive docking prediction of protein–protein complexes and symmetric multimers. Bioinformatics 30(12):1771–1773
49. Alekseenko A, Ignatov M, Jones G et al (2020) Protein–protein and protein–peptide docking with ClusPro server. In: Protein structure prediction. Springer, pp 157–174
50. Nadaradjane AA, Guerois R, Andreani J (2018) Protein-protein docking using evolutionary information. In: Protein complex assembly. Springer, pp 429–447
51. Ragoza M, Hochuli J, Idrobo E et al (2017) Protein–ligand scoring with convolutional neural networks. J Chem Inf Model 57(4): 942–957
52. Wang X, Terashi G, Christoffer CW et al (2019) Protein docking model evaluation by 3D deep convolutional neural networks. Bioinformatics 36(7):2113–2118. https://doi.org/10.1093/bioinformatics/btz870
53. Hwang H, Vreven T, Janin J et al (2010) Protein–protein docking benchmark version 4.0. Proteins: Struct, Funct, Bioinf 78(15): 3111–3114
54. Liu S, Gao Y, Vakser IA (2008) Dockground protein–protein docking decoy set. Bioinformatics 24(22):2634–2635
55. Renaud N, Geng C, Georgievska S et al (2021) DeepRank: a deep learning framework for data mining 3D protein-protein interfaces. Nat Commun 12(1):1–8
56. Eismann S, Townshend RJ, Thomas N et al (2021) Hierarchical, rotation-equivariant neural networks to select structural models of protein complexes. Proteins: Struct, Funct, Bioinf 89(5):493–501
57. Worrall DE, Garbin SJ, Turmukhambetov D et al (2017) Harmonic networks: deep translation and rotation equivariance. In: Proceedings of the IEEE conference on computer vision and pattern recognition, pp 5028–5037
58. Lenc K, Vedaldi A (2015) Understanding image representations by measuring their equivariance and equivalence. In: Proceedings of the IEEE conference on computer vision and pattern recognition, pp 991–999
59. Marcos D, Volpi M, Komodakis N et al (2017) Rotation equivariant vector field networks. In: Proceedings of the IEEE international conference on computer vision, pp 5048–5057
60. Thomas N, Smidt T, Kearnes S et al (2018) Tensor field networks: rotation-and translation-equivariant neural networks for 3d point clouds. arXiv preprint arXiv:1802.08219
61. Wang X, Flannery ST, Kihara D (2021) Protein docking model evaluation by graph neural networks. Front Mol Biosci 8:647915
62. Han Y, He F, Chen Y et al (2021) Quality assessment of protein docking models based on graph neural network. Front Bioinform 1: 693211

63. Zhang M, Cui Z, Neumann M et al (2018) An end-to-end deep learning architecture for graph classification. In: Proceedings of the AAAI conference on artificial intelligence, vol 1
64. Réau M, Renaud N, Xue LC et al (2023) DeepRank-GNN: a graph neural network framework to learn patterns in protein–protein interfaces. Bioinformatics 39(1):btac759
65. Chen X, Morehead A, Liu J et al (2022) DProQ: a gated-graph transformer for protein complex structure assessment. bioRxiv:2022.2005. 2019.492741
66. Dwivedi VP, Bresson X (2020) A generalization of transformer networks to graphs. arXiv preprint arXiv:2012.09699
67. Kabsch W, Sander C (1983) Dictionary of protein secondary structure: pattern recognition of hydrogen-bonded and geometrical features. Biopolymers 22(12):2577–2637
68. Basu S, Wallner B (2016) DockQ: a quality measure for protein-protein docking models. PLoS One 11(8):e0161879
69. Morehead A, Chen X, Wu T et al. (2022) EGR: equivariant graph refinement and assessment of 3D protein complex structures. arXiv preprint arXiv:2205.10390
70. Battaglia PW, Hamrick JB, Bapst V et al (2018) Relational inductive biases, deep learning, and graph networks. arXiv preprint arXiv:1806.01261
71. Feinman R, Lake BM (2018) Learning inductive biases with simple neural networks. arXiv preprint arXiv:1802.02745
72. Neyshabur B, Tomioka R, Srebro N (2014) In search of the real inductive bias: On the role of implicit regularization in deep learning. arXiv preprint arXiv:1412.6614
73. Mariani V, Biasini M, Barbato A et al (2013) lDDT: a local superposition-free score for comparing protein structures and models using distance difference tests. Bioinformatics 29(21):2722–2728
74. Kryshtafovych A, Schwede T, Topf M et al (2021) Critical assessment of methods of protein structure prediction (CASP)—Round XIV. Proteins: Struct, Funct, Bioinf 89(12): 1607–1617
75. Varadi M, Anyango S, Deshpande M et al (2022) AlphaFold Protein Structure Database: massively expanding the structural coverage of protein-sequence space with high-accuracy models. Nucleic Acids Res 50(D1):D439–D444
76. Aderinwale T, Bharadwaj V, Christoffer C et al (2022) Real-time structure search and structure classification for AlphaFold protein models. Commun Biol 5(1):1–12
77. Akdel M, Pires DE, Pardo EP et al (2022) A structural biology community assessment of AlphaFold2 applications. Nat Struct Mol Biol 29:1–12

Part IV

Application of Protein–Protein Docking

Chapter 11

Molecular Docking of Intrinsically Disordered Proteins: Challenges and Strategies

Keyur N. Patel, Dhruvil Chavda, and Moutusi Manna

Abstract

Intrinsically disordered proteins (IDPs) are a novel class of proteins that have established a significant importance and attention within a very short period of time. These proteins are essentially characterized by their inherent structural disorder, encoded mainly by their amino acid sequences. The profound abundance of IDPs and intrinsically disordered regions (IDRs) in the biological world delineates their deep-rooted functionality. IDPs and IDRs convey such extensive functionality through their unique dynamic nature, which enables them to carry out huge number of multifaceted biomolecular interactions and make them "interaction hub" of the cellular systems. Additionally, with such widespread functions, their misfunctioning is also intimately associated with multiple diseases. Thus, understanding the dynamic heterogeneity of various IDPs along with their interactions with respective binding partners is an important field with immense potentials in biomolecular research. In this context, molecular docking-based computational approaches have proven to be remarkable in case of ordered proteins. Molecular docking methods essentially model the biomolecular interactions in both structural and energetic terms and use this information to characterize the putative interactions between the two participant molecules. However, direct applications of the conventional docking methods to study IDPs are largely limited by their structural heterogeneity and demands for unique IDP-centric strategies. Thus, in this chapter, we have presented an overview of current methodologies for successful docking operations involving IDPs and IDRs. These specialized methods majorly include the ensemble-based and fragment-based approaches with their own benefits and limitations. More recently, artificial intelligence and machine learning-assisted approaches are also used to significantly reduce the complexity and computational burden associated with various docking applications. Thus, this chapter aims to provide a comprehensive summary of major challenges and recent advancements of molecular docking approaches in the IDP field for their better utilization and greater applicability.Asp (D)

Key words Intrinsically disordered proteins, Molecular docking, Conformational heterogeneity, Ensemble-based approaches, Fragment-based approaches, AI/ML-based approaches

1 Introduction

With immense advances and expansion in the structural biology field, the recent past has witnessed the emergence of a new and prevalent class of protein, called intrinsically disordered proteins

Agnieszka A. Kaczor (ed.), *Protein-Protein Docking: Methods and Protocols*, Methods in Molecular Biology, vol. 2780, https://doi.org/10.1007/978-1-0716-3985-6_11,
© The Author(s), under exclusive license to Springer Science+Business Media, LLC, part of Springer Nature 2024

(IDPs), that defy the conventional structure-function paradigm [1, 2]. The unusual biophysical features of this distinctive class added another dimension to protein science with new challenges [2–5]. In general, IDPs are defined as proteins lacking a stable, definite structure or conformation in isolation under physiological conditions. Instead, they exist as a structurally heterogeneous ensemble of rapidly interconvertible conformational states. This type of disorder is also seen in various segments of ordered proteins. These segments are referred as intrinsically disordered protein regions (IDRs) [4, 6–10].

Scientific attention on IDPs and IDRs has increased exponentially since their initial discovery. It has been found that a significant predicted disorder is prevalent in eukaryotes than in prokaryotes and archaea [4]. Reports based on 31 different genome analyses indicate 52–67% of eukaryotic proteins contain long IDRs compared to 26–51% in archaea and 16–45% in bacteria [2, 11, 12]. Also, approximately 30% of proteins in the human proteome have IDRs that are more than 30 residues long [13]. Furthermore, multiple comprehensive computational studies covering a considerably large dataset of genomes (3484 in total) from viruses, bacteria, archaea, and eukaryotes, clearly stipulate that the IDPs and IDRs are necessary key players in the complex biological systems [2, 14–18].

In different scenarios, IDPs may undergo disorder-to-order (DOT) transition upon interactions with small-molecule ligands or macromolecular binding partners, or by post-translational modifications [6, 7, 9]. These interactions form the basis of a wide array of functions carried out by these proteins. IDPs and IDRs are majorly associated with molecular recognition, signalling, and regulation [4, 6, 7, 9]. IDPs hold structural advantages in achieving many of these functions over structured proteins [6, 19]. The dynamic disorder endows IDPs with several important attributes, including increased surface area, conformational plasticity for multi-partner interactions, and high association and dissociation rates allowing faster association with reduced binding strength for unique and rapid interactions [19–22]. Due to the elaborative functional role, misfunctions of IDPs and IDRs find their connections in majority of diseases [6], inspiring the "disorder in disorder" (D^2) concept [6, 8]. IDPs are frequently associated with cardiovascular disease, neurodegenerative diseases, cancer, amyloidosis, diabetes, etc. For example, changes in the abundance of IDRs in proteins like c-Myc or p53 may lead to cancer progression, while the misfolding and aggregation of proteins like amyloid-β peptide and α-synuclein are critical for the onset and progression of many neurodegenerative diseases [4, 6, 8, 9, 23, 24]. These unusual functional roles and extensive involvement in diseases, put IDPs and IDRs at the center of attention among the research community.

The unique characteristics of IDPs are encoded in their primary sequence. The first and foremost feature of all IDPs is their propensity to rebel against the ordered folding. Just as the amino acid sequences critically decide the folding in globular proteins, the primary sequences call for the disorder in IDPs and IDRs [2, 9]. IDPs encompass a significant number of polar residues and proline, whereas meagre amounts of hydrophobic residues [2, 4, 13, 25, 26]. Multiple tools have been devised to predict or identify the disorder within proteins based on the amino acid sequences [4, 27]. A comprehensive review and comparison of these IDP and IDR prediction algorithms can be found elsewhere [27–29]. These tools have indicated Pro (P), Glu (E), Ser (S), Gln (Q), and Lys (K) as disorder-promoting amino acids based on their comparatively higher prevalence in disordered structures. Contrastingly, Cys (C), Trp (W), Ile (I), Tyr (Y), Phe (F), and Leu (L) are indicated as structure-promoting amino acids (Fig. 1). At the same time, Gly

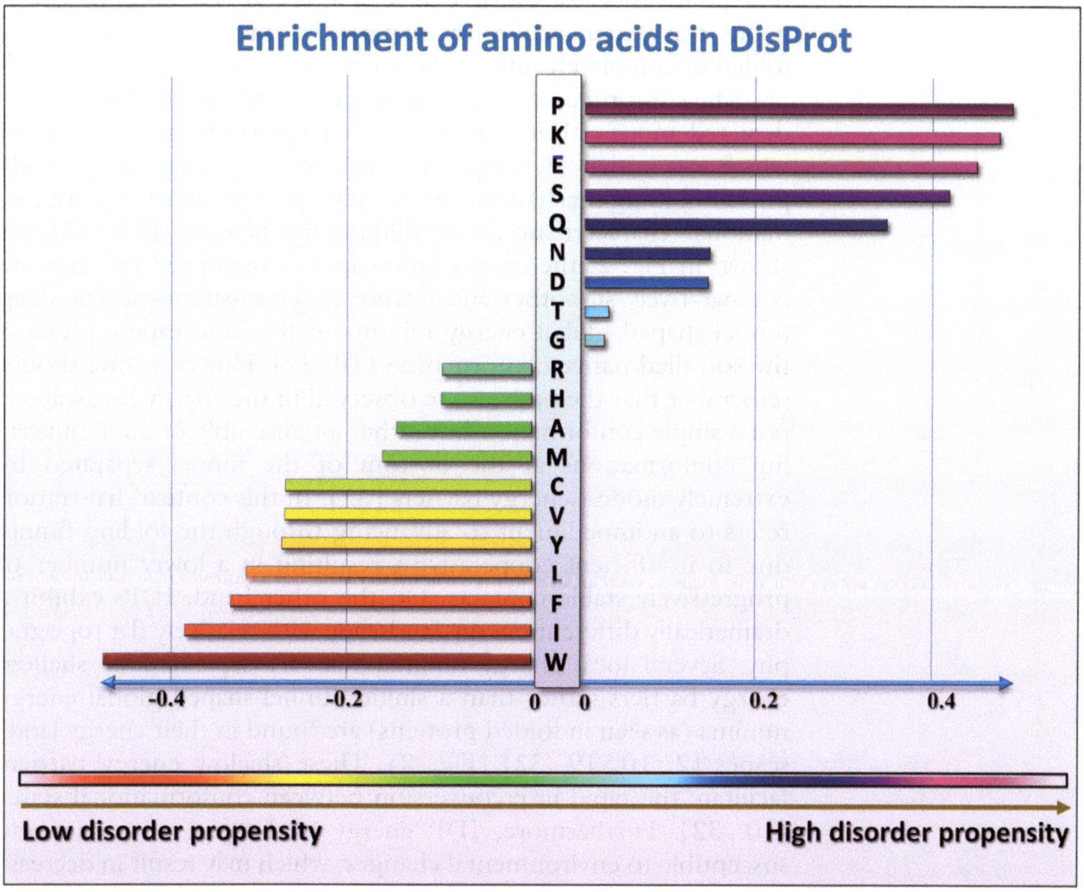

Fig. 1 Enrichment of various amino acids in the IDPs included in DisProt [26] database. The amino acid frequency was calculated for only disordered residues and was further normalized over TrEMBL. (The data represented in the figure was released in March, 2021 and retrieved from https:/www.disprot.org/release-notes on 25/09/2022)

(G), Thr (T), Arg (R), and His (H) mostly show no significant preference for disorder or ordered structure [2, 4, 10]. Usually, IDPs and IDRs are enriched in charged groups leading to relatively high net (usually negative) charge and extreme pI values at neutral pH [2, 30]. In addition, they have comparatively low content of the hydrophobic amino acid residues. Thus, unlike ordered proteins, charge-to-charge repulsion and low or dispersive hydrophobicity resist folding of IDPs. The hydrophobic content in case of IDPs is diversified along the protein length [2, 30]. Additionally, the structural and functional dynamics of IDPs are majorly influenced by the charge patterning due to the peculiar placement of a particularly charged residue into their amino acid sequence [2, 25].

With the cumulative effect of the properties mentioned above, IDPs and IDRs proteins also show remarkable spatiotemporal heterogeneity alongside the substantially reduced structural content. Spatiotemporal heterogeneity is a characteristic in which different parts of the protein show varying degrees of disorder at different time points. Because of this characteristic, IDPs are never homogeneous, but exist as a collection of partially, potentially, or differently folded or completely unfolded conformational states [2, 9, 10, 25].

The conformational heterogeneity of IDPs and IDRs could be depicted more elaborately by their energy landscapes; a concept which was initially developed by Wolynes and colleagues to study protein folding mechanisms by merging protein dynamics, conformational characteristics, and folding mechanisms [31, 32]. As shown in Fig. 2, the energy landscapes in organized proteins are comparatively smoother and feature fewer frustrations. The deep funnel-shaped global energy minima in this landscape represents the so-called native conformation [10, 32]. However, one should remember that the native state observed in the energy landscape is not a single conformation but rather an ensemble of interconverting conformations at the bottom of the funnel separated by extremely modest energy barriers [32]. In this context, frustration refers to an impediment to advancing through the folding funnel due to insufficient cooperativity, resulting in a lower number of progressively stable contacts. On the other hand, IDPs exhibit a dramatically different energy landscape with relatively flat topography. Several local energy minima that are separated by shallow energy barriers rather than a single, funnel-shaped global energy minima (as seen in folded proteins) are found in their energy landscapes [2, 10, 19, 32] (Fig. 2). These shallow energy barriers facilitate the rapid interconversion between conformational states [10, 32]. Furthermore, IDP energy landscapes are particularly susceptible to environmental changes, which may result in decreasing certain energy minima while rising some others [2], thus resulting in extensive conformational plasticity.

As stated by Lermyte, the primary advantage of having conformational plasticity is that it allows a higher binding promiscuity (the

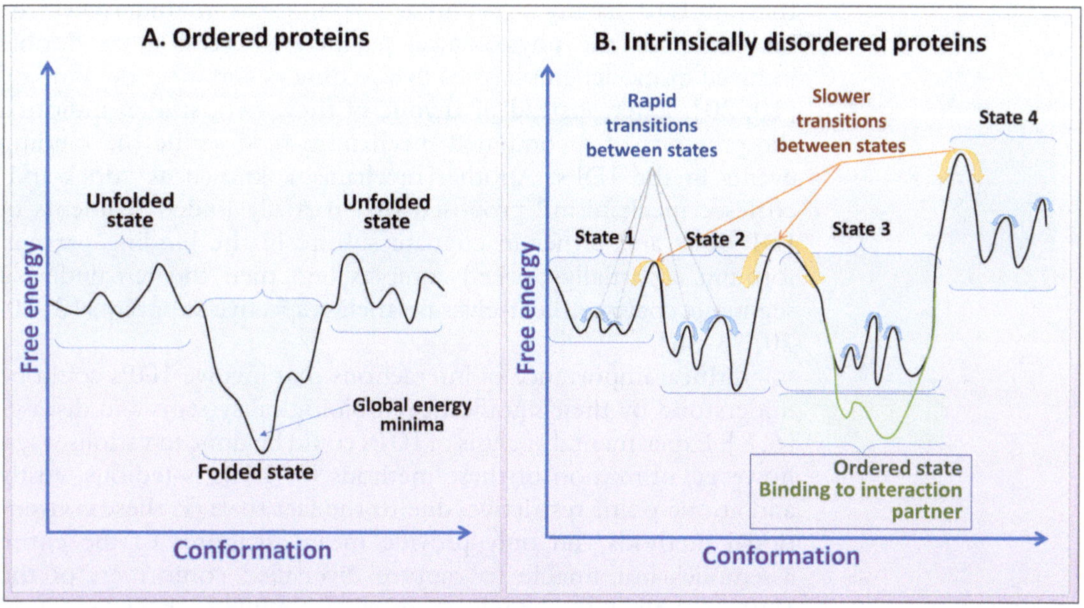

Fig. 2 Schematics of free energy landscape of (**a**) folded proteins and (**b**) intrinsically disordered proteins. The ordered proteins encompass a deep funnel shape energy minimum representing the native state of the protein with minimum frustrations. Contrastingly in IDPs, the energy landscape is considerably rugged with multiple local energy minima separated by both high and low energy barriers

ability to interact with multiple partners) [13], enabling them to act as the "hubs" of protein interactions network [6, 9, 10, 13]. p53 is an excellent example of such an interaction hub which alone can interact with hundreds of binding partners [2]. This huge interactome arises not only due to varied proteoforms but also from the post-translational modifications of 60 out of a total of 393 amino acid residues, many of which are located in IDRs of this protein [4, 13]. Additionally, a protein may have multiple IDRs that may also have various competing binding partners. The protein-protein interactions (PPIs) involving IDPs are highly complex. One-to-many binding, although not prevalent, but has been identified in several IDPs. As a matter of fact, there are examples of IDPs or IDRs with different sequences that bind to and compete for a single structured partner. Similarly, two interacting IDPs may also form mutual synergistic folding [4].

During bindings to interaction partners, IDPs may result in one of two outcomes: (1) they may retain the disorder even after binding with their interaction partners and form "fuzzy complexes"; (2) undergo coupled folding-upon-binding [2, 4, 10]. For the second scenario, different binding mechanisms such as an induced fit or a confirmational selection model were proposed instead of the rigid lock and key model as evident in many structured proteins. The conformational selection model states that an IDP having a pre-folded conformation undertakes the binding with the physiological partner, whereas, more flexible induced fit model emphasizes that folding occurs after the binding [10, 20]. Later, instead of anyone of these two, studies indicated the presence of a combined mechanism to describe the binding events in the IDPs. Another mechanism, known as "dock-and-coalesce mechanism" proposed that partially folded segments of IDP dock and tether in cognate subsite in the binding partner, forming a partially docked complex and then the remainder of segments coalesce themselves on their respective subpoints [3, 10, 20, 33–35].

Critical importance of interactions that involve IDPs could be understood by their significance in biological systems and diseases [6, 8]. Experimental analysis of IDPs could be done in various ways, however, utilization of these methods for IDPs is tedious, costly and at one point restrictive, due to the fact that: (i) these conventional methods can only provide mean attributes of the entire assemblies but unable to capture diversified conformers of the IDPs; (ii) they are unable to provide complete characterization through one method or at one instance; and (iii) due to the noises that take place in the obtained data [8, 10, 20]. In this context, molecular docking can potentially provide more effective and rather direct approach for in-depth analysis and understanding of these features.

Molecular docking is an approach used to predict the atomic-level details of binding interactions between the two molecules in both structural and energetic terms (Fig. 3). To do so, it necessarily involves two important steps: first, predicting the binding modes and second, assessing and ranking the affinity of these modes. Molecular docking is most importantly utilized for protein-ligand interactions, where ligands could be small molecules, peptides, proteins or nucleic acids. Protein-ligand interactions form the basis of different molecular complexes that critically carry out various structural and functional roles in biological systems. Thus, the interaction details of various protein-ligand complexes are significantly important in understanding molecular mechanics of various physiological processes in normal as well as in diseased conditions, for identifying and developing various interaction modifier molecules including various drugs, peptides, proteins, and nucleic acids for therapeutic intervention. Therefore, a varied use of molecular docking approach has been observed in multiple applications including in structural activity studies, in explaining various molecular binding events, in identifying potential leads via virtual screens, in lead optimization, in crystallography studies, etc. The effectiveness of this approach has already been well-established for large variety of ordered proteins. Thus, extending molecular docking studies of IDPs can also open up new doors to understand the IDPs and their interactions.

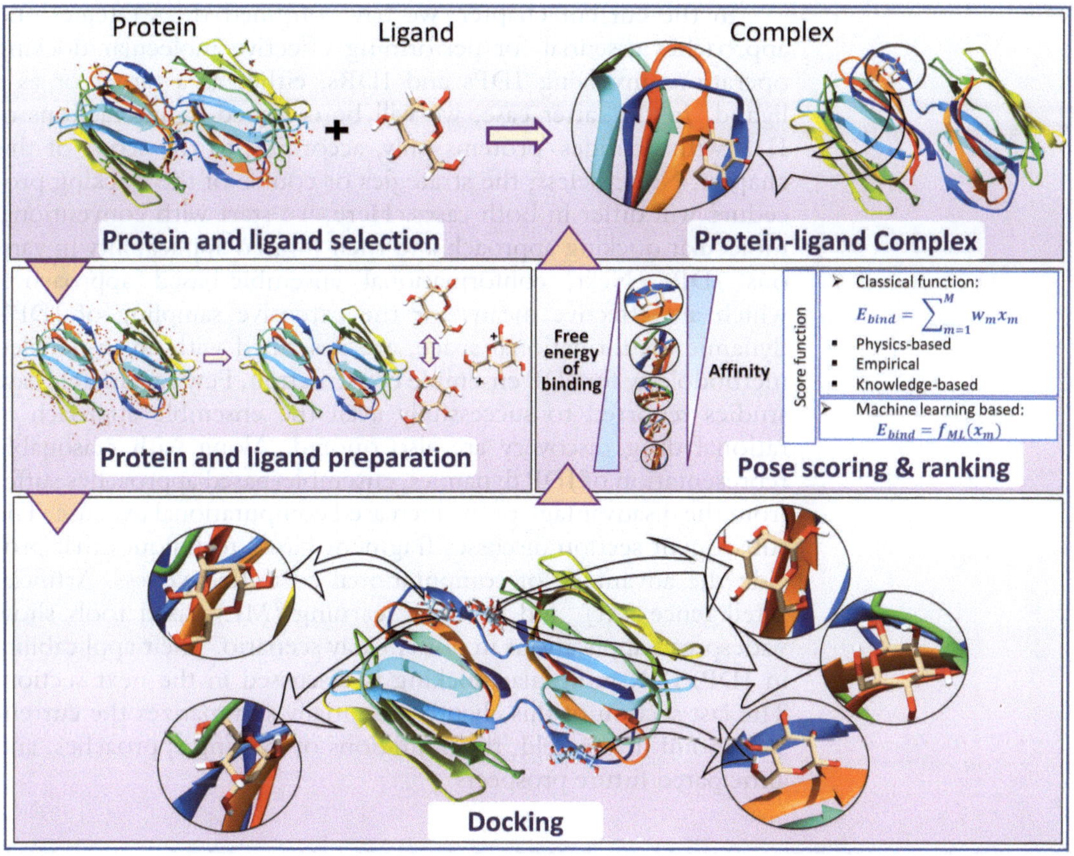

Fig. 3 Steps in conventional molecular docking operation

Unlike the scenario with ordered protein, docking operations face significant hurdles with IDPs and IDRs: (i) molecular docking requires a starting structure to search for the best binding orientation of the ligand molecule, limiting the direct applicability of these docking operations in the IDPs with extensive conformational heterogeneity; (ii) while interacting with various molecules, the IDPs show considerable interaction diversity, which adds on to overall complexity; (iii) the presence of highly dynamic binding pockets (due to dynamic interconverting states) within considerably large structural ensembles, causes pocket-determination challenging. Limited applicability of molecular docking restricts most of the conventional structure-based drug discovery pipelines to be used for IDPs [8, 24, 36–43]. Thus, more unique, tailored and IDP-centric molecular docking strategies are required which can sample the conformational diversity of the IDPs/IDRs and apply it in typical docking operation with reasonable balance between accuracy and computational expense.

In the current chapter, we have outlined the strategies and approaches essential for performing effective molecular docking operations involving IDPs and IDRs, either as a target or as a ligand. In the latter case, we will be confined to interactions of IDP with peptides/proteins only, according to the scope of this chapter. Nonetheless, the strategies or course of the docking procedure will differ in both cases. Here, we start with conventional molecular docking approach and their limited applicability in various IDPs. Next, conformational ensemble-based approaches, which are effective means for the extensive sampling of IDP's dynamic conformational space, are described with an in-detailed methodology for IDP ensemble construction. Few remarkable case studies reported to successfully avail the ensemble approach in rational drug discovery are also quoted. Along with reasonable representation of IDP dynamics, ensemble-based approaches suffer from the disadvantage of the increased computational expense. The subsequent section discusses fragment-based techniques that provide the advantage of computational cost-effectiveness. Artificial intelligence (AI)- and machine learning (ML)-based tools show widespread applications in present-day scenario. Their applicability in IDP/IDR molecular docking is discussed in the next section. The last section of this chapter essentially summarizes the current standpoint in the field, the limitations of existing approaches, and anticipated future prospects.

2 Methods and Application of Molecular Docking in the Field of IDP

The unique features and dynamics enforce the need for specialized approaches to study IDPs. Similarly, molecular docking of IDPs also demands for more customized and unconventional strategies to deal with their complex heterogeneous nature. The following section of the chapter covers these approaches in details.

2.1 Conventional Molecular Docking

Molecular docking aims to model the ligand-target molecular interactions by predicting how these molecules fit together through structural and energetic calculations. Fundamentally, it searches for the dimensional settings of the ligand during its binding to the target. This dimensional setting of the ligand molecule, referred to as "pose", is defined by its conformation, orientation, and position. To find the correct binding pose using molecular docking, possible ligand poses are first generated and then the best pose (which might best represent the original native pose) is selected through ranking system [44–46]. Thus, as shown in Fig. 3, every molecular docking tool consists of two main component features: (i) a tool to generate ensembles of ligand poses (known as "search algorithms") and (ii) a function to evaluate the accuracy of these poses for subsequent ranking (known as "scoring function").

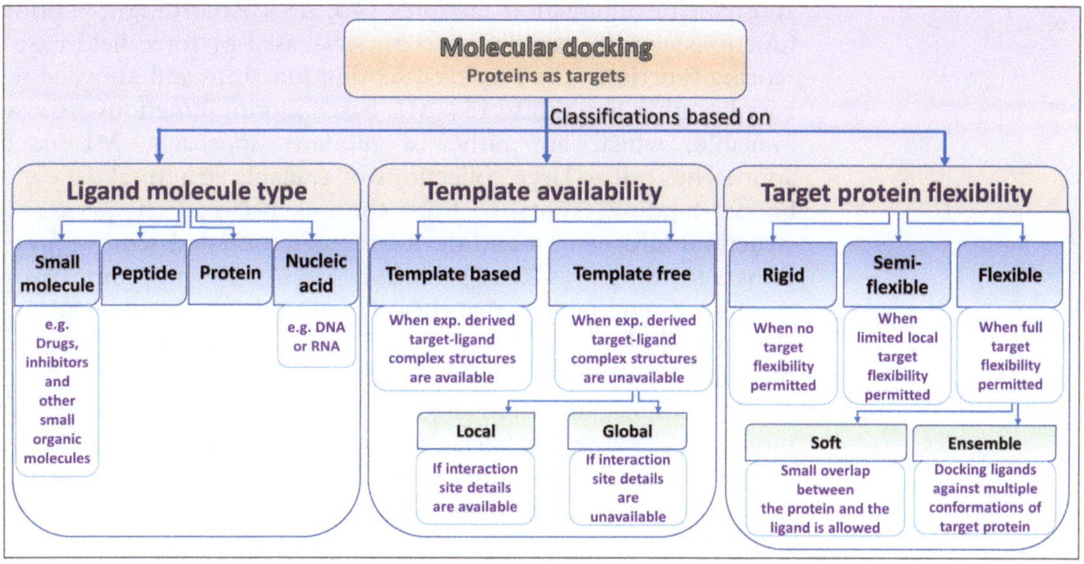

Fig. 4 Types of molecular docking. Here exp. stands for experimentally

The basic aim of the search algorithm is to exhaustively search the conformational space and sample all possible ligand poses with respect to given a target molecule. However, generating all possible conformations and exploring the search space exhaustively might be too expensive in terms of computational resources [44, 45, 47, 48]. Thus, most evolving docking tools adopted a trade-off between molecular flexibility and associated computational cost to model such interactions reasonably. As a part of this trade-off, usual methods allow full flexibility of the ligand while restricting conformational flexibility the target protein fully or partially (Fig. 4). Various search algorithms are available and can be grouped into systematic and stochastic search methods [45, 47, 48]. Systematic search methods are deterministic and involve systematic rotation of all possible bonds of the ligand at predefined intervals. This method is mostly limited to small ligand molecules. Whereas, in stochastic search methods, the ligand conformations and orientations are sampled based on values generated at random at every step. Whether the changes are acceptable or not is determined based on some algorithm-based assessment. These methods are able to produce larger ensembles with broader coverage of energy landscape leading to higher probability of the finding global energy minima.

Once the potential poses are generated, the aim is to determine the pose with the highest proximation to the native pose by using the scoring functions (Fig. 3). The scoring functions rank the poses based on binding affinities or strength of the interactions in the associated protein-ligand complex. The estimation may be based on binding energy, free energy, or a qualitative numerical measure to approximate interaction energies [46, 48]. Accordingly, scoring functions can be classified into physics-based or force field-based scoring functions, the empirical scoring functions and knowledge-based scoring functions [45]. ML-based scoring functions are also available, which are rather a modern approach. ML-based approaches utilize large collections of available structural data and known binding events, to learn the link between energetic and structural information and device the data-oriented scoring functions accordingly [45, 49]. Various assessment and comparison studies strongly suggest that these functions have their own advantages and limitations. None of them is superior in all the tasks but may outperform others in one or few specific ones [45].

In addition to search algorithms and scoring functions, the molecular docking efficiency can also be markedly enhanced if the location of the binding site is provided. The binding site information can be acquired from experimental data and/or various computational approaches [44, 45]. This essentially limit their search within the known binding site leading to a significant reduction in computational power and time [48]. However, blind docking can also be performed in absence of such information (Fig. 4), which analyses the entire conformational space of the target protein to find correct ligand pose fitting appropriately at the putative binding site, at the cost of higher computational burden [44, 45, 47, 48]. Therefore, the availability of the various details can also majorly orchestrate the overall course of molecular docking. Better binding event definition allows for more accurately customized docking, which in turn lowers complexity and computational expense. Thus, based on such customizations, molecular docking can be divided into template-based (when experimentally determined ligand-target complex structure is available), local docking (when information regarding interaction/binding site is available and search space is confined around this site) and global or blind docking (without any prior knowledge) [48, 50] (Fig. 4). Additionally, molecular docking could be rigid, semi-flexible, and flexible based on the flexibility of the target protein and ligand [5, 9] (Fig. 4). A list of docking tools is provided in Table 1.

2.1.1 Applications

The direct applications of conventional molecular docking approaches to study IDPs are challenging due to the lack of their well-defined structures and highly dynamic conformational states. However, in special scenarios IDPs attain partially or fully ordered structures via DOT, such as upon interactions with their binding partner, or due to the formation of fibrillary amyloid aggregates (Fig. 5). In such cases, IDPs could be explored using conventional docking methods where their initial coordinates are derived from experimentally resolved complexes and aggregates. Few of such studies are quoted below.

Table 1
List of majorly utilized molecular docking tools

Docking tool	Description	Major usage[a]	References
GOLD	Based on a genetic algorithm for flexible docking	1, 2	[148]
Surflex-Dock	Based on scoring function from the Hammerhead docking system with surface-based molecular similarity method	1, 2	[149]
pepATTRACT	Proteome-wide in silico protein-peptide docking	2	[138]
HPEPDOCK	Based on a hierarchical algorithm	2	[150]
UNRES	By coarse-grained replica-exchange MD simulations	2, 3	[151]
GalaxyPepDock	Based on interaction similarity and energy optimization	2	[152]
MDockPeP	Based on a statistical potential-based scoring function -ITScorePeP	2	[153]
CABS-dock	Ability to simulate significant backbone flexibility of the entire protein-peptide system	2	[128]
AutoDock Vina	Scoring function is fully empirical, including Gaussian steric interaction terms, a finite repulsion term, piecewise linear hydrophobic and hydrogen-bond interaction terms, and an entropic term proportional to the number of rotatable bonds	1, 2	[70]
AutoDockCrankPep	A novel approach to dock flexible peptides into rigid receptors	2	[154]
DINC 2.0	Peptide-peptide docking based on a meta-docking strategy	2	[155]
ClusPro	Based on fast Fourier transform correlation techniques	3	[66]
PatchDock	Based on shape complementarity principles and symmetry restrictions	2, 3	[156]
SymmDock	Rigid and symmetric docking	3	[156]
GRAMM-X	GRAMM fast Fourier transformation methodology	3	[157]
PIPER	Fast Fourier transform correlation approach with pairwise interaction potentials	3	[158]
RosettaDock	Based on multiscale Monte Carlo-based algorithm for local protein-protein docking	3	[159]
SwarmDock	Based on flexible protein-protein docking algorithm	3	[160]
pyDockWEB	Based on electrostatics and desolation for effective scoring of rigid-body protein-protein docking	3	[161]
HADDOCK	Data driven approach to flexible docking	2, 3	[162]
HDOCK	Based on hybrid algorithm of template-based and template-free docking	3	[163]

(continued)

Table 1
(continued)

Docking tool	Description	Major usage[a]	References
ZDOCK	Interactive docking prediction of protein-protein complexes and symmetric multimers	2, 3	[164]
HawkDock	Based on ATTRACT docking algorithm and MM/GBSA free energy decomposition analysis	3	[165]
FRODOCK	Combine 3D grid-based potentials with the efficiency of spherical harmonics approximations	2, 3	[166]
Pepcrawler	RRT-based algorithm for high-resolution refinement and binding-affinity estimation of peptide inhibitors	2	[167]
PEP_FOLD3	Based on a new hidden Markov model sub-optimal conformation sampling approach	2	[168]
ATTRACT	Based on energy minimization in translational and rotational degrees of freedom of one protein with respect to another protein	2	[169]
HEX	Based on Cartesian-based fast Fourier transform approaches	2	[170]
LEADS_PEP	A benchmark data set for assessment of peptide docking performance	2	[171]

[a]1. Small molecule-protein docking; 2. Peptide-protein docking; 3. Protein-protein docking

Many of the times IDPs form pathogenic aggregates leading to fibril formation that have very rigid structure (Fig. 5). Such aggregates and their interaction studies are of great importance in drug designing. The docking of various ligands and small molecules to such highly ordered conformation of IDPs are often performed using the rigid docking method. A good example of this strategy was demonstrated by Khan et al. [41]. In this study, they targeted intrinsically disordered amyloid-β peptide ($A\beta_{1\text{-}40}$) and its Iowa $D_{23}N$ mutant with number of small molecule inhibitors and their derivatives. These inhibitors were previously shown to have inhibitory effect on amyloid fibrinogenesis. The structures of $A\beta_{1\text{-}40}$ and $D_{23}N$ were retrieved from PDB entry 2M4J and 2LNQ, respectively. Using Discovery Studio 3.5, from both the structures only one chain was extracted and water molecules were removed. Authors have selected total 40 small molecules and their respective structures form PubChem. The target and ligand molecules were prepared by standard protocol using AutoDock tools. Rigid molecular docking was performed using AutoDock 4.2 using Lamarckian genetic algorithm (LGA) [51]. In this study, LGA population size

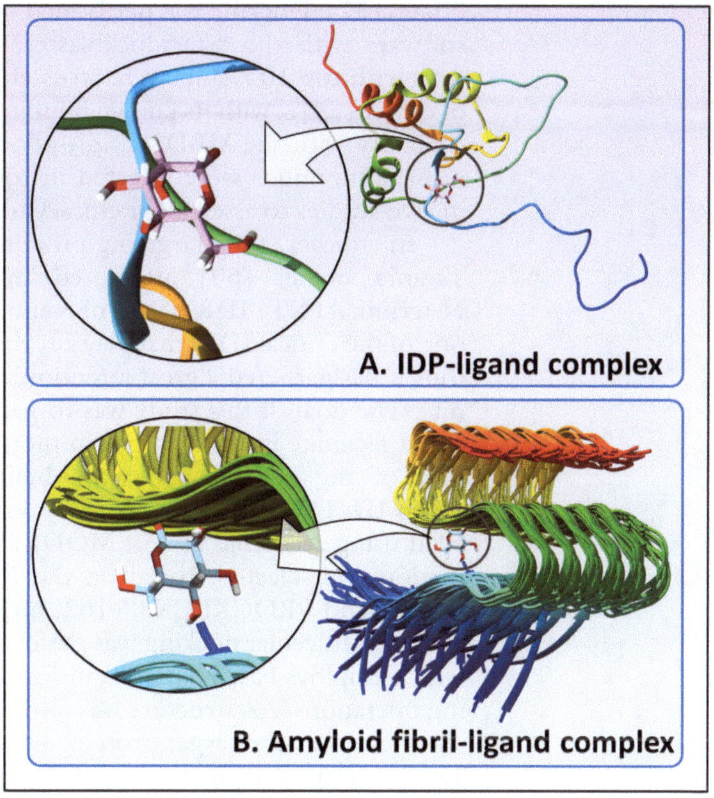

Fig. 5 Schematic representations of docking on (**a**) IDPs that undergo disorder-to-order transition and (**b**) amyloid fibril

was configured to 150 and maximum numbers of energy evolutions were set to 2,500,000. Finally, top 10 ranked compounds were selected and subjected for residue contact determination. In a similar study by Marondedze et al. [52], various polymorphic structures of Aβ were subjected to docking with small ligand molecules using Glide docking tool [71, 72]. Additionally, similar docking studies of IDPs/IDRs with small molecules were also reported in literature [53–55].

In addition to fibrillar aggregates, examples also exist where experimentally-derived folded or partially folded structures of IDPs in complex with respective binding partners were available (Fig. 5), and used to extract IDP structures. These structures were subsequently utilized in classical docking studies. In a recent study by Kim et al. [56], NMR-based conformational ensemble of $MBD2_{360-393}$ complexed with $p66\alpha_{138-178}$ (PDB ID: 2L2L) was retrieved and the lowest-energy conformations were selected for targeted molecular docking. The study was focused to identify antimetastatic agents for IDR of MBD2 protein. Molecular docking-based screen was used to identify most capable interaction partners of MBD2 (at a targeted site) from large ZINC ligand library [57]. Docking was performed using DOCK 3.5.54 [58, 59] software with the "startdockblaster5c" algorithm. The docking runs with top 10 results were cross-checked by using docking run of c-Myc$_{395-430}$ with its known inhibitor 10058-F4, considering its similarity with the MBD2$_{360-393}$-p66α$_{138-178}$ complex. Two topmost compounds were selected in total and pushed forward for in vivo studies to assess their efficacy to in inhibit cancer metastasis.

In another study targeting protein-protein interactions (PPI), Taranta et al. [60] attempted molecular docking between N-terminal (NT) IDR of the p53 and azurin (AZ) protein. p53 is one of the critical IDPs that play key role in cancer progression, and thus it has garnered a great attention among the research community. The goal of this study was to gain insights of the PPI which could facilitate its applications in therapeutic interventions. Structure of the NT domain was obtained from X-ray structures (PDB ID: 1YCQ, 1YCR, 2B3G) and the missing region was modelled using academic 9V1 of MODELLER [61]. The best conformation was selected, based on the Modeler Objective Function scoring and PROCKECK3.5 [62, 63], and subjected for docking studies. Molecular docking was performed using ZDOCK [64, 65] which employs Fast Fourier Transform (FFT) for rigid body docking operations. AZ structure was retrieved from X-ray structure and was treated as the target protein. From top 2000 conformations, 135 structures aligned with mutagenesis experimentation were filtered and further subjected for RMSD-based clustering using ClusPro docking server [66]. Finally, 20 structures were derived and further used for refinement using MD simulations. The best AZ-p53 NT binding mode were selected using free energy calculations and computational mutagenesis approaches.

2.2 Ensemble-Based Approaches

Unlike globular proteins having a well-defined global minimum, the energy landscapes of the IDPs or IDRs are relatively flat with multiple local energy minima, separated by modest energy barriers (Fig. 2). The peculiarities of the primary sequence of IDPs do not allow them to fold into a stable tertiary structure, instead drive the formation of ensemble of the unfolded or partially folded states. These features make modelling of these proteins extremely challenging. Comprehensive modelling of IDPs therefore, necessitates the generation of ensembles of structures that sufficiently reflect the protein's spectrum of accessible conformations.

It is very challenging to provide the reliable description of the conformational ensemble of IDPs or IDRs due to their dynamic nature. Various local and long-range structural organizations of the IDPs and IDRs are frequently analyzed using biophysical techniques like nuclear magnetic resonance (NMR), circular dichroism (CD), single molecule spectroscopy (SMS), fluorescence resonance energy transfer (FRET) microscopy, and small angle X-ray scattering (SAXS) etc. [4, 10, 67]. These conventional experimental

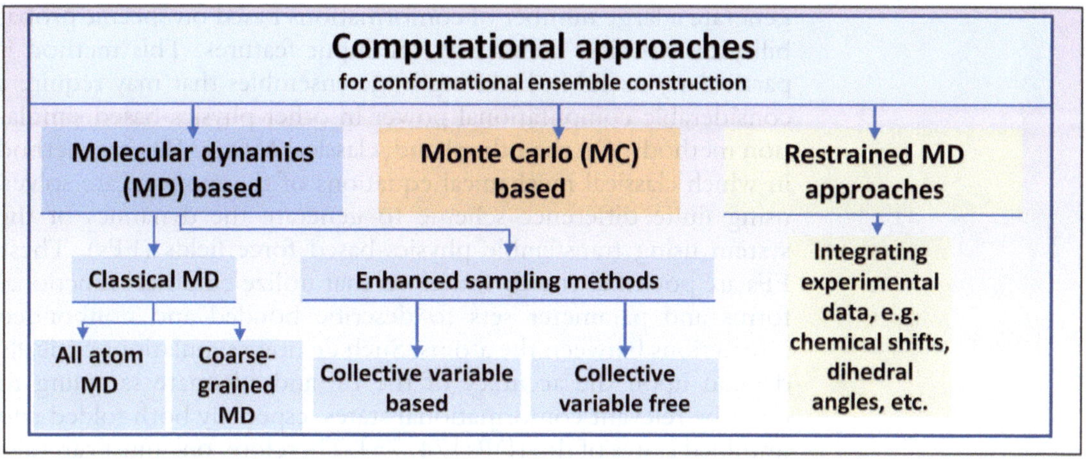

Fig. 6 Methodologies for conformational ensemble construction

methods can only provide ensemble average of the entire assemblies but unable to capture diversified conformers of the IDPs. Alternatively, computational approaches are essential in deriving the heterogeneous confrontational ensemble. The widely employed computational methods for studying IDPs or IDRs include molecular dynamics (MD) simulation, enhanced sampling techniques, experiment-guided restrained MD simulations, and Monte Carlo (MC) simulation (Fig. 6) [4, 10, 67–69].

For docking studies involving IDPs, either as target or as ligand, the first critical need is reliable construction of its conformational ensemble. One must choose computational approaches to sample the range of possible conformations. Once the conformational libraries are generated, typically the next objective is to find the smaller subset of conformations and their relative weights. If available, experimental data is frequently utilized to refine/validate the most representative conformations. Finally, the chosen conformations are submitted to various molecular docking systems such as AutoDock Vina [70], Glide [71, 72], HADDOCK2.2 [73], and others. Here we summarize the computational methods widely utilized to sample IDP conformations (Fig. 6). The in-depth descriptions of these methods and their applications in IDP can be found elsewhere [67–69].

2.2.1 Computational Techniques to Construct IDP Conformational Ensembles

Multiscale Simulations

A particularly attractive approach to generate disordered IDP ensembles is to use MD or MC simulations. MC methods are the computational algorithms that leverage the repeated random sampling to obtain putative results that are otherwise difficult or impossible to obtain using other approaches. During ensemble generation, the same underlying concept is utilized to randomly generate a large number of conformations based on specific probability distribution defined by the input features. This method is particularly useful in handling huge ensembles that may require a considerable computational power in other physics-based simulation methods. On the other hand, classical MD (cMD) is a method in which classical mechanical equations of the motions are solved using finite difference scheme to generate the dynamics of the system using transferable physics-based force fields (FFs). These FFs are potential energy functions that utilize empirical functional forms and parameter sets to describe bonded and nonbonded interactions between the atoms. Such de novo simulations critically depend upon the accuracy of the FF and adequate sampling to describe relevant conformational states, especially both folded and unfolded states of the IDPs [74, 75]. Therefore, one must carefully select the FF and method to warrant the reliable simulation of IDP.

Many non-polarizable FFs that are optimized for folded proteins are not suitable for simulating IDPs, largely due to overstabilization of PPI and poor description of secondary structure content of IDPs [10, 69, 76, 77]. The over-compactness of the IDPs can be alleviated by balancing the protein-protein, protein-water, and water-water interactions [8, 10, 69, 76, 77]. Therefore, various FFs optimized specially for IDPs are used together with a refined water model, such as CHARMM36m/TIP3P*, FF19SB/OPC, and a99SB-disp/TIP4P-D [78–80]. As evident by several benchmark studies, these refined FFs markedly improvised the molecular descriptions of both ordered and disordered proteins [81–85]. However, nonpolarizable FFs have fundamental limitations when it came to accurately describing charged groups that were buried at protein-protein interfaces or exposed to solvent [69, 77]. To overcome this shortcoming, efforts were made to develop polarizable FFs that explicitly approach the electronic polarization by utilizing the empirical models. The fluctuating charge model and the Drude oscillator model in the context of CHARMM FF, the fluctuating charge models in the context of OPLS-AA FF, and detailed multipole expansions and more complicated MM potentials in the context of Amoeba, are a few examples of modern polarizable FFs [69, 77]. These FFs provide better description of the charged and polar protein motifs, but are associated with higher computational cost [86]. Additional studies are required to substantiate their applications in IDP simulations [69, 77].

All-atom (AA) FFs have wide applications in the study of structure and dynamics of biomolecules at the atomistic scale. However, despite of higher accuracy, convergence is a major limitation in atomistic simulations due to the large system size and long simulation time associated with biological condensates [69, 87]. Multiscale approach may facilitate in bridging the gap. Particularly in case of

IDP, large simulation boxes are require to accommodate large and fluctuating protein radius of gyration [22]. Simulations with implicit solvent model can enhance the computational efficiency by removing the solvent degrees of freedom [69, 88]. Coarse-grained (CG) models significantly reduce the degrees of freedom associated with both proteins and solvent by mapping groups of atoms into single beads or particles. CG-MDs are rather quick and can be utilized to sample semiquantitative secondary structures but come at the cost of lower accuracy and resolution. The most popular CG force fields in biomolecular simulation is the MARTINI model [89, 90]. However, MARTINI uses an elastic network to constrain protein to its reference structure and thus does not allow large-scale motion like the conformational transition of protein [91, 92]. MARTINI FF has been previously used to simulate IDP, but with modification in the FF [93]. The recently developed Go-MARTINI model replaces the harmonic potential with Lennard–Jones interactions based on the contact map to preserve the native protein structure [94]. In addition to MARTINI model, there are several CG models, such as self-organized polymer CG model (SOP-CG) [95], hybrid resolution model (HyRes) [96], AWSEM-IDP model [97], etc., that have been specifically developed to study IDP [69]. CG FFs are particularly important in integrative methods where ensembles are constructed or optimized with special considerations of available experimental data. With advancement in the various CG-based methods, multiple examples quoting the usage of CG models for intrinsically disordered could be found in literature [98–103].

Enhanced Sampling Methods

Besides FF accuracy, simulation of the IDP dynamics also demands sufficient sampling of its manifold functionally relevant conformations within a reasonable simulation time. This could be attributed to the large accessible conformational space of IDP where various conformational sub-states may be separated by high-energy barriers. cMD typically limited to nanosecond or microsecond time scale, often fall short in crossing those barriers and exploring full conformational landscape of IDP, even together with the best protein force field. Applications of various enhanced sampling methods can significantly accelerate the conformational sampling of IDP. These advanced techniques facilitate crossing the energy barriers at a faster rate and with a lesser computational cost. These techniques also allow the elucidation of the structural and thermodynamic properties of the system.

Various enhanced sampling strategies are broadly categorized into two groups based on their dependency on predefined collective variables (CV) (Fig. 6). Metadynamics is one of the most used CV-based sampling method [104]. In conventional metadynamics simulations a history-dependent biasing potential is added to the true potential energy. The biasing potential follows the molecular system along predefined CV(s) and accumulates in the attractive basins, which allow crossing the barrier. The amount of bias added gives a measure for the underlying free energy landscape. Over the years, several improved metadynamics variants have been developed to improve the convergence that includes, well-tempered metadynamics and bias-exchanged metadynamics [69, 105]. However, the central limitation of these techniques is that their efficiency is strongly dependent on the quality of the selected CV or CVs [69]. On the other hand, the CV-free enhanced sampling methods do not require any prior knowledge of the CV and appear as a better choice for simulating the high-denominational conformational fluctuations in IDP. The most popular CV-free enchaining technique is replica exchange molecular dynamics (REMD) [106], in which multiple replicas of the system are run in parallel at different temperatures and apply the Metropolis rule to allow exchange of replicas. This technique is widely applied in case of IDP conformational sampling [10, 23, 74, 77, 107]. However, the requirement of parallel simulations of multiple replicas dramatically increases the computational cost. Alternative methods that require lower computational cost are accelerated MD (aMD) or Gaussian accelerated MD (GaMD) [108]. In these techniques a bias potential is added to the true potential in such a way that the potential surface in the vicinity of the minima is raised, while those closer to the barrier or saddle point are not affected. This allows the system to move over the potential barriers with a reduced computational time. In simulating IDP dynamics, a special notation can also be given to the generalized Newton–Euler inverse mass operator (GNEIMO) method [109–111]. This method is particularly useful in simulating lengthy timescale transitions which is rather difficult in AA-MD. The design and functioning mechanism of this method is such that it could be efficiently applied for IDPs and IDRs and can also be combined with other techniques like Temperature Replica Exchange [112]. In addition to the enhanced sampling methods, Markov state model (MSM) provides an approach to determine the kinetically relevant sates and the inter-conversion between the states by analyzing simulation trajectories [113]. However, their effective applicability to IDPs is yet a challenge due to enormous degree of freedom in IDPs [10].

The interested readers may find the detailed information about these methods from the source-papers cited in this chapter as well as from other relevant literature available online.

Information-Driven Ensemble Generation

MD simulations with either classical AA simulations or enhanced sampling methods or with CG models have proved to be crucial foundation pillars to derive good quality conformational ensembles for the IDPs by allowing researchers to look into their energy landscapes. On the other hand, the experimental data of IDPs derived using various biophysical techniques allow to glance into certain important features and provide the good platform to validate the computationally derived observations. However, generating a realistic IDP ensemble solely from the experimental data is rather difficult as they provide ensemble-averaged representation of IDP conformers [68]. In the same way, theoretical models also suffer from the limitations of accuracy and sampling timescales [8, 68]. Therefore, a combined approach using integrative methods serves the purpose and could be efficient enough to construct accurate IDP conformational ensembles congruent with the experimental data [67, 68]. This type of synergic approaches for generating relatively more realistic ensembles could be explored by two main strategies as discussed below.

In the first method, the predetermined experimental information is introduced into the potential function, as a separate term, directing the MD simulations towards sampling the specific region of the conformational space that complies with the experimental data. Importantly, with the case of IDP, the restrains (e.g., distance restraints or dihedral restraints) should be applied to (not just only one conformation but) whole ensemble [67, 68]. However, these methods have their own inherent advantages and disadvantages. Using relatively large number of constrains (devised from experimental data), may lead to more accurate representative ensembles, but how many restrains are necessary is very difficult to predict. Thus, both number of constrains and the type of experimental data used, play crucial role in deriving good quality ensembles.

The second method uses experimental data to choose the appropriate conformations from previously formed set of conformational libraries. The generation of the initial conformational library can be done using the above-mentioned computational techniques. Subsequently, a smaller subset of conformations from the library is selected, so that the average-data calculated over the selected subset matches closely to the experimental data. The conformational selection is often performed by using either MC method like ENSEMBLE tool [114], or evolutionary methods like ASTEROIDS [115], or sample and select methods (SAS). However, one should keep in mind that being in accordance with experimental data does not guarantee that the ensemble is accurate.

2.2.2 Applications

The ensemble approach provides a promising opportunity to efficiently sample the conformational heterogeneity of the IDPs. Therefore, this approach has been extensively used in various IDPs/IDRs to satisfactorily deal with the structural heterogeneity. Some of such chosen studies where ensemble approach has been efficiently employed are cited below.

In IDP-Centric Drug Virtual Screening

The modern drug designing and discovery methods are majorly impacted by the molecular docking-based virtual screening strategies, where numerous docking operations are utilized to screen potential ligands from huge ligand libraries. The conformational ensemble approach put up an important opportunity to target IDPs using such virtual screening operations. In a remarkable study, Ruan et al. [23] proposed a hierarchical computational strategy for Intrinsically Disordered Protein Drug Virtual Screening (IDPDVS) base on multi-conformation-affinity strategy as a novel path for rational drug designing in IDP field. They applied this method to discover effective binding partner of p53 terminal transactivation domain 1 which can activate the p53 pathway as an anticancer strategy. The technique was devised based on the prior observation that IDP drugs should be able to bind with multiple conformations of the IDP. In the first step, two independent MD simulations were conducted for conformational sampling: (i) Classical all-atom MD simulation with explicit solvent (TIP4P-EW), where the initial structure derived from p35 TAD1-MDMX complex and (ii) REMD with implicit solvent using random coil conformation as the starting structure. For both type of simulations AMBERl4SB force field was used [116]. Classical all-atom MD and REMD simulations were performed for 900 ns and 8.5 μs, respectively, and were able to generate similar conformational ensembles. Further, for subsequent derivation of druggable conformations and cavity detections, the authors performed two-step clustering. At first, they performed local clustering based on critical residues involved in p53 TAD1-MDM2 complex. Subsequently, global clustering was performed on p53 backbone atoms using RMSD-based cut-off. In this study, the ligand binding sites were then discovered by CAVITY tool [117, 118]. Eight most druggable conformations were sent forward for docking operations to screen SPECS ligand library with around 20,000 compounds. To balance the computational cost and precision, two-step docking was conducted. To tackle the wide distribution of docking scores among eight conformations, ranking of docking poses (instead of docking score) was approached. First, all the compounds were docked with Glide HTVS model [71, 72] and then 50% of the obtained results were redocked using Glide SP mode. Finally, top 1% compounds populated in at least 3 different conformations were selected as potential drug candidates and considered for further experimental testing.

In Developing Peptide Inhibitor Using Structure-Based Method

Apart from the virtual screening operations, molecular docking is also critically used in designing peptide inhibitors targeting IDPs. In a recent study by Wichapong et al. [119], a structure-driven peptide designing approach was utilized to target the N-terminal region of the histone H4 proteins (NT-H4), by mimicking its interactions with the histone acetyltransferase type B protein (HAtB). NT-H4 is intrinsically disordered and plays active role in cell lysis and tissue damage. In this study, the initial structures of the IDP and the HAtB were extracted from their respective crystal structures. Subsequently, MD simulations, docking, and binding free energy calculations were performed to study the PPI. Guided by the contact residue details at the binding interface, the authors designed a peptide that can disrupt the PPI in question through neutralizing the IDPs. In this study, MD simulations were performed using AMBER14SB force field and TIP3P water model for 100 ns. Docking was performed using HADDOCK2.2 web server [73]. The docking-derived structures were refined using MD runs and binding free energy (BFE) was calculated. For selection of likely poses and/or peptide ligands relative BFE was considered to choose the lowest BFE scoring pose. The selected structure was checked for better stability and residues involved in binding interactions, by extended MD run of 20 ns. This way the sequence EAANYIDETDPS was extracted from HAtB and that served as starting structure for peptide design. Total, nine different cyclic peptides were designed and their binding mode with H4 were predicted using HADDOCK2.2 webserver. Subsequent refinement and BFE calculations were done as like previous step. The same workflow was also utilized in this study for designing inhibitory peptide targeting the IDR of histone H2A protein. Finally, for both cases, lowest BFE-value peptides were sent for further experimental validation to assess their inhibitory activity.

In Selecting Drug Against the "Fuzzy" IDP Interface
During recognition and binding with interaction partner, certain IDPs may undergo extensive conformational transitions, following DOT and the binding may result in a partially or fully folded structures. When IDPs retain a significant disorder even after binding with its physiological partners, these types of IDP-complexes are called "fuzzy complexes". Therefore, the virtual screening strategies as discussed above might not work for such IDP fuzzy complexes and may require alternate strategy. To address this challenge, Neira et al. [120] proposed a synergic approach that utilized biophysical, computational, and biological approach to identify a drug that can inhibit the intrinsically disordered nuclear protein 1 (NUPR1). The overexpression of this protein is associated with the pancreatic ductal adenocarcinoma (PDAC). In this work, the authors started with screening 1120 food and drug administration (FDA)-approved drugs using experimental studies. Fifteen compounds were selected after biophysical analysis and were subjected to in silico analysis. The structures of the both NUPR1 protein and ligands were made by molecular visualization tools. Initial extended

chain structures were collapsed in very short MD simulation through in vacuo condition. Further refinement was done by MD simulations using AMBER ff99SB-ILDN force field. The conformations of the proteins were extracted from the trajectories by cluster analysis and then used for docking studies using AutoDock Vina [70], allowing full flexibility of the ligand. For some special cases, the probability of compounds to interact with multiple protein residues was also explored by both computational and experimental approaches. Finally, those compounds which are seen to interact with the protein were selected and refined through MD simulations. The docking results and experimental data together showed that all the 15 compounds selected for in silico study were found to be forming fuzzy complexes with the target IDP. The results from various biophysical, computational, and biological methods employed in this work were cumulatively used to identify the most effective compound, which completely arrested the PDAC development in mice with induced tumor [120]. Thus, this study provides an excellent example of exploiting molecular docking approach in conjugation with experimental methods to effectively explore the fuzzy complexes using the virtual screening strategy.

In Probing PPI Through MC-Assisted Docking
Alongside the most commonly used MD simulations, MC simulations are also used for ensemble generation. These methods provide distinctive advantages over MD simulations in handling much larger conformational ensemble and are particularly useful in treating IDPs and IDRs [121–125]. Jagannathan et al. [121] in a recent study have employed MC computational modelling of excluded volume to generate conformational ensemble of IDRs of two proteins, Epsin and Eps15, known to be associated with Clathrin-mediated endocytosis (CME). Very large conformational ensembles (3 million) were generated using FoldTraj from TraDES packages [126, 127], following the application of dihedral angle restraints as found from the experimentally determined structures of these proteins with their CME binding partner, alpha subunit of Adaptor protein-2 (AP2α). Subsequent to ensemble generation, rigid docking (i.e., simple docking-by-superposition) approach was employed to dock the IDR conformers to AP2α. *Salign* from TraDES module was used to superimpose the IDR conformers to target protein structures from available complexes. Docking-by-superposition resulted in millions of docked poses. Hard-atom clashes were used as a decisive criterion to determine an "accessible motif". Their results suggest that the PPI may involve one or more motifs and binding-upon expansion of IDRs [121].

In Exploring PPI Using CG-Based Model

The CG models provide a simpler representation of IDPs and assist faster exploration of the elaborated IDP energy landscapes [98–103]. In this context, CABS-Dock, a CG-based protein-peptide docking tool demands special mention [128]. Remarkably, this tool allows simulation of the protein-peptide system with significant backbone flexibility within reasonable simulation time. The tool was originally developed by Kmiecik et al. in 2019 for protein-peptide docking with backbone flexibility [128]. The same group latter shows the potential utilization of this tool in IDPs [128, 129]. CABS-Dock employs a coarser representation in which each residue of proteins is reduced to up to four beads. The basic workflow of the CABS-dock involves usage of MC asymmetric Metropolis sampling scheme. Initially, a total 10,000 frames from 10 different trajectories are selected, out of which 1000 are chosen based on the energy scores and are subjected to clustering by *k*-medoids resulting in 10 final models. These are then reconstructed using MODELLER tool [61] to their respective all-atom representations. The CABS-Dock tool is thus particularly useful to large protein-peptide systems through flexible docking.

All these examples reflect the extensive utilization of the ensemble approaches to sample the huge and dynamic conformational heterogeneity seen in IDPs and IDRs. Altogether, these approaches provide necessary and (although tedious but) viable means to sample and utilize the IDP structural dynamics in rational drug discovery.

2.3 Fragment-Based Approaches

The conventional rigid docking methods suffer from the limitation due to its dependency on the availability of the starting structure, which might be significantly resolved through ensemble approaches. However, the ensemble-based approaches are computationally expensive and tedious. An alternate fragment-based approach can be used which utilizes the underlying idea that although IDP shares a broad interface with its binding partner, the association is mainly governed by the specific regions. Thus, fragments that are long enough to undergo localized interactions, could provide computationally more cost-effective option to determine such specific interactions. A few relevent studies showing the application of fragment-based approaches are discussed below.

2.3.1 Applications

Santofimia-Castaño et al. used the fragmentation approach to understand the PPI interactions between intrinsically disordered NUPR1 and α-chain armadillo unit of Plakophilin 1 (ARM-PKP1) [39]. Both of these proteins are known to be associated with several cancers [38, 39, 120]. Thus, the authors used molecular docking in conjugation with other experimental techniques to model the binding interactions. To do so, the authors retrieved the X-ray crystallography structure of ARM-PKP1 (PDB ID 1XM9) and missing disordered loops in this structure, were predicted using MODELLER [131, 132]. Subsequently, the seven-residue long fragments spanning the whole sequences of NUPR1 were docked onto the target protein using AutoDock Vina [70] and graphical interphase AutoDock tools 1.5.6 [133]. The outcomes of the docking operations were used to determine the interaction hotspots on both target protein and IDP. Along with this study, this fragment-based docking approach was shown to be effective in understanding binding of NUPR1 to multiple type of binding partners [36–38, 40]. Additionally Zhu et al. [43] showed that fragment-based approach can also be extended to search for the binding hotspots in IDPs that could act as interaction sites for the potential drug candidates through its combination with molecular docking.

It is broadly hypothesized that during coupled-folding upon binding mechanism of IDP, partly folded secondary structures that are initially formed may establish the initial contacts, and then the remaining IDP segment coalesces around the binding partner. The fragment-based approaches as described above appear to be useful for sampling the initial contact events, but are only limited to sample the localized interactions of the fragments, instead of the whole IDP. Additionally, it is not possible to extend the resulting binding poses of these fragments to form the full length IDP pose. Therefore, these methods are mainly used to find the so called hot-spot regions of the IDPs which may initiate the association process, but the subsequent coalescing cannot be anticipated.

To deal with this issue, another tool called IDP-LZerD was proposed by Peterson et al. [134], designed specifically based on the dock-and-coalesce model. IDP-LZerD tool can dock IDPs up to 15–69 residue long onto any ordered protein with known structure. This tool particularly follows the dock-and-coalesce model by breaking down the challenge into two sub-problems: anticipating IDP conformations and subsequently docking them onto the target protein [134, 135] (Fig. 7).

The basic workflow of the IDP-LZerD has been shown in Fig. 7. IDP-LZerD begins by fragmenting IDPs into nine-residue long windows with three-residue overlap. These tiny fragments are further predicted for their conformations and probable secondary structures. Subsequently, they are docked globally on the ordered target protein. These steps are referred to as "docking phase". The adjacent docked fragments are then discovered and assembled in the next "coalesce phase" to produce a full IDP conformation. Based on geometric constraints and a scoring system, the optimal fragment combinations are chosen and are refined using MD simulations. In brief, IDP-LZerD divides the whole process into four distinct but manageable steps (Fig. 7). Thus, it is also comparatively

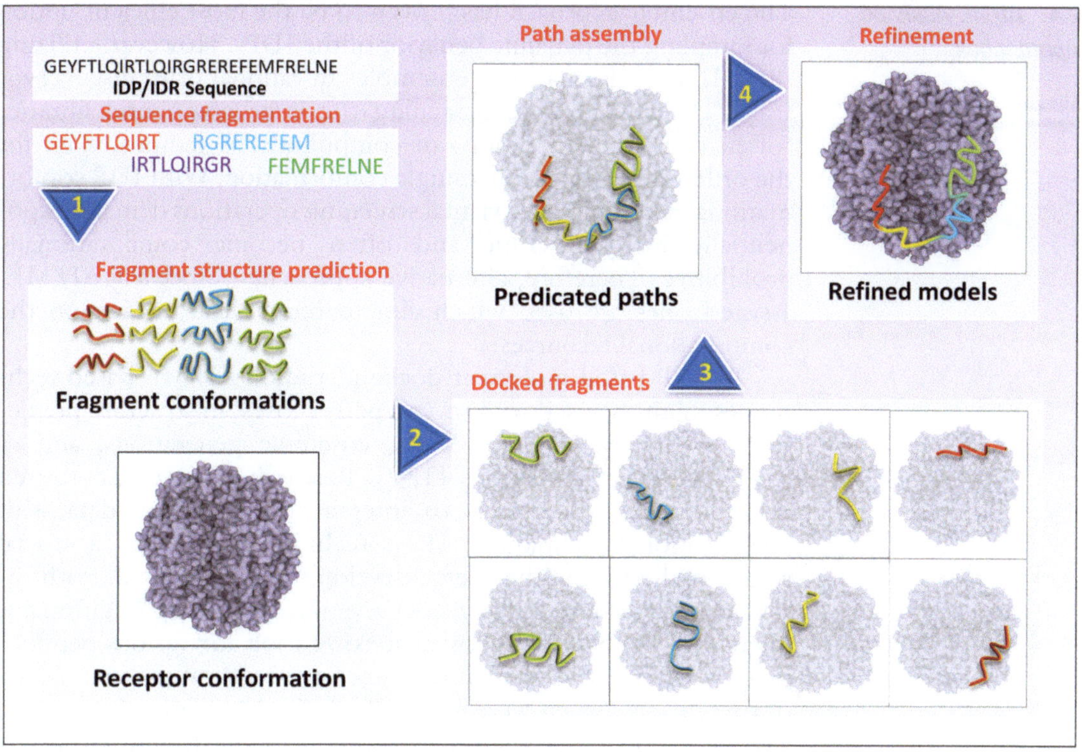

Fig. 7 Steps in IDP-LZerD workflow, adapted from [134]

less expensive in computational terms. Interested readers are encouraged to follow source papers [134–136] for in-depth details of the tool.

The performance of IDP-LZerD tool was tested in various studies [134–137. Studies by Peterson et al. [134] and Christopher et al. [135, 136] showed that IDP-LZerD outperforms conventional docking tools, like pepATTRACT [138] and CABS-dock [128]. In addition, the accuracy of the IDP-LZerD tool was also verified in the CASP-CAPRI protein assembly prediction challenge, 2019 [137]. The results indicated further need to improve the prediction performance of IDP-LZerD for non-globular and more flexible complexes. Another recent study showed that, IDP-LZerD was found to be crossed over by AlphaFold-Multimer [130], particularly in the high-resolution matrix. Overall, IDP-LZerD suffers from two main limitations. First, this method ignores the conformational changes in the target protein and the second, lies in its core assumption that a complete set of its segments can interact with the binding partner. Thus, IDP-LZerD is unable to reliably model the interactions involving the fuzzy complexes. Additionally, various segments of the IDPs are believed to be showing mutual impact on one another which might not be sampled in this fragment-based method.

2.4 AI/ML-Assisted Approaches

The ensemble approach has proven to be the most efficient option for sampling the dynamic behavior of the IDPs. However, utilizing these IDP conformational ensembles in rational drug discovery is still a challenging task. Screening the large compound libraries (often consisting thousands of compounds) is tedious, even for the ordered proteins with a single conformation. With IDP conformational ensembles, the virtual screening operations demand exponentially increased time and often become computationally prohibitive. Therefore, alternative approaches, especially AI/ML-assisted ones are used which aim to reduce the burden on the computational resources.

In IDP-based molecular docking studies, AI/ML-based techniques might be used with several perspectives: in structure prediction, predicting IDPs or IDRs, ensemble generations, and in multiscale IDP simulations [139]. Researchers have also started using AI/ML-based tools to integrate experimental data with computational techniques. These techniques may play a crucial role in orchestrating the overall development of this field. Enthusiastic readers are referred to recent reviews [139, 140]. Additionally, many AI/ML-based molecular docking tools for various peptide-protein and protein-protein docking have been introduced. Many of these tools, which were originally devised for ordered proteins, also show potential to be used for IDP docking studies. However, examples for such applications are rather scarce and often require special care in the context of IDP.

2.4.1 Applications

One classic example of ML-guided molecular docking strategy is the recent work by Baggett and Nath [42]. In this work, molecular docking and an ML-based screening were used to identify small ligands for the intrinsically disordered tau protein, which is known to be associated with Parkinson's disease and dementia. The authors utilized an enhanced sampling technique called Repeated Simulated Annealing (ReSA) MD to first sample the conformational space and generate the conformational ensemble. Subsequently, they adopted the fragment-based approach along with RMSD-based conformational analysis to get 10 most populated clusters from each 25 segments. Altogether, these 250 selected structures are used as docking targets for small molecules from the ChemBridge CNS collection library through virtual screening. To reduce the computational cost of docking large compound libraries (often consisting of $>10^4$ compounds) to the all 250 targets, authors took the help of ML techniques. CACTVS attributes [141] were used as chemical feature descriptors to train the partial least-squares regression (PLSR) analysis model [142]. PLSR was then employed to perform screening operation and increase the overall efficiency using the correlation with the docking scores. In the pilot run, 1000 random compounds were selected to dock with the target conformations using AutoDock Vina [70]. Among these compounds, 990 were randomly selected to construct the training set and remaining 10 were used for validation. Similar sub-sampling cycles were repeated ten times. The overall accuracy of this ML-model was determined using Pearson (Spearman rank-order) correlations scores, indicating the model was reasonably accurate [42]. Subsequently, they fed this trained ML-model with 50,000 compounds of the ChemBridge CNS library and selected 1000 top scoring molecules. These selected molecules were docked with only a druggable segment of the IDP sequence for the sake of computational efficiency. The authors then used the docking results to update the PLSR ML-model. The updated PLSR was again used to predict the best 1000 ligands and the cycle continued until the frequency of finding new best compounds was dropped to ~40%, and docking scores of top-ranked compounds became near constant. At this stage, this model was considered converged and further used to predict docking scores of all the compounds with respect to all the IDP segments and used these scores to select the best candidate compounds for extensive experimental study. The results of the study indicate the ability of this toolkit to adequately deal with the conformational heterogeneity of IDP and its successful utilization in identifying the aggregation inhibitors for tau protein [42].

In another interesting study, Chong et al. [143] have proposed a Reversible Upper Confidence Bound (rUCB) algorithm aiming to solve the bottleneck of virtual screening by significantly reducing the averaged docking number per ligand (as low as 2), while achieving high screening accuracy with low performance loss. In this approach, the aim of virtual screening is to find top m ligands with highest binding affinity from library of N ligands ($m << N$) so that the selected ligands have the highest association constant for the target protein. However, in practice, the ligand is actually docked with only a finite number of conformations, instead of a whole ensemble, which leads to uncertainty in the estimation of binding constant and overall decrease in accuracy. To solve this issue of virtual screening due to the IDP conformational ensemble, Chong et al. has introduced the rUCB algorithm (please find the algorithm details in their article [143]). To benchmark this algorithm, large ligand libraries were docked on the conformational ensemble of c-Myc protein. In a synthetic dataset used in this study (consisting of 283 ligands and 16,716 conformations of c-Myc IDP), rUCB acquired 94% precision when on an average, a ligand was docked only twice. With docking times taken by rUSB method, the usual non-learning algorithm was able to achieve the accuracy of only 47%. Using another dataset (consisting of 828 ligands and 16,716 conformations of c-Myc), the rUCB algorithm shown to reduce the docking time significantly (to T/N = 1.5) with precision of 67% and performance loss of only by factor 0.09. The results suggest the robustness and superiority of rUCB over the uniform non-learning algorithm. Thus, this method enables the virtual screening involving IDPs to achieve high accuracy with computational cost that is in similar scale of ordered proteins. Although, a major limitation of this tool is that it demands fully accurate IDP ensemble and highly efficient molecular docking algorithms for optimal usage.

While emphasizing the AI/ML-based methods for structural biology approaches, AlphaFold2 (AF2) tools cannot pass unnoticed. AF2 came in the limelight during the 14th Critical Assessment of Protein Structure Prediction (CASP14) [144, 145], and afterwards, it left a milestone by achieving structural annotation of more than 98% of human proteome sequences. Most of these studies, however, circled around ordered proteins and their structural complexes, but the utilizations of this tool for various IDPs and IDRs are limited. This tool was broadly used for disorderedness prediction for refinement of protein structure. Although, its applicability for IDP/IDR complex prediction was not very clear [146] until recently when Akdel et al. [145] attempted to assess the AF2 capability for various structural biology applications. As a part of this study, the AF2 was subjected to predict the protein complex structures of 14 IDP-bound complexes. AF2 returned relatively mixed results; it performed remarkably well in fitting short linear motifs into well-defined cavities, but seems to be struggling in case of more challenging long IDRs with tandem motifs arranged in symmetric structure. Additionally, AF2 was failed to predict complex conformations that involved the protein with heavily unusual residue compositions, e.g., in the case of collagen triple helices.

One more important study has shown direct applicability of AF2 in predicting IDR-protein complexes [147]. In this study, AF2 was shown to produce high-confidence conformations of a large number of human IDRs based on the predicted structures of AlphaFold Database (AFDB). Also, AF2 was able to identify conditionally foldable IDRs. However, this was somewhat unexpected because: (i) IDRs with predicted high confidence structures did not seem to have sequence similarity with PDB entries which is the source of data upon which AF2 was trained; (ii) the IDRs were minimally present in the training data set. Thus, AF2's such predicting ability is both surprising and under-studied, leaving many more questions wide open. Nonetheless, it should be noted that AF2 is less effective for IDPs/IDRs forming fuzzy complexes. It also cannot provide the functionally relevant structural plasticity of IDPs/IDRs and thus unable generate realistic ensemble representation of conditionally folded IDPs/IDRs. Furthermore, in a recently published comparison study [130] where an extension of

AF2, AlphaFold-Multimer (AF-M), were assessed for its docking efficiency along with two other docking tools: CABS-DOCK and IDP-LZerD. Although the performances of all three were quite nearer with low- and medium-resolution metrices, AF-M appeared to outperform the other two in instances with high-resolution metrics. However, it was concluded that due to the highly dynamic nature of the IDPs, the efficacy of this tool may not be so much clear, promoting the utilization of conventional physics-based tools. Nonetheless, future approaches may demand hybrid strategies involving AI/ML-, physics-, and MD-based methods.

3 Summary and Future Perspective

The milestone advancements during the recent years have revolutionized the workflow of the various structural biology approaches. Molecular docking also being one of them has evolved to a great extent. Thus, extending the reach of molecular docking to IDPs may open up new opportunities to answer various unsolved mysteries of these novel class of proteins.

In this chapter, we presented a comprehensive review of methodologies, utilized in various molecular docking operations involving IDPs. The dynamic conformational heterogeneity in IDPs/IDRs presents a significant challenge while using them in conventional molecular docking operations. Thus, applications of conventional docking tools in case of IDPs are limited. Interestingly, ensemble approaches can be introduced as an additional step to successfully sample the conformational ensembles of the IDPs. These approaches emphasize the application of MD- and MC-based simulations and enhanced-sampling methods to generate the IDP conformational ensemble and afterwards, these conformations are utilized for performing docking operations. Noticeably, these methods show added complexity and are computationally expensive due to increased number of input structures and subsequent docking events. Therefore, alternative fragment-based approaches are also available, that reduces the complexity and computational cost by segmenting the IDPs in manageable fragments during the docking runs. On the other hand, they also suffer from major limitations due to the assumption that all the segments of the IDPs can interact with their binding partners. These methods disregard the mutual effects of local segments and are unable to sample fuzzy interactions. Additionally, certain AI- and ML-assisted methods are presented to assist extensive docking operations during virtual screening for IDP ligands. Nevertheless, AI/ML-based approaches are still in the preliminary stage to be used for generalized application in IDP docking.

Therefore, current standpoint delineates an enormous number of molecular docking-based applications of IDPs in structural biology and therapeutic fields. However, this may require many hurdles to be crossed first. More effective pipelines for docking involving IDPs/IDRs are required which can sample their conformational dynamics with a critical balance between accuracy and computational expense. Capturing the dynamic conformational heterogeneity of IDPs using various ensemble-based methods demands for the accurate simulations of these molecules with reasonable computational resources. Therefore, better force fields and improvised sampling techniques that are designed specifically for IDPs are required to provide more accurate IDP ensembles. More robust strategies are required that can utilize the experimental observations for customization, refinement, and validation of docking operations in IDPs. Further adaptation of AI/ML-based approaches can extensively aid the hurdles of IDP docking. Finally, in order to ameliorate the limitations of individual approaches, synergic approaches are also shown to be of great importance.

Acknowledgments

KNP and DC thank CSIR-UGC and DBT for JRF fellowships. The high-performance computer resources of CSIR Fourth Paradigm Institute, Bangalore, India, is duly acknowledged for providing computational facility. This manuscript has PRIS registration number 189/2022.

Author Contributions Conceptualization and design: KP and MM; literature survey and analysis: KP, MM, and DC; manuscript preparation: KP and MM; figure preparation: DC and KP; proofreading and finalization: KP, MM, and DC. All authors reviewed and approved the final version of the manuscript.

References

1. Chouard T (2011) Breaking the protein rules: if dogma dictates that proteins need a structure to function, then why do so many of them live in a state of disorder? Nature 469(7337):151–154
2. Uversky VN (2019) Intrinsically disordered proteins and their "mysterious"(meta) physics. Front Phys 7:10
3. Malagrinò F, Diop A, Pagano L, Nardella C, Toto A, Gianni S (2022) Unveiling induced folding of intrinsically disordered proteins–protein engineering, frustration and emerging themes. Curr Opin Struct Biol 72:153–160
4. Oldfield CJ, Dunker AK (2014) Intrinsically disordered proteins and intrinsically disordered protein regions. Annu Rev Biochem 83(1):553–584
5. Uversky VN (2013) Unusual biophysics of intrinsically disordered proteins. Biochimica et Biophysica Acta (BBA)-Prot Proteom 1834(5):932–951
6. Uversky VN, Oldfield CJ, Dunker AK (2008) Intrinsically disordered proteins in human diseases: introducing the D2 concept. Annu Rev Biophys 37(1):215–246

7. Zhou J, Oldfield CJ, Yan W, Shen B, Dunker AK (2019) Intrinsically disordered domains: sequence → disorder → function relationships. Protein Sci 28(9):1652–1663
8. Ruan H, Sun Q, Zhang W, Liu Y, Lai L (2019) Targeting intrinsically disordered proteins at the edge of chaos. Drug Discov Today 24(1):217–227
9. Uversky VN (2020) Intrinsically disordered proteins: targets for the future? In: Structural biology in drug discovery: methods, techniques, and practices, pp 587–612
10. Bhattacharya S, Lin X (2019) Recent advances in computational protocols addressing intrinsically disordered proteins. Biomol Ther 9(4): 146
11. Dunker AK, Romero P, Obradovic Z, Garner EC, Brown CJ (2000) Intrinsic protein disorder in complete genomes. Genome Inform 11:161–171
12. Oldfield CJ, Cheng Y, Cortese MS, Brown CJ, Uversky VN, Dunker AK (2005) Comparing and combining predictors of mostly disordered proteins. Biochemistry 44(6): 1989–2000
13. Lermyte F (2020) Roles, characteristics, and analysis of intrinsically disordered proteins: a minireview. Life 10(12):320
14. Xue B, Dunker AK, Uversky VN (2012) Orderly order in protein intrinsic disorder distribution: disorder in 3500 proteomes from viruses and the three domains of life. J Biomol Struct Dyn 30(2):137–149
15. Ward JJ, Sodhi JS, McGuffin LJ, Buxton BF, Jones DT (2004) Prediction and functional analysis of native disorder in proteins from the three kingdoms of life. J Mol Biol 337(3):635–645
16. Xue B, Williams RW, Oldfield CJ, Dunker AK, Uversky VN (2010) Archaic chaos: intrinsically disordered proteins in Archaea. BMC Syst Biol 4(1):1–21
17. Walsh I, Giollo M, Di Domenico T, Ferrari C, Zimmermann O, Tosatto SC (2015) Comprehensive large-scale assessment of intrinsic protein disorder. Bioinformatics 31(2): 201–208
18. Burra PV, Kalmar L, Tompa P (2010) Reduction in structural disorder and functional complexity in the thermal adaptation of prokaryotes. PLoS One 5(8):e12069
19. Turoverov KK, Kuznetsova IM, Uversky VN (2010) The protein kingdom extended: ordered and intrinsically disordered proteins, their folding, supramolecular complex formation, and aggregation. Prog Biophys Mol Biol 102(2-3):73–84
20. Mollica L, Bessa LM, Hanoulle X, Jensen MR, Blackledge M, Schneider R (2016) Binding mechanisms of intrinsically disordered proteins: theory, simulation, and experiment. Front Mol Biosci 3:52
21. Yang J, Gao M, Xiong J, Su Z, Huang Y (2019) Features of molecular recognition of intrinsically disordered proteins via coupled folding and binding. Protein Sci 28(11): 1952–1965
22. Clerc I, Sagar A, Barducci A, Sibille N, Bernadó P, Cortés J (2021) The diversity of molecular interactions involving intrinsically disordered proteins: a molecular modeling perspective. Comput Struct Biotechnol J 19: 3817–3828
23. Ruan H, Yu C, Niu X, Zhang W, Liu H, Chen L, Xiong R, Sun Q, Jin C, Liu Y (2021) Computational strategy for intrinsically disordered protein ligand design leads to the discovery of p53 transactivation domain I binding compounds that activate the p53 pathway. Chem Sci 12(8):3004–3016
24. Uversky VN (2016) p53 proteoforms and intrinsic disorder: an illustration of the protein structure–function continuum concept. Int J Mol Sci 17(11):1874
25. Ghosh K, Huihui J, Phillips M, Haider A (2022) Rules of physical mathematics govern intrinsically disordered proteins. Annu Rev Biophys 51:355–376
26. Quaglia F, Mészáros B, Salladini E, Hatos A, Pancsa R, Chemes LB, Pajkos M, Lazar T, Peña-Díaz S, Santos J (2022) DisProt in 2022: improved quality and accessibility of protein intrinsic disorder annotation. Nucleic Acids Res 50(D1):D480–D487
27. Liu Y, Wang X, Liu B (2019) A comprehensive review and comparison of existing computational methods for intrinsically disordered protein and region prediction. Brief Bioinform 20(1):330–346
28. Katuwawala A, Ghadermarzi S, Kurgan L (2019) Computational prediction of functions of intrinsically disordered regions. Prog Mol Biol Transl Sci 166:341–369
29. Zhao B, Kurgan L (2021) Surveying over 100 predictors of intrinsic disorder in proteins. Expert Rev Proteomics 18(12): 1019–1029
30. Ahmed SS, Rifat ZT, Lohia R, Campbell AJ, Dunker AK, Rahman MS, Iqbal S (2022) Characterization of intrinsically disordered regions in proteins informed by human genetic diversity. PLoS Comput Biol 18(3): e1009911

31. Frauenfelder H, Sligar SG, Wolynes PG (1991) The energy landscapes and motions of proteins. Science 254(5038):1598–1603
32. Wei G, Xi W, Nussinov R, Ma B (2016) Protein ensembles: how does nature harness thermodynamic fluctuations for life? The diverse functional roles of conformational ensembles in the cell. Chem Rev 116(11):6516–6551. https://doi.org/10.1021/acs.chemrev.5b00562
33. Qin S, Pang X, Zhou H-X (2011) Automated prediction of protein association rate constants. Structure 19(12):1744–1751
34. Dogan J, Gianni S, Jemth P (2014) The binding mechanisms of intrinsically disordered proteins. Phys Chem Chem Phys 16(14): 6323–6331
35. Wu D, Zhou H-X (2019) Designed mutations alter the binding pathways of an intrinsically disordered protein. Sci Rep 9(1):1–10
36. Neira JL, Correa J, Rizzuti B, Santofimia-Castaño P, Abian O, Velázquez-Campoy A, Fernandez-Megia E, Iovanna JL (2019) Dendrimers as competitors of protein–protein interactions of the intrinsically disordered nuclear chromatin protein NUPR1. Biomacromolecules 20(7):2567–2576
37. Neira JL, Rizzuti B, Jiménez-Alesanco A, Palomino-Schätzlein M, Abián O, Velázquez-Campoy A, Iovanna JL (2020) A phosphorylation-induced switch in the nuclear localization sequence of the intrinsically disordered NUPR1 hampers binding to importin. Biomol Ther 10(9):1313
38. Santofimia-Castaño P, Rizzuti B, Abián O, Velázquez-Campoy A, Iovanna JL, Neira JL (2018) Amphipathic helical peptides hamper protein-protein interactions of the intrinsically disordered chromatin nuclear protein 1 (NUPR1). Biochimica et Biophysica Acta (BBA)-Gen Subj 1862(6):1283–1295
39. Santofimia-Castaño P, Rizzuti B, Pey AL, Fárez-Vidal ME, Iovanna JL, Neira JL (2021) Intrinsically disordered protein NUPR1 binds to the armadillo-repeat domain of Plakophilin 1. Int J Biol Macromol 170: 549–560
40. Santofimia-Castaño P, Rizzuti B, Pey ÁL, Soubeyran P, Vidal M, Urrutia R, Iovanna JL, Neira JL (2017) Intrinsically disordered chromatin protein NUPR1 binds to the C-terminal region of Polycomb RING1B. Proc Natl Acad Sci 114(31):E6332–E6341
41. Khan RH, Siddiqi MK, Uversky VN, Salahuddin P (2019) Molecular docking of Aβ1–40 peptide and its Iowa D23N mutant using small molecule inhibitors: possible mechanisms of Aβ-peptide inhibition. Int J Biol Macromol 127:250–270
42. Baggett DW, Nath A (2018) The rational discovery of a tau aggregation inhibitor. Biochemistry 57(42):6099–6107
43. Zhu M, De Simone A, Schenk D, Toth G, Dobson CM, Vendruscolo M (2013) Identification of small-molecule binding pockets in the soluble monomeric form of the Aβ42 peptide. J Chem Phys 139(3):07B609_601
44. Meng X-Y, Zhang H-X, Mezei M, Cui M (2011) Molecular docking: a powerful approach for structure-based drug discovery. Curr Comput Aided Drug Des 7(2):146–157
45. Stanzione F, Giangreco I, Cole JC (2021) Use of molecular docking computational tools in drug discovery. Prog Med Chem 60: 273–343
46. Makhouri FR, Ghasemi JB (2018) Combating diseases with computational strategies used for drug design and discovery. Curr Top Med Chem 18(32):2743–2773
47. Xu X, Huang M, Zou X (2018) Docking-based inverse virtual screening: methods, applications, and challenges. Biophys Rep 4(1):1–16
48. Torres PH, Sodero AC, Jofily P, Silva-Jr FP (2019) Key topics in molecular docking for drug design. Int J Mol Sci 20(18):4574
49. Singh S, Baker QB, Singh DB (2022) Molecular docking and molecular dynamics simulation. In: Bioinformatics. Elsevier, pp 291–304
50. Ciemny M, Kurcinski M, Kamel K, Kolinski A, Alam N, Schueler-Furman O, Kmiecik S (2018) Protein–peptide docking: opportunities and challenges. Drug Discov Today 23(8):1530–1537
51. Morris GM, Goodsell DS, Halliday RS, Huey R, Hart WE, Belew RK, Olson AJ (1998) Automated docking using a Lamarckian genetic algorithm and an empirical binding free energy function. J Comput Chem 19(14):1639–1662
52. Marondedze EF, Govender KK, Govender PP (2020) Computational investigation of the binding characteristics of β-amyloid fibrils. Biophys Chem 256:106281
53. Bieschke J, Herbst M, Wiglenda T, Friedrich RP, Boeddrich A, Schiele F, Kleckers D, Lopez del Amo JM, Grüning BA, Wang Q (2012) Small-molecule conversion of toxic oligomers to nontoxic β-sheet–rich amyloid fibrils. Nat Chem Biol 8(1):93–101
54. Chia S, Faidon Brotzakis Z, Horne RI, Possenti A, Mannini B, Cataldi R, Nowinska M, Staats R, Linse S, Knowles TP, Habchi J, (2022) Structure-based discovery of small molecule inhibitors of the autocatalytic

proliferation of α-synuclein aggregates. Mol Pharm 20(1):183–193

55. Wärmländer S, Tiiman A, Abelein A, Luo J, Jarvet J, Söderberg KL, Danielsson J, Gräslund A (2013) Biophysical studies of the amyloid β-peptide: interactions with metal ions and small molecules. Chembiochem 14(14):1692–1704
56. Kim MY, Na I, Kim JS, Son SH, Choi S, Lee SE, Kim J-H, Jang K, Alterovitz G, Chen Y (2019) Rational discovery of antimetastatic agents targeting the intrinsically disordered region of MBD2. Sci Adv 5(11):eaav9810
57. Irwin JJ, Sterling T, Mysinger MM, Bolstad ES, Coleman RG (2012) ZINC: a free tool to discover chemistry for biology. J Chem Inf Model 52(7):1757–1768
58. Lorber DM, Shoichet BK (2005) Hierarchical docking of databases of multiple ligand conformations. Curr Top Med Chem 5(8): 739–749
59. Chen Y, Shoichet BK (2009) Molecular docking and ligand specificity in fragment-based inhibitor discovery. Nat Chem Biol 5(5): 358–364
60. Taranta M, Bizzarri AR, Cannistraro S (2009) Modeling the interaction between the N-terminal domain of the tumor suppressor p53 and azurin. J Mol Recog Interdiscipl J 22(3):215–222
61. Webb B, Sali A (2021) Protein structure modeling with MODELLER. In: Structural genomics. Springer, pp 239–255
62. Morris AL, MacArthur MW, Hutchinson EG, Thornton JM (1992) Stereochemical quality of protein structure coordinates. Proteins 12(4):345–364
63. Laskowski RA, MacArthur MW, Moss DS, Thornton JM (1993) PROCHECK: a program to check the stereochemical quality of protein structures. J Appl Crystallogr 26(2): 283–291
64. Chen R, Weng Z (2002) Docking unbound proteins using shape complementarity, desolvation, and electrostatics. Proteins 47(3): 281–294
65. Chen R, Weng Z (2003) A novel shape complementarity scoring function for protein-protein docking. Proteins 51(3):397–408
66. Comeau SR, Gatchell DW, Vajda S, Camacho CJ (2004) ClusPro: a fully automated algorithm for protein–protein docking. Nucleic Acids Res 32(suppl_2):W96–W99
67. Fisher CK, Stultz CM (2011) Constructing ensembles for intrinsically disordered proteins. Curr Opin Struct Biol 21(3):426–431
68. Thomasen FE, Lindorff-Larsen K (2022) Conformational ensembles of intrinsically disordered proteins and flexible multidomain proteins. Biochem Soc Trans 50(1):541–554
69. Gong X, Zhang Y, Chen J (2021) Advanced sampling methods for multiscale simulation of disordered proteins and dynamic interactions. Biomol Ther 11(10):1416
70. Trott O, Olson AJ (2010) AutoDock Vina: improving the speed and accuracy of docking with a new scoring function, efficient optimization, and multithreading. J Comput Chem 31(2):455–461
71. Friesner RA, Banks JL, Murphy RB, Halgren TA, Klicic JJ, Mainz DT, Repasky MP, Knoll EH, Shelley M, Perry JK (2004) Glide: a new approach for rapid, accurate docking and scoring. 1. Method and assessment of docking accuracy. J Med Chem 47(7):1739–1749
72. Halgren TA, Murphy RB, Friesner RA, Beard HS, Frye LL, Pollard WT, Banks JL (2004) Glide: a new approach for rapid, accurate docking and scoring. 2. Enrichment factors in database screening. J Med Chem 47(7): 1750–1759
73. Van Zundert G, Rodrigues J, Trellet M, Schmitz C, Kastritis P, Karaca E, Melquiond A, van Dijk M, De Vries S, Bonvin A (2016) The HADDOCK2. 2 web server: user-friendly integrative modeling of biomolecular complexes. J Mol Biol 428(4):720–725
74. Kasahara K, Terazawa H, Takahashi T, Higo J (2019) Studies on molecular dynamics of intrinsically disordered proteins and their fuzzy complexes: a mini-review. Comput Struct Biotechnol J 17:712–720
75. Manna M, Róg T, Vattulainen I (2014) The challenges of understanding glycolipid functions: an open outlook based on molecular simulations. Biochimica et Biophysica Acta (BBA)-Mol Cell Biol Lipids 1841(8): 1130–1145
76. Chong S-H, Chatterjee P, Ham S (2017) Computer simulations of intrinsically disordered proteins. Annu Rev Phys Chem 68: 117–134
77. Mu J, Liu H, Zhang J, Luo R, Chen H-F (2021) Recent force field strategies for intrinsically disordered proteins. J Chem Inf Model 61(3):1037–1047
78. Tian C, Kasavajhala K, Belfon KA, Raguette L, Huang H, Migues AN, Bickel J, Wang Y, Pincay J, Wu Q (2019) ff19SB: amino-acid-specific protein backbone parameters trained against quantum mechanics energy surfaces in solution. J Chem Theory Comput 16(1):528–552

79. Huang J, Rauscher S, Nawrocki G, Ran T, Feig M, De Groot BL, Grubmüller H, MacKerell AD (2017) CHARMM36m: an improved force field for folded and intrinsically disordered proteins. Nat Methods 14(1): 71–73
80. Robustelli P, Piana S, Shaw DE (2018) Developing a molecular dynamics force field for both folded and disordered protein states. Proc Natl Acad Sci 115(21):E4758–E4766
81. Song D, Liu H, Luo R, Chen H-F (2020) Environment-specific force field for intrinsically disordered and ordered proteins. J Chem Inf Model 60(4):2257–2267
82. Rahman MU, Rehman AU, Liu H, Chen H-F (2020) Comparison and evaluation of force fields for intrinsically disordered proteins. J Chem Inf Model 60(10):4912–4923
83. Abriata LA, Dal Peraro M (2021) Assessment of transferable forcefields for protein simulations attests improved description of disordered states and secondary structure propensities, and hints at multi-protein systems as the next challenge for optimization. Comput Struct Biotechnol J 19:2626–2636
84. Piana S, Robustelli P, Tan D, Chen S, Shaw DE (2020) Development of a force field for the simulation of single-chain proteins and protein–protein complexes. J Chem Theory Comput 16(4):2494–2507
85. Song D, Luo R, Chen H-F (2017) The IDP-specific force field ff14IDPSFF improves the conformer sampling of intrinsically disordered proteins. J Chem Inf Model 57(5): 1166–1178
86. Jing Z, Liu C, Cheng SY, Qi R, Walker BD, Piquemal J-P, Ren P (2019) Polarizable force fields for biomolecular simulations: recent advances and applications. Annu Rev Biophys 48:371
87. Chen J, Liu X, Chen J (2020) Targeting intrinsically disordered proteins through dynamic interactions. Biomol Ther 10(5):743
88. Das RK, Ruff KM, Pappu RV (2015) Relating sequence encoded information to form and function of intrinsically disordered proteins. Curr Opin Struct Biol 32:102–112
89. Marrink SJ, Tieleman DP (2013) Perspective on the Martini model. Chem Soc Rev 42(16): 6801–6822
90. Marrink SJ, Risselada HJ, Yefimov S, Tieleman DP, De Vries AH (2007) The MARTINI force field: coarse grained model for biomolecular simulations. J Phys Chem B 111(27): 7812–7824
91. Marrink SJ, Corradi V, Souza PC, Ingolfsson HI, Tieleman DP, Sansom MS (2019) Computational modeling of realistic cell membranes. Chem Rev 119(9):6184–6226
92. Corradi V, Mendez-Villuendas E, Ingólfsson HI, Gu R-X, Siuda I, Melo MN, Moussatova A, DeGagné LJ, Sejdiu BI, Singh G (2018) Lipid–protein interactions are unique fingerprints for membrane proteins. ACS Cent Sci 4(6):709–717
93. Ashbaugh HS, Hatch HW (2008) Natively unfolded protein stability as a coil-to-globule transition in charge/hydropathy space. J Am Chem Soc 130(29):9536–9542
94. Poma AB, Cieplak M, Theodorakis PE (2017) Combining the MARTINI and structure-based coarse-grained approaches for the molecular dynamics studies of conformational transitions in proteins. J Chem Theory Comput 13(3):1366–1374
95. Baul U, Chakraborty D, Mugnai ML, Straub JE, Thirumalai D (2019) Sequence effects on size, shape, and structural heterogeneity in intrinsically disordered proteins. J Phys Chem B 123(16):3462–3474
96. Liu X, Chen J (2017) HyRes: a coarse-grained model for multi-scale enhanced sampling of disordered protein conformations. Phys Chem Chem Phys 19(48):32421–32432
97. Wu H, Wolynes PG, Papoian GA (2018) AWSEM-IDP: a coarse-grained force field for intrinsically disordered proteins. J Phys Chem B 122(49):11115–11125
98. Lin X, Roy S, Jolly MK, Bocci F, Schafer NP, Tsai M-Y, Chen Y, He Y, Grishaev A, Weninger K (2018) PAGE4 and conformational switching: insights from molecular dynamics simulations and implications for prostate cancer. J Mol Biol 430(16):2422–2438
99. Bereau T, Deserno M (2009) Generic coarse-grained model for protein folding and aggregation. J Chem Phys 130(23):06B621
100. Rutter GO, Brown AH, Quigley D, Walsh TR, Allen MP (2015) Testing the transferability of a coarse-grained model to intrinsically disordered proteins. Phys Chem Chem Phys 17(47):31741–31749
101. Rutter G, Brown A, Quigley D, Walsh T, Allen M (2018) Emergence of order in self-assembly of the intrinsically disordered biomineralisation peptide n16N. Mol Simul 44(6): 463–469
102. Kurcinski M, Kolinski A, Kmiecik S (2014) Mechanism of folding and binding of an intrinsically disordered protein as revealed by ab initio simulations. J Chem Theory Comput 10(6):2224–2231
103. Ciemny MP, Debinski A, Paczkowska M, Kolinski A, Kurcinski M, Kmiecik S (2016) Protein-peptide molecular docking with

large-scale conformational changes: the p53-MDM2 interaction. Sci Rep 6(1):1–8

104. Barducci A, Bonomi M, Parrinello M (2011) Metadynamics. Wiley Interdiscipl Rev Comput Mol Sci 1(5):826–843
105. Bussi G, Laio A (2020) Using metadynamics to explore complex free-energy landscapes. Nat Rev Phys 2(4):200–212
106. Sugita Y, Okamoto Y (1999) Replica-exchange molecular dynamics method for protein folding. Chem Phys Lett 314(1-2): 141–151
107. Shrestha UR, Juneja P, Zhang Q, Gurumoorthy V, Borreguero JM, Urban V, Cheng X, Pingali SV, Smith JC, O'Neill HM (2019) Generation of the configurational ensemble of an intrinsically disordered protein from unbiased molecular dynamics simulation. Proc Natl Acad Sci 116(41): 20446–20452
108. Miao Y, Feher VA, McCammon JA (2015) Gaussian accelerated molecular dynamics: unconstrained enhanced sampling and free energy calculation. J Chem Theory Comput 11(8):3584–3595
109. Jain A, Vaidehi N, Rodriguez G (1993) A fast recursive algorithm for molecular dynamics simulation. J Comput Phys 106(2):258–268
110. Vaidehi N, Jain A, Goddard WA (1996) Constant temperature constrained molecular dynamics: the Newton– Euler inverse mass operator method. J Phys Chem 100(25): 10508–10517
111. Vaidehi N, Jain A (2015) Internal coordinate molecular dynamics: a foundation for multiscale dynamics. J Phys Chem B 119(4): 1233–1242
112. Gangupomu VK, Wagner JR, Park I-H, Jain A, Vaidehi N (2013) Mapping conformational dynamics of proteins using torsional dynamics simulations. Biophys J 104(9): 1999–2008
113. Pande VS, Beauchamp K, Bowman GR (2010) Everything you wanted to know about Markov State Models but were afraid to ask. Methods 52(1):99–105
114. Krzeminski M, Marsh JA, Neale C, Choy W-Y, Forman-Kay JD (2013) Characterization of disordered proteins with ENSEMBLE. Bioinformatics 29(3):398–399
115. Nodet G, Salmon L, Ozenne V, Meier S, Jensen MR, Blackledge M (2009) Quantitative description of backbone conformational sampling of unfolded proteins at amino acid resolution from NMR residual dipolar couplings. J Am Chem Soc 131(49):17908–17918
116. Maier JA, Martinez C, Kasavajhala K, Wickstrom L, Hauser KE, Simmerling C (2015) ff14SB: improving the accuracy of protein side chain and backbone parameters from ff99SB. J Chem Theory Comput 11(8): 3696–3713
117. Valdés JJ, Schwarz A, Cabeza de Vaca I, Calvo E, Pedra JH, Guallar V, Kotsyfakis M (2013) Tryptogalinin is a tick Kunitz serine protease inhibitor with a unique intrinsic disorder. PLoS One 8(5):e62562
118. Yuan Y, Pei J, Lai L (2013) Binding site detection and druggability prediction of protein targets for structure-based drug design. Curr Pharm Des 19(12):2326–2333
119. Wichapong K, Silvestre-Roig C, Braster Q, Schumski A, Soehnlein O, Nicolaes GA (2021) Structure-based peptide design targeting intrinsically disordered proteins: Novel histone H4 and H2A peptidic inhibitors. Comput Struct Biotechnol J 19:934–948
120. Neira JL, Bintz J, Arruebo M, Rizzuti B, Bonacci T, Vega S, Lanas A, Velázquez-Campoy A, Iovanna JL, Abián O (2017) Identification of a drug targeting an intrinsically disordered protein involved in pancreatic adenocarcinoma. Sci Rep 7(1):1–15
121. Jagannathan NS, Hogue CW, Tucker-Kellogg L (2021) Computational modeling suggests binding-induced expansion of Epsin disordered regions upon association with AP2. PLoS Comput Biol 17(1):e1008474
122. Cragnell C, Durand D, Cabane B, Skepö M (2016) Coarse-grained modeling of the intrinsically disordered protein Histatin 5 in solution: Monte Carlo simulations in combination with SAXS. Proteins 84(6):777–791
123. Cragnell C, Rieloff E, Skepö M (2018) Utilizing coarse-grained modeling and monte carlo simulations to evaluate the conformational ensemble of intrinsically disordered proteins and regions. J Mol Biol 430(16). 2478–2492
124. Fagerberg E, Månsson LK, Lenton S, Skepö M (2020) The effects of chain length on the structural properties of intrinsically disordered proteins in concentrated solutions. J Phys Chem B 124(52):11843–11853
125. Staneva I, Huang Y, Liu Z, Wallin S (2012) Binding of two intrinsically disordered peptides to a multi-specific protein: a combined Monte Carlo and molecular dynamics study. PLoS Comput Biol 8(9):e1002682
126. Feldman HJ, Hogue CW (2000) A fast method to sample real protein conformational space. Proteins 39(2):112–131

127. Feldman HJ, Hogue CW (2002) Probabilistic sampling of protein conformations: new hope for brute force? Proteins 46(1):8–23
128. Kurcinski M, Pawel Ciemny M, Oleniecki T, Kuriata A, Badaczewska-Dawid AE, Kolinski A, Kmiecik S (2019) CABS-dock standalone: a toolbox for flexible protein–peptide docking. Bioinformatics 35(20): 4170–4172
129. Ciemny MP, Badaczewska-Dawid AE, Pikuzinska M, Kolinski A, Kmiecik S (2019) Modeling of disordered protein structures using monte carlo simulations and knowledge-based statistical force fields. Int J Mol Sci 20(3):606
130. Verburgt J, Zhang Z, Kihara D (2022) Multi-level analysis of intrinsically disordered protein docking methods. Methods 204:55
131. Fiser A, Do RKG (2000) Modeling of loops in protein structures. Protein Sci 9(9): 1753–1773
132. Šali A, Blundell TL (1993) Comparative protein modelling by satisfaction of spatial restraints. J Mol Biol 234(3):779–815
133. Morris GM, Huey R, Lindstrom W, Sanner MF, Belew RK, Goodsell DS, Olson AJ (2009) AutoDock4 and AutoDockTools4: automated docking with selective receptor flexibility. J Comput Chem 30(16): 2785–2791
134. Peterson LX, Roy A, Christoffer C, Terashi G, Kihara D (2017) Modeling disordered protein interactions from biophysical principles. PLoS Comput Biol 13(4):e1005485
135. Christoffer CW, Kihara D (2019) Modeling protein-protein interactions with intrinsically disordered proteins. In: Intrinsically disordered proteins. Elsevier, pp 189–206
136. Christoffer C, Kihara D (2020) IDP-LZerD: software for modeling disordered protein interactions. In: Protein structure prediction. Springer, pp 231–244
137. Lensink MF, Brysbaert G, Nadzirin N, Velankar S, Chaleil RA, Gerguri T, Bates PA, Laine E, Carbone A, Grudinin S (2019) Blind prediction of homo-and hetero-protein complexes: the CASP13-CAPRI experiment. Proteins 87(12):1200–1221
138. Schindler CE, de Vries SJ, Zacharias M (2015) Fully blind peptide-protein docking with pepATTRACT. Structure 23(8): 1507–1515
139. Ramanathan A, Ma H, Parvatikar A, Chennubhotla SC (2021) Artificial intelligence techniques for integrative structural biology of intrinsically disordered proteins. Curr Opin Struct Biol 66:216–224
140. Kumar P, Bhardwaj A, Uversky VN, Tripathi T, Giri R (2022) Computational methods to study intrinsically disordered proteins. In: Advances in protein molecular and structural biology methods. Elsevier, pp 489–504
141. Kim S, Thiessen PA, Bolton EE, Chen J, Fu G, Gindulyte A, Han L, He J, He S, Shoemaker BA (2016) PubChem substance and compound databases. Nucleic Acids Res 44 (D1):D1202–D1213
142. Abdi H (2003) Partial least square regression (PLS regression). Encyclopedia Res Methods Soc Sci 6(4):792–795
143. Chong B, Yang Y, Wang Z-L, Xing H, Liu Z (2021) Reinforcement learning to boost molecular docking upon protein conformational ensemble. Phys Chem Chem Phys 23(11):6800–6806
144. Jumper J, Evans R, Pritzel A, Green T, Figurnov M, Ronneberger O, Tunyasuvunakool K, Bates R, Žídek A, Potapenko A (2021) Applying and improving AlphaFold at CASP14. Proteins 89(12): 1711–1721
145. Akdel M, Pires DE, Pardo EP, Jänes J, Zalevsky AO, Mészáros B, Bryant P, Good LL, Laskowski RA, Pozzati G, Shenoy A (2022) A structural biology community assessment of AlphaFold 2 applications. Nat Struct Mol Biol 29(11):1056–1067
146. Nussinov R, Zhang M, Liu Y, Jang H (2022) AlphaFold, Artificial Intelligence (AI), and Allostery. J Phys Chem B 126(34):6372
147. Alderson TR, Pritišanac I, Kolarić Đ, Moses AM, Forman-Kay JD (2023) Systematic identification of conditionally folded intrinsically disordered regions by AlphaFold2. Proceedings of the National Academy of Sciences. 1;120(44):e2304302120
148. Jones G, Willett P, Glen RC, Leach AR, Taylor R (1997) Development and validation of a genetic algorithm for flexible docking. J Mol Biol 267(3):727–748
149. Jain AN (2003) Surflex: fully automatic flexible molecular docking using a molecular similarity-based search engine. J Med Chem 46(4):499–511
150. Zhou P, Jin B, Li H, Huang S-Y (2018) HPEPDOCK: a web server for blind peptide–protein docking based on a hierarchical algorithm. Nucleic Acids Res 46(W1):W443–W450
151. Krupa P, Karczyńska AS, Mozolewska MA, Liwo A, Czaplewski C (2021) UNRES-Dock—protein–protein and peptide–protein docking by coarse-grained replica-exchange

MD simulations. Bioinformatics 37(11): 1613–1615

152. Lee H, Heo L, Lee MS, Seok C (2015) GalaxyPepDock: a protein–peptide docking tool based on interaction similarity and energy optimization. Nucleic Acids Res 43(W1): W431–W435

153. Xu X, Yan C, Zou X (2018) MDockPeP: an ab-initio protein–peptide docking server. J Comput Chem 39(28):2409–2413

154. Zhang Y, Sanner MF (2019) AutoDock CrankPep: combining folding and docking to predict protein–peptide complexes. Bioinformatics 35(24):5121–5127

155. Antunes DA, Moll M, Devaurs D, Jackson KR, Lizée G, Kavraki LE (2017) DINC 2.0: a new protein–peptide docking webserver using an incremental approach. Cancer Res 77(21):e55–e57

156. Schneidman-Duhovny D, Inbar Y, Nussinov R, Wolfson HJ (2005) PatchDock and SymmDock: servers for rigid and symmetric docking. Nucleic Acids Res 33 (suppl_2):W363–W367

157. Tovchigrechko A, Vakser IA (2006) GRAMM-X public web server for protein–protein docking. Nucleic Acids Res 34 (suppl_2):W310–W314

158. Kozakov D, Brenke R, Comeau SR, Vajda S (2006) PIPER: an FFT-based protein docking program with pairwise potentials. Proteins 65(2):392–406

159. Lyskov S, Gray JJ (2008) The RosettaDock server for local protein–protein docking. Nucleic Acids Res 36(suppl_2):W233–W238

160. Torchala M, Moal IH, Chaleil RA, Fernandez-Recio J, Bates PA (2013) SwarmDock: a server for flexible protein–protein docking. Bioinformatics 29(6):807–809

161. Cheng TMK, Blundell TL, Fernandez-Recio J (2007) pyDock: electrostatics and desolvation for effective scoring of rigid body protein–protein docking. Proteins 68(2): 503–515

162. De Vries SJ, Van Dijk M, Bonvin AM (2010) The HADDOCK web server for data-driven biomolecular docking. Nat Protoc 5(5): 883–897

163. Yan Y, Tao H, He J, Huang S-Y (2020) The HDOCK server for integrated protein–protein docking. Nat Protoc 15(5):1829–1852

164. Pierce BG, Wiehe K, Hwang H, Kim B-H, Vreven T, Weng Z (2014) ZDOCK server: interactive docking prediction of protein–protein complexes and symmetric multimers. Bioinformatics 30(12):1771–1773

165. Weng G, Wang E, Wang Z, Liu H, Zhu F, Li D, Hou T (2019) HawkDock: a web server to predict and analyze the protein–protein complex based on computational docking and MM/GBSA. Nucleic Acids Res 47(W1): W322–W330

166. Ramírez-Aportela E, López-Blanco JR, Chacón P (2016) FRODOCK 2.0: fast protein–protein docking server. Bioinformatics 32(15):2386–2388

167. Donsky E, Wolfson HJ (2011) PepCrawler: a fast RRT-based algorithm for high-resolution refinement and binding affinity estimation of peptide inhibitors. Bioinformatics 27(20): 2836–2842

168. Lamiable A, Thévenet P, Rey J, Vavrusa M, Derreumaux P, Tufféry P (2016) PEP-FOLD3: faster de novo structure prediction for linear peptides in solution and in complex. Nucleic Acids Res 44(W1):W449–W454

169. Zacharias M (2005) ATTRACT: protein–protein docking in CAPRI using a reduced protein model. Proteins 60(2):252–256

170. Macindoe G, Mavridis L, Venkatraman V, Devignes MD, Ritchie DW (2010) HexServer: an FFT-based protein docking server powered by graphics processors. Nucleic acids research, 38(suppl_2), W445-W449

171. Hauser AS, Windshügel B (2016) LEADS-PEP: a benchmark data set for assessment of peptide docking performance. J Chem Inf Model 56(1):188–200

Chapter 12

Challenges of Protein-Protein Docking of the Membrane Proteins

Yusra Sajid Kiani and Ishrat Jabeen

Abstract

Despite the recent advances in the determination of high-resolution membrane protein (MP) structures, the structural and functional characterization of MPs remains extremely challenging, mainly due to the hydrophobic nature, low abundance, poor expression, purification, and crystallization difficulties associated with MPs. Whereby the major challenges/hurdles for MP structure determination are associated with the expression, purification, and crystallization procedures. Although there have been significant advances in the experimental determination of MP structures, only a limited number of MP structures (approximately less than 1% of all) are available in the Protein Data Bank (PDB). Therefore, the structures of a large number of MPs still remain unresolved, which leads to the availability of widely unplumbed structural and functional information related to MPs. As a result, recent developments in the drug discovery realm and the significant biological contemplation have led to the development of several novel, low-cost, and time-efficient computational methods that overcome the limitations of experimental approaches, supplement experiments, and provide alternatives for the characterization of MPs. Whereby the fine tuning and optimizations of these computational approaches remains an ongoing endeavor.

Computational methods offer a potential way for the elucidation of structural features and the augmentation of currently available MP information. However, the use of computational modeling can be extremely challenging for MPs mainly due to insufficient knowledge of (or gaps in) atomic structures of MPs. Despite the availability of numerous in silico methods for 3D structure determination the applicability of these methods to MPs remains relatively low since all methods are not well-suited or adequate for MPs. However, sophisticated methods for MP structure predictions are constantly being developed and updated to integrate the modifications required for MPs. Currently, different computational methods for (1) MP structure prediction, (2) stability analysis of MPs through molecular dynamics simulations, (3) modeling of MP complexes through docking, (4) prediction of interactions between MPs, and (5) MP interactions with its soluble partner are extensively used. Towards this end, MP docking is widely used. It is notable that the MP docking methods yet few in number might show greater potential in terms of filling the knowledge gap. In this chapter, MP docking methods and associated challenges have been reviewed to improve the applicability, accuracy, and the ability to model macromolecular complexes.

Key words Membrane proteins (MPs), Membrane protein docking, Membrane protein scoring functions, MP docking software, CAPRI benchmarks

Agnieszka A. Kaczor (ed.), *Protein-Protein Docking: Methods and Protocols*, Methods in Molecular Biology, vol. 2780, https://doi.org/10.1007/978-1-0716-3985-6_12,
© The Author(s), under exclusive license to Springer Science+Business Media, LLC, part of Springer Nature 2024

1 Introduction

1.1 Membrane Proteins (MPs)—Functional Significance

Biomembranes form an integral component of complex biological systems, thus defining the cellular boundaries and facilitating various chemical and physical processes that are essential for life [1]. In addition, the embedded proteins or membrane-associated proteins are the indispensable mediators for a wide array of biological and cellular processes [1]. The complete genomic sequence analysis of numerous organisms has revealed that membrane proteins (MPs) are encoded by ~20–30% of all genes [2]. Whereby the MPs constitute a large family representative of approximately 60% of the approved drug targets that span a wider therapeutic range from antibiotics to cancer, but the available structural and functional information still remains relatively meagre [3, 4].

Additionally, the occurrence of mutations or inappropriate folding of MPs is reported to be concomitant to a number of diseases including heart disease, cystic fibrosis, depression, obesity, and cancer [5]. Although elusive, MPs play a central role in various biological processes such as signal transduction, molecule-transport (i.e., ion, drug, peptides, and water), photosynthesis, respiration, catalysis, and channeling/transport of molecules across membranes [6–8]. Therefore, the explication of the structural and functional significance through structure determination at atomic level/resolution is extremely incumbent [1]. Furthermore, the investigation of bimolecular behavior and interaction of MPs at the molecular level is highly necessitated to understand the biological complexity of life [9].

1.2 Challenges Associated with Structure Determination of MPs

Despite the recent advances in the determination of high-resolution MP structures, the structural and functional characterization remains extremely challenging, mainly due to the hydrophobic nature, low abundance, poor expression, purification, and crystallization difficulties associated with MPs [1]. Whereby the major challenges/hurdles for MP structure determination are associated with the expression, purification, and crystallization procedures [10].

Generally, MP surfaces are hydrophobic in nature, therefore require extraction from the membrane through detergents which adds difficulty at all phases (i.e., from expression to structure solution) [8]. Although the utility of artificial/non-natural protein environment offers partial aggregation protection but it can also have a considerable effect on the overall conformation and stability of MPs [11]. The major challenges associated with the structural determination of MP are discussed as follows.

1.2.1 Expression

Escherichia coli serves as a rapid, inexpensive, and convenient expression system for the prokaryotic proteins. Whereas, the expression of eukaryotic proteins requires a eukaryotic expression system such as yeast or insect cells [12, 13]. Whereby the successful outcomes of an expression system are dependent upon a number of factors [14].

Particularly, for MPs the natural low level expression renders the separation of MPs (i.e., in sufficient quantity) from different biological mediums a challenging and arduous exertion [15, 16]. The expression of MPs requires targeting MPs to the cell membrane of the host prior to correct folding which requires the utilization of specific system (i.e., SRP-Sec61) for the MP insertion into the eukaryotic endoplasmic reticulum [17]. In addition, the lipid composition for embedded proteins varies significantly which can have a considerable impact on the stability and the possibility of crystallization. However, in particular, for eukaryotic proteins high level eukaryotic cell lines are extremely necessitated to offer an essential machinery for the mediation of post-translational modifications (i.e., glycosylation) [18]. Moreover, experimentation with a range of expression systems for MPs is largely indispensable for the expression of MPs [8].

1.2.2 Extraction/ Solubilization and Purification

Prior to purification, the MPs require extraction and solubilization from the membrane while concurrently maintaining stability, which remains a persevering challenge in the structural biology of MPs. Conventionally, surfactants (i.e., detergents) are the most commonly used methods for solubilization of MPs [16]. Whereby the choice of detergents for MP extraction is a key component in the purification process as it provides coverage to the hydrophobic surface of the protein, which in turn allows solubilization [19]. Detergents as amphipathic agents allow the replacement of the lipid bilayer through micelle formation which enables the MPs to become soluble in aqueous mediums [19]. It is very unfortunate that numerous detergents do not offer the preservation and maintenance of MP structure and function. This is due to the fact that lack of lateral pressure allows MPs to become misfolded after their extraction from the native lipid environment [16].

Furthermore, it has also been demonstrated that some powerful detergents (i.e., FOS-Choline) facilitate efficient extraction of MPs, but do not guarantee the solubilization of stable MPs. Generally, detergents with shorter chains are not very effective for solubilization but are appropriate to be used for crystallization as compared to the long chain detergents that might impede/intervene with the protein-protein interactions within the packed crystals. Most evidently some detergents require a detergent exchange before purification because they might impede with affinity purification [19]. Largely, detergents display highly variable characteristics which makes testing and final selection of detergents a challenging endeavor, that is governed by a number of factors such as the quantity of active, soluble, stable, homogeneous MPs produced, and the overall cost of the detergent used [8, 19]. Although detergents assist in the solubilization and purification of MPs, but it is extremely challenging to evade aggregation and allow the maintenance of the functional and structural integrity [8].

1.2.3 Crystallization

In order to achieve well-diffracting crystals, the testing of numerous crystallization reagents followed by further optimization is extremely necessary. The initial screening of the crystallization reagents is performed using the 96-well plates and the vapor diffusion-based method. For soluble proteins more than 30 different (96-well sparse matrix) screening systems are available but they provide certain conditions in which MP crystallization is extremely unlikely. In addition, an important feature of the crystallization process is the screening of the detergent used for crystallization [19]. The quality of crystals produced for MPs can be improved using additional detergents with either lipidic cubic, sponge phases or bicelles as an alternative for detergent micelle [20, 21], because these methods allow dispensing of viscous lipidic cubic phase samples in nanodrop quantities [22]. Another prominent technique is the utility of microfluidics for combining minute quantities of sample proteins (i.e., 10 nL of protein and crystallization agent). This often leads to the production of highly anisotropic crystals refraining diffraction above 5 Å [23, 24]. Although optimization through the utilization of additives can improve resolution but it still remains challenging for a large number of crystals for which optimization is strenuous. Whereby such instances require the extension of the available surfaces for crystal contacts which in some cases has been achieved through the formation of stable complexes with Fab and Fv antibody regions [25, 26].

1.2.4 Data Collection

For soluble proteins the crystallization in a high-throughput setting has become a common practice, whereby the crystals are mounted with the sample changer, data is gathered automatically and structure resolution is executed in a semi-automated manner [27]. However, particularly for MPs this is extremely challenging as the crystals have a high solvent content due the usage of detergent micelle that covers the hydrophobic region of the protein. As a result, the crystals obtained are fragile, complex to handle, diffract to lower resolution, sustain radiation damage and the quality may vary considerably amongst the crystals produced from the same drop. This points towards the screening of a larger number of crystals at synchrotron during the initiation of data collection [28].

1.2.5 Structure Solution

The factors governing data collection can have a considerable impact on the structure solution process. Mainly, the damages caused due to radiations can arise difficulties in the collection of data with adequate quality for further usage in phase by multi-wavelength or single-wavelength anomalous diffraction (MAD/-SAD). Additionally, the non-isomorphism in the crystals can impede the structure solution through isomorphous replacement process [8].

1.3 Approaches to Overcome the Challenges Associated with Structure Determination of MPs

The overexpression of MPs brings about aggregation in the cytoplasm as well as metabolic changes in the cells [29]. Moreover, the extraction and purification of MPs is also an onerous procedure that results in MPs with different outcomes in context of MP state, variability and stability, particularly to be used for structure determination [30]. In addition to the aforementioned factors the actual 3D structure determination of purified MPs through the three commonly used methods including NMR, X-ray crystallography, and cryo-electron microscopy is also extremely challenging [31]. However, during the recent years concerted efforts have been made to overcome numerous hurdles/difficulties associated with the determination of 3D structures of the purified MPs through these methods, mainly for the structural and functional explication of MPs [32–34].

This has been achieved through the exploration of recombinant MP overexpression in different expression hosts [32–34] followed by the development of more efficiently solubilizing and crystalizing detergents and lipids [19, 35, 36]. Moreover, many novel approaches have also been developed for improving the stability of MPs which has facilitated their structure determination. A few methods based on *E. coli* strains that are less affected by protein overexpression, i.e., Walker strains have been developed to circumvent cytotoxicity problems associated with the overexpression of MPs [32]. Additionally, many fast screening assays based on green fluorescent protein (GFP) fusion constructs have also been utilized for the optimization of conditions that lead to stabilization [37].

Other approaches such as inclusion of mutations, deletions, fusion engineering, and monoclonal antibodies have also been used to enhance protein stability and quality of diffraction crystals [38–41]. Furthermore, the augmentation or improvement of the automation, miniaturization, integration of initial and optimization crystallization strategies [42], developments in synchrotron radiation and beamline are also amongst the approaches that have assisted in overcoming the challenges associated with the structure determination of MPs [43].

1.4 Computational Modeling/Methods to Overcome the Challenges Associated with Structure Determination of MPs

The difficulties associated with the experimental solubilization and stability hinders the structure determination of MPs. Despite significant advances in the experimental determination of MP structures only 4611 MP structures or sub-domains can be identified amongst a total of 190,404 deposited protein structures at the PDB (statistic from May 15th 2022) (https://www.rcsb.org/stats/growth/growth-released-structures) [18], that constitute a limited number of unique MP structures. This indicates that approximately less than 1% of all structures deposited in the PDB belong to MPs and this number might comprise multiple entries of the same MP determined under different experimental conditions. Contrary to a small number of MP structures deposited in the PDB there exists a larger number (~24,127) of MP gene clusters in the UniProt UniRef (accessed May 15th 2022) [18].

Therefore, the structures of a large number of MPs still remain unresolved, which leads to the availability of widely unplumbed structural and functional information related to MPs. As a result, recent advancements in the drug discovery realm and the significant biological/pharmaceutical contemplation have led to the development of several novel, low-cost, and time-efficient computational methods that: (1) overcome the limitations of experimental approaches, (2) supplement experiments, and (3) provide alternatives for the characterization of MPs. Whereby the fine tuning and optimizations of these computational approaches remains an ongoing endeavor [17].

Computational methods offer a potential way for the elucidation of structural features for the augmentation of currently available MP information. However, the use of computational modeling can be extremely challenging for MPs mainly due to insufficient knowledge of (or gaps in) atomic structures of MPs. Despite the availability of numerous in silico methods for 3D structure determination the applicability of these methods to MPs remains relatively low since all methods are not well-suited or adequate for MPs. However, sophisticated methods for MP structure predictions are constantly being developed and updated to integrate the modifications required for MPs [44]. Currently, different computational methods for (1) MP structure prediction, (2) stability analysis of MPs through molecular dynamics simulations, (3) modeling of MP complexes through docking, (4) prediction of interactions between MPs, and (5) MP interactions with its soluble partner are extensively used [31].

1.4.1 In Silico Prediction of MP Structures

Generally, different in silico methods have been developed that are well suited to the solvent environment or can be used for structure prediction of soluble proteins. However, particularly for MPs these methods require modification to provide the actual representation of a hydrophobic layer within a hydrophilic environment [31]. The sequence-based secondary and tertiary structure prediction methods form a challenging component within the structural bioinformatics domains, whereby considerable advancement has been achieved over the last two decades. The secondary structure prediction methods in combination with large datasets and improved/optimized alignment methods has led to the improvement in the overall accuracy and utility of tertiary structure prediction methods [45].

De novo, fold recognition (threading), and the homology-based methods are the most commonly used methods for the prediction of tertiary structures. Where, the de novo methods can rely on the previously reported secondary structure features and topology, which can have a significant impact on the computational costs that remain high otherwise [21]. The de novo methods do not require templates as they are based on simulated annealing algorithms that are governed by general principles of protein energetics and structures. ROSETTA [46], PROTINFO [47], FILM [48], and FRAGFOLD [49] are some of the most prominent ab initio methods used for the prediction of MP structures.

Both threading and homology-based methods rely on structural template that requires selection of one or more homologs. Although homology-based methods provide better and fast results but their applicability to MPs remains limited due to the paucity of the available 3D homologues. A number of fold recognition and homology-based methods have now been developed for the modeling of MPs. Some of the most important methods are Rosetta-Membrane [50], MODELLER [51], MEMOIR (membrane protein modeling pipeline) [52], PROSPECT [53], RAPTOR [54], TMFoldRec [55], TMFoldWeb [55], pGenTHREADER [56], pro-sp3-TASSER [57], I-TASSER [58], and iMembrane [59]. It is clearly evident that each method displays a unique set of specificities, scoring functions and advantages, however, for structure prediction the level of accuracy lies in a similar range. However, several successful applications of ab initio and homology-based methods for the prediction of membrane protein families have been reported in literature [60–62].

In addition to the aforementioned methods, the use of hybrid approaches combining diverse experimental data and computational modeling have also been reported for structure determination and macromolecular assembly of MPs (i.e., bilitranslocase (BTL)), for which no structural information or homolog is available. Where, such an integrative approach might assist in increasing the efficiency and accuracy of structure determination when a single approach either computational or experimental is inadequate for the elucidation of MPs structures [44]. However, in absence of any 3D structural information and homologs different ligand-based approaches including pharmacophore and structure-activity modeling (i.e., quantitative structure activity relationships (QSAR)) can be used for the investigation of structural features, as demonstrated for BTL with lacking structural information [44, 63, 64].

Due to the recent advancements in MP structure prediction, molecular dynamics (MD) has emerged as an indispensable tool for the investigation of MPs. Towards this, a number of force fields (i.e., CHARMM36 [65] and AMBER Lipid14 [66]) have been developed particularly for membranes. Additionally, performing MD simulations of membrane-embedded proteins requires the careful modeling of protein-lipid interactions which can have a significant impact on the entire system, i.e., hydrophobic mismatch. A thicker bilayer in comparison to the transmembrane protein will allow exposure of hydrophobic region to water which will ultimately lead to slight tilting of the protein and energy forfeit. Whereas, a thinner bilayer will lead to compression, thus preventing the interactions between the hydrophilic residues and the membrane [67].

For MD an explicit (i.e., all atom/coarse-grained) or implicit (i.e., mathematical function for some part and all atom for the remaining part of the system) representation of a membrane can be used. An explicit representation with very little approximations is presented by the all-atom models. However, different approaches such as replica exchange MD (REMD) have been used to increase sampling of the system in a parallel fashion with a different Hamiltonian and temperatures [68]. The REMD approach has been effectively applied to investigate the mechanism of α-helical peptide insertion, demonstrating that folding is more likely to occur within the membrane rather than at the membrane/water interface [69].

Other techniques such as steered MD (SMD) that are based on external driving forces for simulation, have been successfully used to explore the gating mechanism of mechanosensitive channel of large conductance (MscL) through the application of external forces (i.e., 35–70 Pn) to residues in the vicinity of the membrane water interface [70]. Metadynamics is yet another advance sampling technique that allows fast and efficient depiction of the full energy landscape through the calculation of several state functions including free energy and the addition of positive Gaussian potential to already explored search spaces [71]. The metadynamics approach has been applied to MPs for the effective characterization of ion permeation through the ion channels [72].

Regardless of the hardware and software advancements achieved over the recent years, running explicit MD simulations of MPs remain challenging for a large number of systems. The coarse-grained models provide a better alternate by providing the lipid and protein representation through the approximation of atom ensembles as single particles. MARTINI is one of the finest and computationally less intensive coarse-grained force field that relies on the concept of four to one mapping (i.e., considers four heavy atoms with their respective hydrogen atoms as a single particle) [73].

In addition to explicit modeling of membranes the implicit representation can also be used which leads to increased computational efficiency. In the implicit representation the membrane is modeled on the basis of solvent free energy. Whereby the solvent free energy is dependent on the free energies of solute-solute, solute-solvent, solute-solvent vdW interactions and the cavity formation cost due to the insertion of protein in the membrane. Membranes can be implicitly modeled through different mathematical equations such as the Poisson-Boltzmann (PB) equation that describes the electrochemical potential of a solution perpendicular to a charged surface of the membrane [74].

The implementation of which is hard to establish in MD simulations due to the associated high computational costs [75]. Therefore, approximations through generalized Born (GB) method have been utilized for reproducing PB-based model [76]. Two other implementations of the GB are GB with a simple SWitching (GBSW) [77] and the heterogeneous dielectric generalized Born (HDGB) [78]. The former method considers identical dielectric constants for the membrane and protein, and a smoothing function for the modeling of interactions between the membrane-water and solute-water, whereby the later method takes into account the fact that membrane has several layers with different dielectric constants [77, 78].

Furthermore, it is well known that MPs interact with wide array of molecules and particles to perform vital cellular functions, whereby the interactions are typically between MPs and its soluble partners or between the MPs. Therefore, over the years a number of computational approaches have also been developed to predict interactions between the MP and its soluble partners as well between MPs. Particularly, for the prediction of these interactions the identification of the binding interfaces remains useful. However, for the identification of binding interfaces a number of approaches have been reported including the sequence-based methods that have gained considerable importance [79]. PS-HomPPI and NPS-HomPPI are some of the methods that use interfacial residue data from homologs for the interface prediction [80]. PPIPP [81] and PSIVER [82] are yet other sequence-based methods that rely on the calculation of propensity score of residues at the interface and Naïve Bayes classification with kernel density estimation, respectively.

Some structure-based methods including ProMate [83] and SPPIDER [84] have been implemented, that are based on the estimation of different secondary structure features and fingerprints of protein-protein interactions (PPIs), respectively. Additionally, machine learning (ML) methods, various meta servers (i.e., CPORT [85]) and hybrid approaches (both sequence- and structure-based, i.e., PAIRpred [86]) have also been developed to provide information about the protein interfaces. As far as the performance is of these methods is considered, it is generally reported that the use of hybrid approaches with partner information from other tools significantly improves the interface predictions of the proteins [87].

It is noticeable that the aforementioned methods have been specifically implemented for soluble proteins but might display limited applicability when used for prediction of MP binding sites. Since MPs are also capable of interaction and association, therefore for prediction of MP interfaces different MP interface prediction methods have been developed over the recent years. Amongst these the ML-trained web server MemBrain [88] predicts the interaction interfaces for helical MPs. Some of the other most prominent tools are the transmembrane strand interaction potential (TMSIP) [89] and BTMX (β-barrel TransMembrane eXposure) [90] that have been developed for β-barrel MPs. The TMSIP is designed particularly for the prediction of oligomer interfaces [89], whereas the BTMX web server performs the prediction of β-barrel interfaces [90].

2 Modeling of MP Complexes Through Docking

Docking is yet another computational method that is geared towards modeling MP complexes. MPs being located within a lipid membrane play vital roles in a wide array of functions that are essential for life, mainly through their compartmentalization in two distinct biological systems. MPs are mainly involved in sensing, signaling, motility, endocytosis, etc., whereby two or more MPs interact to form complexes that are important for both the functioning and malfunctioning of the biological processes [91]. Markedly, a wide range of biological processes are mediated through PPIs but due to scarcity of knowledge regarding MP structures, the utilization of highly accurate in silico tools might assist in filling the information gap and providing better insights into the underlying biological mechanism.

Particularly, for molecular biology and the drug discovery realm, the prediction of protein 3D structures is of paramount importance. Molecular docking is one computational method that can be applied for assembling known constituents into a biologically relevant complex [92]. Broadly, docking is based on three steps that include searching, sampling, and scoring. Searching involves the generation of possible conformations of a molecular complex. Initially, during the search phase a large number (tens of thousands) of complexes are generated by utilizing the currently available biological information relevant to the location on the interface. This information can ultimately assist in achieving docking poses that are nearly similar to the native structure [31]. Moreover, sampling refers to clustering of these conformations. Once searching and sampling is performed, the next step is to rank the poses using scoring functions (SF). This helps identify highly representative native conformations from a set of biologically unrelated conformations [93]. Several methods are used for sampling (i.e., fast Fourier transformation (FFT)-based methods), whereby scoring can be performed on the basis of statistical potentials or physicochemical properties [91, 94].

Additionally, a number of sampling and scoring methods applicable to soluble proteins, each with its own set of advantages and limitations, have been reported in literature. However, these methods do not provide accurate identification of native conformations particularly for MPs. The details of the docking benchmarks, sampling, scoring methods, and available tools for MP docking are elaborated in the upcoming sections.

2.1 Protein-Protein Interfaces in Membrane Versus Aqueous Solution and the Effect of Environment on the Docking Algorithms

A plethora of algorithms, approaches, and programs are currently available for the structure prediction of protein complexes, that rely on the already available individual monomer structures [95, 96]. It is notable that due to the availability of supplementary experimental data of soluble proteins, these approaches are limited to the protein complexes found in aqueous solution, for which parameterization of scoring functions and theoretical testing of predictions is convenient. Yet, the structure prediction of protein complexes in other environments, i.e., membrane-embedded is also of utmost concern. Since the protein-protein interfaces are different in the aqueous and membrane environment, therefore the docking methods are highly dependent on the environment.

For instance, during docking the generated poses are ranked using different scoring functions that are based on geometric complementarity, solvent exclusion from the interface, and interactions such as electrostatic, van der Waals, and hydrogen bonding [31]. These scoring functions, however, display limited accuracy in identifying nearly similar to native solutions, particularly in the case of MPs. This is due to the fact that already developed docking programs primarily cater the structure prediction of soluble proteins and are not suitable to deal with the lipidic environment in which the MPs are embedded. In addition, the search algorithms employed in docking do not take into account the additional restraints that might arise due to the possible orientations of the components of MP within the lipid membrane. Regardless of this, a few methods have been developed specifically to perform MP docking, which are discussed in detail in the proceeding sections [31].

2.2 Factors Determining the Success of Docking Algorithms When Applied to Different Environments

The docking algorithms developed so far for water-soluble proteins are highly parameterized on the basis of assumptions and the available information, as a result they might not be appropriate for transmembrane or MPs [97]. Therefore, the success of docking algorithms directly depends upon the suitability to a particular environment. Generally, a number of accurate energy functions have been optimized for soluble proteins, however, considerable improvements are required particularly for MPs because the inclusion of lipid bilayer brings along a number of challenges for the development of energy functions [98]. Owing to this fact, the biomolecular modeling through docking requires effective benchmarking with macromolecular complexes as well as thermodynamic data. In addition, these modeling approaches perform calculations on the basis of an implicit membrane representing solvent as the continuous medium. This obscures the comparison of the predictions with the actual lipid bilayer-associated experimental data [98]. Therefore, the choice of the scoring method is yet another determining factor for assessing the success of the docking approach when applied to different environments.

Moreover, it is highly necessitated to use appropriate and highly efficient searching methods for the exploration of the possible conformational landscape that ultimately leads to arrangements for the minimization of the scoring function. Thus, the selected searching method might also contribute towards the success of docking performed in different environments. Apart from this, a number of other factors also govern the success of a protein-protein docking algorithm when applied to different environments. These include the magnitude of conformational change upon protein binding, the area and hydrophobicity of the interface, respectively [37]. The success of docking algorithms can become comprised at higher levels of docking difficulty mainly due to higher conformational changes upon binding/complex formation (C_a RMSD>2), small surface interface (<700 (<1400) Å), and lower hydrophobicity as demonstrated by Vajda et al. [99].

2.3 Role of Membrane and Membrane-Water Interface in Protein-Protein Interactions

The structure of biological membrane constitutes a complex combination of lipids, proteins, and glycans that are capable of involvement in membrane protein-protein interactions (M-PPIs) [100]. Herein, the lipid/protein interactions play important roles in the folding, assembly, stability, and functioning of MPs [98]. Furthermore, membranes provide anchorage to different MPs that are broadly divided into transmembrane proteins on the basis of topology. These are either permanently assimilated to the membrane interior, non-covalently associated (peripheral MPs) or covalently attached (lipid-anchored proteins) [101]. Amongst these, the peripheral MPs play crucial roles in different cellular processes including signal transduction, transport, protein/enzyme activation, etc. [98].

It is avowed that the over- or under-activation, and abnormal attachment of peripheral MPs can lead to the dysregulation of cellular pathways and disease [101, 102]. Therefore, the modulation of M-PPIs embodies promising therapeutic options for different disease conditions and the targeting of previously declared undruggable MPs (i.e., the membrane-anchored KRAS protein, and PI3Kα lipid kinases) [102]. Besides this, membranes provide facilitation in receptor orientation and increasing binding avidity through strengthening of individually weak PPIs via clustering [103, 104].

As far as the role of the membrane in PPIs is concerned, it is notable that MPs have unfolded a very intricate interaction network mainly with the lipids, extracellular environment, and other membrane- or cytoplasm-associated proteins. Conclusively, M-PPIs play a pivotal role in the direct regulation of the communication between cells that range from high affinity (covalent) to low affinity (transient) interactions. The quantification of which is extremely challenging due to the complexities associated with structure determination of MPs and intricate lipid composition of the membrane [105].

In addition to the membrane interface, the MPs experience a complex environment that contains aqueous medium on either side of the lipid bilayer. The lipid water interface comprises a relatively larger area (approximately 10 Å) of the overall thickness of the bilayer [106]. However, in comparison to the hydrophobic membrane core, the lipid water interfacial region embodies a complex chemical environment that supports non-covalent interactions with protein side chains, thus allowing the stable insertion of proteins or peptides into the bilayer. This indicates that the interfacial region plays a crucial role in the association of proteins with membrane [107].

Furthermore, there exists a polarity gradient across the interfacial region, i.e., from the highly polar head groups near aqueous medium to the non-polar hydrocarbon tail regions of the membrane. The positioning of the water molecules around the lipid head groups allows for occurrence of different types of interactions including dipole-dipole, hydrogen bonding, and electrostatic interactions. The interactions of MPs with the membrane-water interfacial region are therefore highly dependent upon the properties (i.e., polarity, charge, hydrophobicity, etc.) of the side chain amino acids [107].

Largely, for soluble proteins and peptides the role of membrane-water interfacial region in influencing interactions, promoting binding, folding, localization, and orientation is well demonstrated [108]. However, the MPs might show interaction with this interfacial region on either side of the membrane due to the residues flanking the hydrophobic membrane spanning segments, which determines the exact positioning of these interfacial segments or influences the orientation within the membrane [107].

3 CAPRI (Critical Assessment of Predicted Interactions) Docking Benchmarks for Protein-Protein Complexes

During recent years, the increased availability of structural data has led to the development of structure prediction approaches that remained focused to soluble proteins. Despite the crucial roles of MPs in cellular pathways the experimental structure determination of MPs continues to remain challenging due to the associated complexities. However, the utility of in silico approaches has proven to be valuable particularly for investigating MPs, their associated assemblies, and PPIs. Notably, PPIs are crucial for a wide array of cellular functions that are mediated through the physical interactions between two proteins. Therefore, the elucidation of comprehensive atomic-level details of these interactions points towards the utility of protein-protein docking, which is by far amongst the most well-established computational approaches used for structure prediction of protein complexes. However, particularly for MPs the general docking methods are not ideal and, thus, require optimization to tailor different physicochemical environments and spatial constraints levied due to the inclusion of membranes [109]. Towards this, considerable efforts have been made to develop and benchmark MP docking methods, using different high-quality datasets of MP complexes [109].

Largely inspired from CASP (Critical Assessment of Methods of Protein Structure Prediction), the CAPRI is a community-wide protein docking-based experiment that is implemented for the assessment of a docking method particularly for its predictability of PPIs. Generally, in the CAPRI experiments different groups develop docking protocols to predict the three-dimensional structures of protein complexes from already known component structures. Most prominently, since its inception in the year 2001, CAPRI experiment has played a pivotal role in augmenting the development of different docking algorithms [110, 111]. Therefore, in order to allow the fair assessment of different docking algorithms for further improvements, the docking community extensively relies on the utility of docking benchmarks. Which not only assists in the blind docking experiments (i.e., CAPRI) but also aids in the comparison of various other methods used for small molecule, protein-peptide, and protein-protein docking [111, 112]. However, amongst the available benchmarks the protein-DNA [113], protein-ligand [114], protein-peptide [115], and protein-protein [112] benchmarks are the highly cited.

Currently, CAPRI experiment counts a total of 54 docking rounds conducted so far, whereby the 54th round of CAPRI was initiated to be run through August 2022. A total of 230 targets spanning diverse organizations of complexes including enzyme-inhibitor, single protein with multiple domains, protein homo- and hetero-oligomers, protein-peptide, protein-polysaccharides, protein-lipid, and protein-nucleic acids have been included in these rounds (https://www.ebi.ac.uk/pdbe/complex-pred/capri/). As far as the CAPRI targets are concerned, the most appropriate ones are the unpublished X-ray crystallographic structures of protein complexes (i.e., known structures of component proteins), that are put forward for unbound docking predictions. However, due to scarcity of such structures both bound and unbound complexes with known and novel protein structures are also accepted for consideration [116]. The information regarding all CAPRI rounds and associated target proteins is provided in Table 1.

Particularly, for the CAPRI rounds 1–2, a total of seven protein-protein targets (Table 1) were submitted for evaluation through blind structure predictions. Overall, 19 groups were involved in these rounds. Highly accurate and acceptable predictions were made for a total of four and five targets respectively [116]. Additionally, a total of 10 targets (i.e., enzyme-inhibitor, antigen-antibody, cellular signaling complexes, homo-oligomers, bacterial cellulose components (complex), Table 1) were considered for the rounds 3–5. Largely, these rounds were based on both bound and unbound components (i.e., known or unknown structures), which led to considerable conformational differences, thus signifying a major challenge for the protein-protein docking methods. A total of 1866 predictions were completed, whereby the predictions of one target could not lead to completion. Most prominently, absolutely novel docking algorithms and scoring functions addressing the challenges such as side chain and backbone flexibility were applied by 1/3 of the groups. The assessments were made on the basis of comparison with experimental structures. Notably, 24 out of 30 groups provided acceptable to highly accurate predictions for the selected targets, whereby substantial differences were observed for the bound versus unbound structures [117].

Furthermore, a total of nine targets (Table 1) with majority unbound-unbound category were assessed during the CAPRI rounds 6–12, amongst which the two targets (targets 22–23) submitted for round 8 were later on canceled. The target set was inclusive of protein types which were not often included in the docking benchmarks, thus they were representative of the more challenging and realistic docking difficulties. A total of 63 groups and eight automatic docking servers were included for the assessment of the targets. Overall, for the considered targets the

Table 1
The details of CAPRI rounds and the targets investigated so far

Round no	Targets	Assessment perspective	References
CAPRI Round 1	*Target 1*: (unbound/unbound) Lactobacillus HPr kinase—*B. subtilis* HPr *Target 2*: (unbound/bound) Bovine rotavirus VP6—Fab *Target 3*: (unbound/bound) Flu hemagglutinin—Fab HC63	Protein-protein docking predictions	[116, 180]
CAPRI Round 2	*Target 4*: Alpha-amylase complex camelide antibody VH domain 1 *Target 5*: Alpha-amylase complex camelide antibody VH domain 2 *Target 6*: Alpha-amylase complex camelide antibody VH domain 3 *Target 7*: Receptor/toxin complex	Protein-protein docking predictions	[116, 180]
CAPRI Round 3	*Target 8*: (unbound/bound) Nidogen-G3/laminin complex *Target 9*: (unbound/unbound) Wild type LicT homodimer	Protein-protein docking predictions	[116, 117]
CAPRI Round 4	*Target 10*: TBE virus E trimer *Target 11*: Cohesin/dockerin (unbound) *Target 12*: Cohesin/dockerin (bound) *Target 13*: SAG1/Fab	Protein-protein docking predictions	[116, 117]
CAPRI Round 5	*Target 14*: Phosphatase 1δ/MYPT1 *Target 15*: Colicin D/Imm D *Target 18*: Xylanase/TAXI *Target 19*: Ovine prion/Fab Note: *Targets 16* and *17* have been cancelled from round 5.	Protein-protein docking predictions	[116, 117]
CAPRI Round 6	*Target 20*: HemK/eRF1 Complex (*E. coli* peptidyl tRNA release factor 1 (RF1)—Methyl transferase complex)	Docking and scoring protein complexes	[122]

CAPRI Round 7	*Target 21*: Orc1/Sir1 complex (yeast Orc1 BAH-Sir1p OIP domains complex)	Docking and scoring protein complexes	[122]
CAPRI Round 8	Note: *Target 22* and *Target 23* have been cancelled from round 8.	Docking and scoring protein complexes	[122]
CAPRI Round 9	*Target 24*: Mouse Arf1-GTP/model built ArfBD of ArhGAP10 (unbound) *Target 25*: Mouse Arf1-GTP/bound ArfBD of ArhGAP10 (unbound)	Docking and scoring protein complexes	[122]
CAPRI Round 10	*Target 26*: *E. coli* proteins TolB (PDB 1c5k) in complex with Pal (PDB 1oap)	Docking and scoring protein complexes	[122]
CAPRI Round 11	*Target 27*: Hip2/UBC9 complex (unbound)	Docking and scoring protein complexes	[122]
CAPRI Round 12	*Target 28*: Catalytic domain of NEDD4L (dimer, residues 569–946)	Docking and scoring protein complexes	[122]
CAPRI Round 13	*Target 29*: *S. cerevisiae* Trm8/Trm82 tRNA guanine-N(7)-methyltransferase (Trm82 Subunit is bound) Pdb ID: 2VDU	Near-native pose selection	[119]
CAPRI Round 14	*Target 30*: Rnd1-GTP in complex to the RBD dimer (crystal structure) *Target 31*: Rac1-GTP (1mh1) bound with the RBD monomer (NMR structure)	Near-native pose selection	[119]
CAPRI Round 15	*Target 32*: (unbound/unbound) Protein-protein complex of savinase and barley α-amylase/subtilisin inhibitor (BASI) Pdb ID: 3BX1 *Target 33*: (homology/unbound) Methyltransferase and an RNA transcript complex (protein-RNA complex)	Near-native pose selection	[119]

(continued)

Table 1 (continued)

Round no	Targets	Assessment perspective	References
	Target 34: (homology/bound) Methyltransferase and an RNA transcript complex (protein-RNA complex) *Target 35*: (homology/homology) Intramolecular complex of Xyn10B A pair of covalently linked modules Complex between two domains of Xyn10B including the catalytic domain and the carbohydrate-binding module *Target 36*: (homology/homology with one bound) Intramolecular complex of Xyn10B A pair of covalently linked modules Complex between the catalytic domain and the carbohydrate-binding modules of Xyn10B		
CAPRI Round 16	*Target 37*: (unbound/homology) Human G-protein ARF6 and LZ2, the second leucine zipper motif of JIP4 (JNK-interacting protein 4) in complex Protein-protein complex Pdb ID: 2W83	Near-native pose selection	[119]
CAPRI Round 17	*Target 38*: (unbound/homology) Centaurin-α1 and the Forkhead-associated (FHA) domain of KIF13B in complex Protein-protein complex Pdb ID: 3FM8 *Target 39*: (with a bound component) Centaurin-α1 and the Forkhead-associated (FHA) domain of KIF13B in complex Protein-protein complex Pdb ID: 3FM8	Near-native pose selection	[119]

CAPRI Round 18	*Target 40*: (bound/unbound) Bovine trypsin and the double-headed arrowhead protease inhibitor (API-A) in complex Protein-protein complex Pdb ID: 3E8L	Near-native pose selection	[119]
CAPRI Round 19	*Target 41*: (unbound/unbound) DNase domain of colicin E9 and the Im2 immunity protein complex Protein-protein complex Pdb ID: 2WPT *Target 42*: Designed TPR repeat homodimer Protein-protein complex (unbound) Pdb ID: 2WQH	Near-native pose selection	[119]
CAPRI Round 20	*Targets 43 and 44*: Designed protein-protein complexes Provided by D. Baker and collaborators (University of Washington, Seattle)	Docking, scoring, and affinity prediction	[120, 121]
CAPRI Round 21	*Target 45*: Designed protein-protein complexes Provided by Sarel Fleishman and David Baker (University of Washington)	Docking, scoring, and affinity prediction	[120, 121]
CAPRI Round 22	*Target 46*: Methyltransferase MTq2 in complex with Trm12, an activator protein from *E. cuniculi* (PDB ID: 3Q87) A protein-protein complex (homology/homology)	Docking, scoring, and affinity prediction	[120, 121]
CAPRI Round 23	*Target 47*: DNase domain of colicin E2 and the cognate Im2 immunity protein (PDB ID: 3U43) in complex Protein-protein complex (unbound/homology) *Target 48*: T4moC/T4moH mono-oxygenase complex Protein-protein complex (unbound/unbound) *Target 49*: T4moC/T4moH mono-oxygenase complex Protein-protein complex (unbound/unbound)	Docking, scoring, and affinity prediction	[120, 121]

(continued)

Table 1
(continued)

Round no	Targets	Assessment perspective	References
CAPRI Round 24	*Target 50*: Spanish influenza virus hemagglutinin (HA) with HB36.3 (a de novo designed protein) Protein-protein complex *Target 51*: Xylanase CtXyl5A A multi-domain protein	Docking, scoring, and affinity prediction	[120, 121]
CAPRI Round 25	*Target 52*: NS1B-ISG15 complex Note: Cancelled	Docking, scoring, and affinity prediction	[120, 121]
CAPRI Round 26	*Target 53*: A protein-protein complex (homology/unbound) Two designed alpha repeats Rep2 and Rep4 (PDB ID: 4JW2) Initiated from the unbound Rep4 structure (PDB ID: 3LTJ) *Target 54*: A protein-protein complex (homology/unbound). Rep16 protein (modeled through homology) in complex with neocarzinostatin (PDB ID: 4JW3) *Target 55*: Designed complexes *Target 56*: Designed complexes	Docking, scoring, and affinity prediction	[120, 121]
CAPRI Round 27	*Target 57*: A protein-polysaccharide complex. BT4661 protein in complex with a heparin (six-sugar unit) molecule (PDB ID: 4AK2) *Target 58*: A protein-protein complex, with unbound, SAXS data PliG/SalG: Lysozyme-inhibitor complex (PDB ID: 4G9S)	Docking, scoring, and affinity prediction	[120, 121]
CAPRI Round 28	*Target 59*: Rps28b/LSm-Edc3 A protein-protein complex *Target 60*: m-Importin-α/Gu-α Protein-peptide complexes PDB ID: 3zin *Target 61*: m-Importin-α/a28 Protein-peptide complexes	Modeling protein-protein and protein-peptide complexes	[123]

	PDB ID: 3zio *Target 62*: m-Importin-α/a58 Protein-peptide complexes PDB ID: 3zip *Target 63*: m-Importin-α/b6 Protein-peptide complexes PDB ID: 3ziq *Target 64*: m-Importin-α/b141 Protein-peptide complexes PDB ID: 3zir		
CAPRI Round 29	Round 29 of CAPRI has three targets: *Target 65*: RNase-H/SSB-Ct Protein-peptide complex PDB ID: 4z0u *Target 66*: PriA DNA helicase/SSB-Ct Protein-peptide complex PDB ID: 4n18 *Target 67*: Nedd4 WW3/ PPxY of ARRDC3 Protein-peptide complex PDB ID: 4n7h	Modeling protein-protein and protein-peptide complexes	[123]
CAPRI Round 30	*Target 68*: Plectin 1 and 2 repeats of the human periplakin *Target 69*: Putative esterase (*Parabacteroides distasonis*) *Target 70*: Modulator protein MzrA (*Klebsiella pneumoniae* subsp.) *Target 71*: Leucine-rich repeat protein (*Bacteroides capillosus*) *Target 72*: SusD homolog (*Bacteroides thetaiotaomicron*) *Target 73*: Putative glycosyl hydrolase (*Parabacteroides distasonis*) *Target 74*: Hypothetical protein (*Bacteroides vulgatus*) *Target 75*: Putative GDSL-like lipase (*Parabacteroides merdae*) *Target 77*: Conserved hypothetical protein (*Streptococcus pneumoniae* TIGR4) *Target 78*: Hypothetical protein (*Bacillus cereus*) *Target 79*: OSKAR-N *Target 80*: Sugar aminotransferase WecE (*Escherichia coli* K-12) *Target 81*: cGMP-dependent protein kinase II leucine zipper Rab11b protein	Protein docking and template-based modeling of homo- and hetero-protein complexes CASP-CAPRI experiment	[124]

(continued)

Table 1
(continued)

Round no	Targets	Assessment perspective	References
	Target 82: Nitro-reductase rv3368 *Target 84*: Triose phosphate isomerase *Target 85*: Cyclohexadienyl dehydrogenase from *Sinorhizobium meliloti* in complex with NADP *Target 86*: Putative polyketide cyclase (*Sinorhizobium meliloti*) *Target 87*: Histidinol-phosphate aminotransferase (*Sinorhizobium meliloti* in complex with pyridoxal-5′-phosphate) *Target 88*: WRAP-5 *Target 89*: RON receptor tyrosine kinase subunit Macrophage stimulating protein subunit (MSP) *Target 90*: Ats13 *Target 91*: Human Bj-Tsa-9 *Target 92*: Glutathione S-transferase domain (*Haliangium ochraceum*) *Target 93*: Cals8 (*Micromonospora echinospora*, P294S mutant) *Target 94*: APC103154 uncharacterized 2-domain protein (putative esterase in accordance with Pfam, *Capnocytophaga ochracea*)		
CAPRI Round 31	*Target 95*: PRC1 ubiquitylation module/nucleosome A challenging Enzyme-Nucleosome complex. PDB ID: 4r8p *Target 96*: eGFP-binder A/eGFP Protein-protein complex PDB ID: 4x15 *Target 97*: eGFP-binder C/eGFP Protein-protein complex PDB ID: 4xvp	Modeling protein-protein and protein-peptide complexes	[123]

CAPRI Round 32	*Target 98*: ubiquitin Ct hydrolase L5/RPN13 Protein-protein complex PDB ID: 4uem *Target 99*: ubiquitin Ct hydrolase L5/Ubiquitin/RPN13 Protein-protein complex PDB ID: 4uel *Target 100*: ubiquitin Ct hydrolase L5/Ubiquitin/Ino80G Protein-protein complex PDB ID: 4uf6 *Target 101*: ubiquitin Ct hydrolase L5/Ino80G Protein-protein complex PDB ID: 4uf5	Modeling protein-protein and protein-peptide complexes	[123]
CAPRI Round 33	*Target 102*: It was at first offered as target 102. But erroneously without the unbound structure of the large component (HxuA protein), which was very challenging from the modeling perspective. Therefore, the target 102 was represented and offered later as the target 107 Protein-protein complex. *Target 103*: UBE2Z/FAT10 Protein-protein complex.	Modeling protein-protein and protein-peptide complexes	[123]
CAPRI Round 34	*Target 104*: PyoAP41/ImAP41 Interface water and side chain predictions Protein-protein complex. PDB ID: 4uhp *Target 105*: PyoS2/ImS2 Interface water and side chain predictions Protein-protein complex. PDB ID: 4qko The targets represent complexes of DNase domains with immunity proteins provided by Colin Kleanthous from the Oxford University, UK The major area of focus was the: Precise and correct modeling of interface side chains and water. Assessment of classical docking and scoring.	Modeling protein-protein and protein-peptide complexes	[123]

(continued)

Table 1 (continued)

Round no	Targets	Assessment perspective	References
CAPRI Round 35	*Target 107*: HxuA/hemopexin PDB ID: 4rt6	Modeling protein-protein and protein-peptide complexes	[123]
CAPRI Round 36	*Targets 108–109*: Two protein-peptide complexes (protein UFM1 in complex with the peptide segment from the UBA5 protein (E1 of UFM1)) were used in the round 36 The two targets representing the same protein-peptide complex were provided by Prof. Reuven Wiener (Hebrew University of Jerusalem) Note: The CAPRI round 36 was cancelled due to description of the complex in an article.		https://www.ebi.ac.uk/pdbe/complex-pred/capri/
CAPRI Round 37	*Target 110*: Tri-meric viral (raptor siadenovirus 1) fiber head domains *Target 111*: Fiber 1 protein from Lizard adenovirus 2 *Target 112*: Trimeric assembly was to be built from the models of C-terminal domain (residues 258 to 461) of goose adenovirus 4 (UNIPROT ID I3PMP2) *Target 113*: Docking performed using models of two distinct proteins including the small C-terminal part (residues 3395–3469) of a very large exoprotein (UNIPROT B3R1C1) and a relatively small 116 residue protein CPX19-CDII2 (UNIPROT B3R1C2) from *Cupriavidus taiwanensis* bacteria *Target 114*: Homodimeric assembly from the models of LV4 2A2 protein, which is a novel domain (residues 823–944) of a very large 2254 residue genome polyprotein from Ljungan virus *Target 115*: Cancelled *Target 116*: Histidine kinase CckA from *Caulobacter crescentus* *Target 117*: Assembly of tetramer from models of two *Drosophila melanogaster* proteins: TPR receptor region (residues 25–406 of	Modeling protein assemblies	[125, 181]

	658 residue PINS protein) and the ligand (residues 283–623 of 859 residue INSC protein) *Target 118*: Octameric fructose biphosphatase (T118/T0906) *Target 119*: Dimeric Red Sea protein (T119/T0917) *Target 120*: Cohesin-Dockerin heterodimer		
CAPRI Round 38	*Target 121*: *P. aeroginosa* TolAIII domain/N-terminus *P. aeruginosa* TolB complex The target 121 is complex formed between the C-terminal domain (TolAIII domain) of TolA periplasmic protein and the N-terminal peptide of TolB from *P. aeruginosa* Protein-peptide NMR complex made accessible by Prof. Colin Kleanthous (University of Oxford UK) Difficulty level high	Prediction of binding modes and the 3D conformational modeling of bound peptides and protein subunits	[127]
CAPRI Round 39	*Target 122*: Human cytokine receptor complex (IL23-IL23R) Pdb ID: 5mzv and Resolution: 2.8 Å The target was provided by Savvas Savvides, Dept. Bochemistry & Microbiology, Ghent University, Belgium Difficulty level high *Target 123*: PorM-Nt/nb (pdb id:6ey0) *Target 124*: PorM-Nt/nb (pdb id:6ey6) Targets 123–124 are the complexes of PorM component (from the bacterial type 9 secretion system) with Nter and Cter domains respectively Where PorM component is with a camelid nanobody The targets were presented by Alain Roussel, Université Aix Marseille The difficulty level was high for both targets 123–124	Prediction of binding modes and the 3D conformational modeling of bound peptides and protein subunits	[127]
CAPRI Round 40	*Target 125*: The hetero-hexamer LLT1/NKR-P1 Hetero-hexamer constituting the extracellular domains LLT1 in complex with the extracellular domain of its inhibitor NKR-P1 Provided by Dr. Ondrej Vanrk, Charles University in Prague Difficulty level high	Prediction of binding modes and the 3D conformational modeling of bound peptides and protein subunits	[127]

(continued)

Table 1 (continued)

Round no	Targets	Assessment perspective	References
CAPRI Round 41	*Target 126*: Arabino-oligosaccharide binding protein, G stearothermophilus, with AbnE/A6(ABC transporter-bound periplasmic protein from *Geobacillus stearothermophilus/ arabinohexaose*, 6 sugar units) Difficulty level high *Target 127*: Arabino-oligosaccharide binding protein, *G. stearothermophilus*, with AbnE/A5 (ABC transporter-bound periplasmic protein from *Geobacillus stearothermophilus/ arabinopentaose*, 5 sugar units) Difficulty level high *Target 128*: Arabino-oligosaccharide binding protein, *G. stearothermophilus*, with AbnE/A4 (ABC transporter-bound periplasmic protein from *Geobacillus stearothermophilus/ arabinotetraose*, 4 sugar units) Difficulty level medium *Target 129*: Arabino-oligosaccharide binding protein, *G. stearothermophilus*, with AbnE/A3 (ABC transporter-bound periplasmic protein from *Geobacillus stearothermophilus/ arabinohexaose/arabinotriose*, 3 sugar units) Difficulty level medium *Target 130*: Mutant (E201A) of Arabino-oligosaccharide binding protein, *G. stearothermophilus*, with AbnB/A5 This target is a complex of the intracellular arabinanase AbnB mutant, an enzyme involved in the degradation of intra-cellular arabino-oligosaccharides, bound to A5 (arabinopentaose, 5 sugar units) Difficulty level easy Targets 126–130 made accessible by Ms. Shifra Lansky and Prof. Gil Shoham of the HUJI, Israel	Prediction of binding modes and the 3D conformational modeling of bound peptides and protein subunits	[127]

CAPRI Round 42	*Target 131*: Human CEACAM1/HopQ-Type-I *H. pylori* X-ray structure, Pdb id: 6gbg Difficulty level high *Target 132*: Human CEACAM1/hopQ-Type-II *H. pylori* X-ray structure, Pdb id: 6gbh Difficulty level medium Target 131–132 are host-pathogen hetero complexes (protein-protein) of human CEACAM1 cell adhesion protein with other cell adhesion proteins, i.e , HopQ Type I, and HopQ Type II respectively Target provided by Prof. Han Remaut (the Free University of Brussels, and the Vlaamse Institute voor Biotechnologie)	Prediction of binding modes and the 3D conformational modeling of bound peptides and protein subunits	[127]
CAPRI Round 43	*Target 133*: Colicin E2 DNase/Im2 complex Redesigned Both components are redesigned relative to the wild-type complex at the level of (1) sequence, (2) backbone conformation, and (3) rigid body orientation X-ray structure, Pdb id: 6ere Difficulty level easy Made accessible by Ravit Netzer and Sarel Fleishman (Weizmann Institute, Israel)	Prediction of binding modes and the 3D conformational modeling of bound peptides and protein subunits	[127]
CAPRI Round 44	*Target 134*: Dynein light chain subunit 8 (DLC8 dimer)/MAG 50-residue fragment (cognate protein) Pdb id: 6gzj Difficulty level easy *Target 135*: DLC8 dimer (Rat)/MAG 12-residue fragment (cognate protein) Pdb id: 6gzl Difficulty level easy Targets offered by Petri Kursula from the University of Bergen, Norway The major challenge for the associated targets is the production of greatly accurate models for the aforementioned protein-peptide complex	Prediction of binding modes and the 3D conformational modeling of bound peptides and protein subunits	[127]

(continued)

Table 1 (continued)

Round no	Targets	Assessment perspective	References
CAPRI Round 45	*Target 136*: Novel bacterial lysine decarboxylase LdcA Homo-decamer Large Cryo-EM complex Determined through cryo-EM The target provided by Irina Gutsche (Institut de Biologies Structurale, Grenoble, France)	Prediction of binding modes and the 3D conformational modeling of bound peptides and protein subunits	[127]
CAPRI Round 46	***Difficult targets***: *Target 137*: NADP-dependent reductase Pdb id: 6D2V *Target 138*: RasRap1 site-specific endopeptidase Pdb id: 5W6L *Target 141*: Rhodanese-like family protein, bacteria Pdb id: 2MXV *Target 146*: Lipid-transport, bacterial outer membrane *Target 149*: Pentafunctional AROM polypeptide: five main enzymes of the shikimate pathway *Target 150*: Idem, with SAXS data *Target 151*: Idem, with crosslinking data *Target 154*: Alpha-xylosidase Pdb id: 6DRU *Target 155*: CDI_213 protein, bacteria *Target 156*: 201_INDD4 protein, *E. coli* *Target 157*: CDI207t protein, *E. coli* *Target 159*: 18-mer heterocomplex; EM ***Easy targets*** *Target 139*: Acyl-CoA dehydrogenase from *Bdellovibrio bacteriovorus* *Target 140*: Bacteriophage ESE058 coat protein *Target 142*: Repressor-antirepressor complex (lysogeny switch)	Blind predictions complexes (i.e., of homo- and hetero-proteins)	[128]

	Target 143: Cals10 protein *Target 144*: Two-pore calcium channel protein; EM Pdb id: 6NQ1 *Target 147*: Cyanide dihydratase (*B. pumilus*); EM *Target 152*: ALAS2, 5′-Aminolevulinate synthase 2 Pdb id: 6HRH *Target 153*: Putative membrane transporter (*C. desulfamplus*) Pdb id: 6QEK *Target 158*: SLAC1 protein		
CAPRI Round 47	*Target 159–160*: Assembly domain of SAP, the surface layer (S-layer) protein from *Bacillus anthracis* 3D structure of the isolated SAP assembly domain in complex with two (different) nanobodies (Nb684 and Nb694) Solved to high resolution by X-ray diffraction Povided by Han Remote and Antonella Floravanti from the VIB-VUB Center for Structural biology (Brussels, Belgium) Atomic coordinates of backbone atoms in in random orientation and translation were also provided in order to facilitate the modeling task of the SAP 3D structure	Prediction of 3D structure of the protein complexes by taking into consideration the intra- domain-domain interactions and the interfaces and contacts with the 2 nanobodies	https://www.ebi.ac.uk/pdbe/complex-pred/capri/
CAPRI Round 48	*Targets 161–162*: Both target systems are toxin-antitoxin complexes of *E. coli* The target hetero-complexes constitute a toxin protein homodimer which is bound to two molecules of antitoxin The challenges for the target T161 and T162 were the prediction of the bound form (toxin homodimer) and full structure of toxin-antitoxin complex respectively	Prediction of the bound form of the toxin homodimer and the structure of full toxin-antitoxin complex	https://www.ebi.ac.uk/pdbe/complex-pred/capri/
CAPRI Round 49	*Targets 163*: A heterocomplex with experimentally confirmed A2B2 stoichiometry of two small helical protein components of the synaptonemal complex	Modeling protein-protein and complexes	https://www.ebi.ac.uk/pdbe/complex-pred/capri/

(continued)

Table 1 (continued)

Round no	Targets	Assessment perspective	References
CAPRI Round 50	*Target 164 (T1032)*: Easy target *Target 165 (H1036)*: Easy target *Target 166 (H1045)*: Easy target *Target 167 (T1050)*: Cancelled *Target 168 (T1052)*: Easy target *Target 169 (T1054)*: Difficult target *Target 170 (H1060)*: Difficult target *Target 171 (T1063)*: No structure *Target 172 (H1066)*: No structure *Target 173 (H1069)*: No structure *Target 174 (T1070)*: Difficult target *Target 175 (T1073)*: Cancelled *Target 176 (T1078)*: Difficult target *Target 177 (H1081)*: Easy target *Target 178 (T1083)*: Difficult target *Target 179 (T1087)*: Difficult target *Target 180 (T1099)*: Easy target *Target 181 (H1103)*: No structure		https://www.ebi.ac.uk/pdbe/complex-pred/capri/
CAPRI Round 51	First round of CAPRI COVID-19 Open Science initiative *Target 182, Target 183, Target 184, and Target 185* Three SARS-COV-2 complexes with human host proteins	Modeling of SARS-COV-2 complexes with human host proteins	https://www.ebi.ac.uk/pdbe/complex-pred/capri/
CAPRI Round 52	*Target 186*: A large protein complex which is the part of a stable sub-complex of the *E. coli* respiratory complex constituting NADH: Ubiquinone oxidoreductase (6 polypeptides) Determined at high resolution by cryo-EM, by Prof. Rouslan Efremov, at the VIB-VUB center for structural biology, Brussels, Belgium The main challenge of this target is the correct modeling of the bound conformation (~100 residue loop) in the *E. coli* complex	Correct modeling of bound conformations	https://www.ebi.ac.uk/pdbe/complex-pred/capri/

CAPRI Round 53	*Targets 187 and 188*: The system of both targets is the same important bacterial protein-DNA complex, whose high-resolution structure is solved for the first time by single particle cryo-EM T187 is the more challenging problem, where predictors are provided only with sequence information for the protein and the DNA construct used for the cryo-EM study T188 represents an easier modeling exercise where predictors are also provided with the 3D structure of the bound DNA moiety from the complex Scoring challenges are offered for both targets independently	https://www.ebi.ac.uk/pdbe/complex-pred/capri/
CAPRI Round 54	In progress: Target and other associated data is currently not available	https://www.ebi.ac.uk/pdbe/complex-pred/capri/

predictions appeared to be superior in situations when (1) no conformational changes were observed in at least one of the subunits upon association and (2) the components side chains were correctly positioned. Additionally, the assessments indicate that only half of the groups produced models with acceptable or medium accuracy. For all targets only one highly accurate submission was observed excluding the homodimer.

Likewise, a total of 14 targets were included in the CAPRI rounds 13–19 (as shown in Table 1) for the assessment of the docking protocols. These rounds were focused on the assessment of the ability of the protein-protein docking algorithms to single out the near-native poses in the 14 selected targets, after searching the translational and conformational space. Although the scoring predictions have achieved greater progress over the previous CAPRI rounds but the identification of a near-native or best model remains a challenging endeavor which has an adverse effect on the model ranking. Therefore, the aforementioned CAPRI rounds focused around these aspects. However, in order to achieve success in the predictions, the conformational changes as well as the model-associated inaccuracies were brought into consideration. Assessments were made for a total of 4420 docking predictions made by 64 groups and 12 web servers. Remarkably, 67% of the groups produced acceptable models for one target while many other groups submitted several medium to high accuracy models for 2–6 targets [118]. Moreover, on the basis of the performance of different assessment groups it was identified that the inclusion of biochemical as well as structural information can lead to a considerable improvement in the quality of predicted models through protein-protein docking. However, in the absence of biochemical or structural information the interface information can still be utilized [119].

Additionally, during the years 2010–2012, a total of 8 CAPRI rounds (20–27) were conducted using a set of 15 targets. Amongst all only five targets displayed protein assemblies similar to the targets used in the previous CAPRI rounds. Excluding two targets, the remaining targets were specifically designed protein complexes obtained through the engineering experiments. The major focus of the predictions were the engineered structures or the assessment of the binding affinities of the newly designed ligands. Most prominently affinity predictions were for the very first time initiated in the rounds 20–21. Affinity prediction as a new module in CAPRI experiments remained challenging for the participants, as it is based on the assessment of the models not in terms of atomic details but rather on the basis of interaction thermodynamics.

The outcomes of these CAPRI experiments invigorated the development of novel procedures and scoring functions that will ultimately assist in: (1) improving the overall performance of the docking protocols, (2) the design of binders, and (3) providing improved binding free energy estimates (i.e., structure-based) of the natural assemblies [120, 121]. Although the rounds were run for the targets 21–27 but the final results are reported for the targets 25–27 mainly due to the cancellation of the targets 22–23 and uploading of incorrect models for the target 24. Lately, all docking protocols failed to reproduce the huge conformational adjustment that is incumbent for the correct assembly which points towards the fact that managing protein flexibility remains a key challenge. Additionally, it was observed that currently available scoring methods are perhaps not sensitive to an adequate level, as the best models could not be identified in most cases [122].

The sixth CAPRI report focusses on the assessment of the seven rounds including 28–29 and 31–35 respectively, whereby the results of round 30 have been presented as first joint CASP-CAPRI experiment. Overall, the sixth CAPRI report focusses on a total of 20 targets ranging from very complex/challenging protein-protein complexes (i.e., 12 targets) to short peptide motifs-protein (i.e., 8 targets). These targets have gained considerable interest over the recent years mainly due to the crucial roles of these peptides in the regulatory processes. Most evidently two of the protein-protein complexes were assessed for the predictions of side chain conformations and positioning of interfacial water molecules. Whereas, seven other complexes from this class were principally very challenging due to the multi-component behavior and conformational changes upon binding. These assessment rounds revealed the submission of acceptable quality models for 14 targets, high-quality models for 8 and medium-quality models for a total of 13 targets. Thus, indicating the progress of the computational approaches and an overall improvement in the accuracy when applied to protein complex modeling. The major attributing factors for this progress are the better amalgamation of docking protocols with different modeling tools along with model scoring on the basis of erudite evolutionary information. Despite the progress achieved, modeling of the conformational flexibility in interacting proteins continues to remain the essential component that necessitates improvement [123].

As far as the joint CASP-CAPRI experiment (round 30) is concerned the assessments were based on protein structure and protein docking predictions pertaining to a total of 25 targets that include a number of homodimers, a few homo-tetramer, and heterodimers. Whereby the included targets could be modeled readily through extraction of templates from the Protein Data Bank. Overall, 24 CAPRI groups focused on docking predictions of the targets, whereas 12 CAPRI groups were involved in the scoring experiment. Greater than 9500 models were assessed and it was revealed that the homodimer assembly predictions based on homology modeling and docking approaches appeared to be fairly successful for the targets that represent stable associations mainly due to adequately large interfaces of the subunits. Poor predictions were observed in case of the targets with ambiguous oligomeric assignments. However, the worst/poorest performance was achieved for the tetrameric targets with smaller pair-wise interfaces and inaccurately built homology-based subunits. It was also shown that the docking methods overperform the standard homology modeling approaches, and the assemblage of highly accurate proteins components is not always necessitated for the identification of the sufficiently accurate association modes [124].

Particularly, for the combined CAPRI 37-CASP 12 round a total of 12 homo-oligomers and 3 hetero-complexes were considered, where the evaluations were based on the challenge of modeling protein assemblies. The major part of the analysis was focused on 10 targets that included 3 hetero-complexes, as well as the protein chains or the full assemblies that could be easily modeled from PDB structural templates. Due to the availability of templates for complete assembly, a number of high-quality models were obtained for a total of 6 targets. Whereas, for the remaining targets poor results were obtained mainly due to the availability of low-quality templates for complete complex or the individual protein chains. This points towards the fact that the challenges associated to the modeling of structural adjustments of protein components upon binding or template-based modeling still remain unmet [125].

Furthermore, for the CAPRI rounds 38–45, a total of 18 target (i.e., 8 protein-protein, 3 protein-peptide, and 5 protein-oligosaccharide, Table 1) complexes were considered for which the observations were based on the prediction of binding modes and 3D conformational modeling of different bound peptides and protein subunits, etc. It is notable that each set of targets represented distinctive challenges for modeling. However, for the respective rounds the challenging aspects were (1) the level of template availability for different subunits (i.e., for heterodimers), (2) the aptitude to correctly identify the binding mode, and (3) the interface optimizations. As far as the targets are concerned the associated challenges were (1) inter-domain flexibility within the heterotrimer target 122 and its homologs, (2) unavailability of templates for PorM moieties in target 123–124, (3) difference in the interfacial region of the target versus templates for both targets 131 and 132 and the occurrence of several loops regions in the target 31 interface that were missing in the templates [126, 127].

Furthermore, a total of 10 docking servers including CLUSPRO, HDOCK, HADDOCK, LZERD, MDOCKPP, GALAXYPPDOCK, SWARDOCK, PYDOCKWEB, INTERPRED, and ZDOCK were used for server-based ranking in addition to human-based ranking of complexes. Amongst all CLUSPRO reserved its high rank presentation as shown in the previous assessments of CAPRI round. Additionally, HDOCK server displayed a performance very near to CLUSPRO. Similar to the last CAPRI assessments, the overall prediction performance of the servers for rounds 38–45 remained lower in comparison to the human (predictor) groups. This might be due to lower performance of servers on target complexes, i.e., protein-peptide and protein-sugar. Whereby the protein-sugar targets represent a reasonably recent CAPRI target category that requires optimization of automatic servers.

Overall, as per target, the predictive performance remained poor for the targets including 122, 123, 124, 131, and 132 (Table 1). Moreover, for the discussed rounds it was notable that the groups showing better performance for such difficult targets essentially integrated the sophisticated template-based modeling approaches not only at the individual subunit level but also at the complex level [127]. However, in cases where the template-based modeling was not adequate for handling modeling problems the strategy was augmented by utilizing biochemical-guided information, binding affinity approximations, quality score assessments, and redocking of subunits modeled so far [127]. Another limiting factor was that the crystal structures of similar hetero-complexes are susceptible to exhibiting different binding modes due to the effect of other component proteins and their transitory characteristics [114].

Additionally, a total of 20 targets divided into two difficulty levels (i.e., easy and difficult) were included in the CAPRI round 46. These include 14 homo-oligomers and 6 hetero-complexes (Table 1) for which the predictions were assessed in terms of representation as tractable modeling problems. The modeling of 8 homo-oligomer was readily achievable using PDB templates, however, for the remaining 11 heterodimer targets the difficulty level was high mainly due to the ab initio docking of subunits performed on the basis of distantly related templates and higher-order assemblies. The prediction performance of this round was assessed by 30 groups and 9 automatic servers, whereby medium- and higher-quality predictions were achieved for 3 models only. However, the other 11 difficult targets displayed poor predictions. Conclusively, the major gap in this round was identified which indicates that modeling conformational changes in protein components upon binding should be brought into consideration and must also be accounted for in the homology-based modeling. Moreover, it was highlighted that in comparison to the previous rounds the binding interfaces of the targets in round 46 appeared to be not as much well predicted [128]. Refer to Table 1 to get additional information on the targets used during the CAPRI rounds 47–53.

4 Challenges in Prediction of Interaction Sites of Membrane Proteins

Generally, the physiochemical properties and the lack of structural information makes the MPs a difficult target. MPs are generally integrated in lipid-bilayer and presence of suitable lipids governs the functionality of MPs, whereby the crucial cellular and biochemical processes are mediated by the interactions between these proteins [129]. A number of interacting protein pairs have been identified through the utility of genome-wide experiments but the residues involved in interactions are largely not known which demands the structural characterization of majority of these interactions. Since the experimental determination of protein-protein complexes is challenging, resource and time intensive, therefore comparative modeling (limited cases) and protein-protein docking is used. The docking method takes into account the surface complementarity and electrostatics to predict structural complexes through the interacting surfaces. Even though significant advancements have been achieved but the methods are still hindered due to vague understanding of the involved forces and conformation changes upon ligand binding. Furthermore, due to an increase in the numbers of proteins with atomic resolutions various groups have performed the extraction of basic features including shape and chemical complementarity or a combination of both [130].

However, certain challenges still remain in the prediction of the interaction sites of MPs. MPs form complexes to perform their biological roles similar to other non-membrane protein; however, due to the scarcity of high-resolution crystallographic structures (<1%) the structural basis of predicting protein-protein interaction sites is relatively challenging. The prediction of interaction sites in MP is complex because the residue frequencies on the surface of MPs differ significantly from soluble proteins, mainly due to the hydrophobic and hydrophilic environments respectively [131]. The hydrophobic nature of MPs is a major challenge for classical approaches to characterize the interaction with partners. Therefore, distinct predictors are required for to each particular class of proteins. Additionally, genetic and biochemical techniques have been developed to determine PPIs of MPs. However, these high-throughput approaches may generate high number of false-positive results [132]. Moreover, the computational prediction methods used for the interaction sites training show limited reliability due to lack of complex structures and the large degree of redundancy in the PDB. Hence, the reliability of testing of new methods also becomes difficult [130]. Another challenge is the misclassification of membrane proteins due to presence of signal peptides. Signal peptides contain hydrophobic regions that are mistaken by transmembrane segments [133].

Another challenging aspect is the presence of multiple interfaces and how to treat these interfaces. Since it is well established that proteins might be involved in a wide range of interactions either directly or transiently with many different protein complexes in different or same pathways respectively. Therefore, the lack of validated experimental complexes in the PDB hinders the understanding of different interactions in diverse magnitudes, since some proteins might be involved in multiple interactions due to the presence of multiple and overlapping interaction surfaces [130]. Another important aspect that needs to be brought into consideration is the overall importance of disordered regions in PPIs. It is revealed that these disordered regions allow for more interaction partners and modification sites [134]. The disordered regions have shown to be important in many interactions. It is also well-known the per-residue surface and interface areas of complexes can be used to distinguish ordered structures (monomers) from disordered structures. This further adds to the complexity of predicting interaction sites [130]. Furthermore, the homodimers are easy to discriminate since these are generally large and hydrophobic in nature. Moreover, on binding these dimers tend to bury large areas of non-polar residues. Whereas, hetero-complexes are generally difficult to predict because they include different transient complexes where the interaction sites have to be predicted blindly from unbound structure [130].

5 Challenges in Conformational Sampling and Refinement

PPIs form a central component in the molecular mechanisms of the wide range of biological processes which highlights the significance of the protein-protein docking in deciphering mechanistic insights. This points towards the rational that proteins are quite suitable for rigid body docking approximation mainly due to their exclusivity and limited variation in the global fold. Particularly, for docking protocols such approximation is of utmost concern as it allows considerable reduction in the global search space (i.e., to six coordinates of two rigid body system) mainly by the exclusion of massive collection of internal degrees of freedom generated for the interacting molecules. Another landmark is that to a larger extent the protein shape reflecting the protein fold determines the docking solution also at the coarse-grained depiction level [135].

Particularly, protein and all other macromolecules are subjected to conformational changes upon binding. However, the structural knowledge is required to study the functions of macromolecular assembly and their constituent parts along with the interactions formed. PPI interfaces are necessary for the formation of macromolecular assemblies, highlighting the difficulty in making accurate interface predictions. Since various protein complexes may prefer different conformational states and alter upon binding, these dynamical surface structures should be well described by experimental methods to maximize the success rate of the assembly structure prediction. Proteins may change between different conformations, which makes it challenging to find a new complex because of the computational challenges associated with conformational sampling and energy weighing. However, integrative modeling, that integrates computational and experimental techniques to model, score, and evaluate structures, is a potent strategy to increase modeling accuracy and effectiveness. Integrative modeling has previously demonstrated its value in determining the structures of assemblies, some of which were quite enormous and couldn't have been obtained any other way [136].

Moreover, global flexibility may not be an issue for protein docking but it is often considered as a key factor in docking methods. Evidence supports that to precompute the conformational ensemble of the interacting molecules and dock the individual conformers with minimal conformational search may resolve this issue [137–139]. Whereas, determining the adequate conformational ensemble is more often complicated for molecules with inherent flexibility (such as lipids, peptides, etc.) and/or for cases where a significant contribution of induced fit mechanism is observed in binding. The local conformational refinement in addition to the global flexibility remains challenging. For many biological systems pertaining to the prediction of binding interfaces for functional assessment, the global search-based approximate docking prediction may be adequate. Whereas, for various other cases related to inhibition, affinity assessment, etc., the atomic resolutions of predicted matches are highly necessitated. It is notable that in comparison to the global conformational search the local search takes into account lower degree of freedom and, hence, is more amenable. Towards this, significant progress has been made but the problem remains challenging particularly for the backbone conformational changes [140–143].

6 Scoring of Membrane Protein Complexes

The prediction of protein structures is highly coupled with the development of accurate scoring functions that are capable of discriminating between the native and near-native protein models. Scoring functions broadly lie under the categories of physics-, knowledge-, learning-based or a meld of the aforementioned types. The scoring functions for membrane proteins are discussed in detail as follows.

6.1 Physics-Based Scoring Functions

This particular type of scoring functions generally takes into account the energy terms used for describing the bonds, angles, dihedral angles, electrostatics, and van der Waals interactions. Whereby the solvent interactions can be modeled either explicitly by modeling each interaction pair individually, or implicitly by representing solvent as a continuous field with average solvent properties. As far as the membrane bilayer is considered its modeling as an implicit solvent was extended by Lazaridis [144]. The EEF1 solvation potential was extended by its application to the lipid environment where the parameterization of the solvation term was performed on the basis of experimental energy transfer observations from water to cyclohexane. The resultant IMM1111, potential was combined with the Rosetta knowledge-based scores in case of soluble proteins, mainly for the creation of RosettaMembrane [145], i.e., a high-resolution scoring function. This scoring function was used as a complement to the low-resolution Rosetta-Membrane scoring function that relies solely on the knowledge-based potentials based on modeling of residue pairs, solvation, etc. [145].

In other scoring functions, extensions have been made in the generalized Born theory of solvation particularly for the inclusion of membrane as a planar hydrophobic region. This approach was originally developed by Spassov et al., and led to 1–2 orders of magnitude improvement in comparison to the explicit bilayer simulations [146]. When combined with replica exchange techniques this work has facilitated for the first time the unguided protein folding simulations of small membrane-spanning polypeptide segments through the utility of classical MD or MC methods and the CHARMM, OPLS, and GROMOS standard molecular mechanics force fields [147]. The limitation of such methods is that they oversimplify the chemical complexity of the lipid bilayer interface and are inadequate in capturing the deformations caused in the membrane due to charged side chain partitioning. MARTINI is yet another force field that was initially developed for the modeling of lipid bilayers but was later on extended for protein modeling in the membrane [148]. This works by taking into account four lipid or amino acid atoms as a super atom bundle which considerably increases the speed (i.e., by 3–4 orders of magnitude). Furthermore, more recently MARTINI has been combined with the atomistic force field GROMOS for the modeling of lipid bilayer in coarse-grained mode while maintaining the ability to atomistically model proteins [149].

6.2 Knowledge-Based Scoring Functions

Largely for helical bundles, the knowledge-based scoring functions have been developed using a small number of the determined MP structures. The interacting residues within trans-membrane helices were elucidated and it was shown that the most probable interactions were observed between pairs of aromatic and basic residues (i.e., WR, WH, and YK) [150]. Additionally, the investigation of packing revealed the presence of larger residues in the pockets (where small residues were extremely unlikely), and a correlation between residue size and number of contacts was also observed [150]. Furthermore, Fleishman & Ben-Tal developed a packing score for the modeling of helix-helix interactions [151]. This scoring works by penalizing larger residues and rewarding smaller residues that are present within the interface region. Particularly, for the oligomeric helix-helix interfaces a distance-dependent scoring function was developed by Park et al. [152].

To guide the modeling or screening of erroneous models the knowledge-based potentials probing the interactions with the solvent can be used effectively. MPRAP191, a sequence-based SASA predictor is based on the support vector machines which was trained using a dataset of both soluble and MPs [153]. This method was developed to overcome the limitations associated with other scoring methods; however, the overall performance remains better in either of the forms, i.e., solution or membrane but not in both. In addition, a lipid propensity scale was utilized by Liangs group for the training of the LIPS server, which is capable of predicting lipid accessible surfaces solely from the protein sequences [154, 155]. Another fine-grained scoring function describing the PH-dependent binding of peptides to membrane bilayer was proposed by Lazaridis et al. [156]. Herein, the binding free energy components were calculated using MD simulation results. Later on, at the water-bilayer interface region the potentials of mean force for ionizable side chains were derived by the same group [157].

6.3 Comparison of the Available Scoring Functions

It is important to mention that the energy landscape is highly dependent on the granularity of modeling a protein; therefore, appropriate and correct scoring functions must be chosen in different scenarios. In order to determine and discriminate native-like conformations a number of physics-based potentials were compared by Forrest & Woolf who studied the loop conformations in MPs [158]. It was established that the continuum-solvent models and finite-difference PB models were amongst the most powerful ones. It was also notable that under a homogeneous environment the effective energy functions could distinguish the native-like conformations. The loop conformations in MP were also investigated through the testing of many high-resolution potentials on homology modeled decoys [159]. It was shown that PLOP, Rosetta and DFIRE are the scoring functions with superior discrimination power. These were followed by other scoring functions such as AMBER/GBSA, RAPDF, and MODELLER [160].

Moreover, RosettaMembrane with dependence on the membrane thickness was used to assess the discrimination power of a number of physics-based scoring function decoys [145]. It was identified that the best results at a membrane thickness of 25.4 Å were obtained when IMM1 was used in combination with CHARMM [160]. ProQM197, a learning-based scoring function, was developed by the Wallner group using support vector machine training on structural characteristics such as (1) atom-atom contacts, (2) residue-residue contacts, (3) SASA, secondary structure, (4) TM spans, (5) membrane depth, and (6) evolutionary information [161]. It was evident that this particular method produced higher correlation coefficients in comparison to other methods. Moreover, a sevenfold enrichment in near-native models was achieved when used in combination with RosettaMembrane [161]. Later on this method was outperformed by a more recent and simple method that is based on the known residue-residue interactions [162]. Based on the number of interactions and their distributions the IQ-score was calculated and it was observed that it shows a superior performance (on average by ~7%) in comparison to Pro QM [160].

7 Algorithms/Available Software Comparison and Challenges

During biological processes, a significant portion of signaling and regulatory proteins partake in brief protein-protein interactions. However, only a few membranes protein-protein complexes with known structures are available. This is primarily because of the difficulties of solidifying such proteins or performing NMR spectroscopy examinations. Therefore, the generation of computational methods for docking membrane proteins is imperative [163]. Various strategies for docking globular proteins exist, yet only a few cater to membrane proteins and intend to address docking inside the lipid bilayer environment.

Moreover, membrane proteins have unique properties identified via studying their backbone and exploring their hydrophobic nature. Since the layer has areas of strength for force imperatives, the polar gatherings of the polypeptide spine of the membrane protein partake in intra-atomic hydrogen bonds. This is consistently achieved through two primary themes: the layer traversing the alpha-helix pack as well as the beta-barrel. The more confined conformational search area that encompasses those two overarching themes is one of the merits of membrane protein demonstration [164]. Membrane protein modeling has advanced significantly during the past 10 years. Amongst the plethora of increasingly sophisticated protein-protein docking approaches, some of the most important MP softwares are enlisted as follows:

Memdock is an α-helical membrane protein docking algorithm introduced by Naama et al. [163]. The developed software not only performs the lipid bilayer docking but also takes into consideration the refinement and ranking of the docked proteins. Memdock performs a three-step algorithm for docking, whereby in the initial step calculation is dependent on the PatchDock algorithm [163]. PatchDock is an effective unbound delicate docking algorithm, which positions docking mainly by mathematical shape complementarity between the docked proteins [165]. In the subsequent step, a refinement algorithm is utilized using the FiberDock algorithm [166], which performs flexible refinement of docked proteins.

The technique models both side chain and backbone, and performs rigid body optimization of the ligand orientation. Lastly, the refined models are rescored and positioned using the energy functions formulated explicitly for membrane proteins. Additionally, the application of Memdock to a dataset of proteins with α-helical layers demonstrates that it works well and is much faster when docking membrane proteins. The calculation was applied to a set of 21 bound protein designs and 11 unbound protein structures. In 33% of the bound set and 36% of the unbound set, an adequate docking competitor was positioned at the top. In contrary, among the top 20 placement results, a satisfactory docking was projected in 71% of the bound benchmark set and 45% of the unbound benchmark sets respectively [163].

Furthermore, Lucas et al., proposed the transmembrane protein docking algorithm *JabberDock* for docking membrane proteins which is illustrative of the membrane protein design and dynamics in their biphasic climate [91]. The capacity of JabberDock to yield precise expectations was checked by applying it to 20 transmembrane dimers, returning a triumph pace of 75.0%, thus making the software exceptionally aggressive among accessible layer protein-protein docking tools [91]. JabberDock docks transmembrane proteins by means of a multistage cycle that involves: (1) fixing of missing deposits and particles, (2) drenching pre-arranged protein into pope bilayer, solvation and neutralization of the framework, (3) running MD reproductions, i.e., energy minimization, 60 ns equilibration, 10 ns creation, (4) conversion of MD results into special and temporal influence density (STID) map, (5) placement of guides of the ligands and receptors at the beginning, (6) characterization of the ligand space, followed by (7) docking run. JabberDock requires the input proteins to be aligned with the proteins center of mass for their transmembrane region at $z = 0$, where the z-pivot is located opposite to the bilayer plane [91]. Prior to being immersed in a POPE bilayer by utilizing the PACKMOL-memgen tool, the structure is fixed where necessary in order to use the Model [167]. GROMACS [168] is then used to create the Amber14SB [169] and SLipid [170] simulation information, which empowers the formation of STID maps. The maps of both the docking proteins are then converted into iso-surfaces using a predetermined end and docked to increase the complementarity of their surfaces. This surface-based scoring capability is compelling on the grounds that it sidesteps the need to explicitly deal with the pressing of interfacial particles, thereby, yielding smoother and gentler inclinations compared to the commonplace atomistic portrayals. Here, the space is explored by utilizing the molecule swarm advancement calculation executed in the POWer improvement motor [171].

LightDock, a new multi-scale way to deal with protein docking was presented by Garcı'a et al. [172]. The LightDock protein docking procedure is equipped to take into account the conformational adaptability and scoring capabilities at various goal levels. A significant challenge in membrane protein docking is the productive testing of the primary landscape and relationship at various resolution levels, considering that each scoring capability is frequently and profoundly coupled to a particular kind of search strategy [172]. LightDock depends on the Glowworm Swarm Optimization (GSO) [173] calculation for inspecting the translational and rotational space of protein docking, and anisotropic network mode (ANM) [167] portrayal for the incorporation of adaptability. LightDock intends to serve as a publicly available structural tool for developing and testing novel scoring systems for protein docking. Furthermore, when combined with other features, such as DFIRE [174], the use of pyDockLite's scoring function yields accomplishment rates that are nearly similar to those of cutting-edge conventions. This multi-scale docking system has the flexibility to use a broad variety of scoring capabilities (alone or in combination) and to take adaptation into account at different objective levels. However, the major disadvantage of GSO over Particle Swarm Optimization (PSO) is the calculation time, which will often be one significant degree higher [175].

HADDOCK is an integrative biomolecular docking stage that utilizes trial information or bioinformatics expectations to drive the docking system. Normally, this data is transformed into distance constraints that enable docking. Docking comprises three phases: (1) rigid body energy minimization, (2) semi-adaptable refinement by simulated annealing in torsional space, and (3) refinement in explicit solvent. For the principal stage, the accomplices are arbitrarily arranged and interpreted away from one another followed by inflexible body energy minimization. During this stage, the complex's connection point buildups to demonstrate flexibility, initially just at the edge chains followed by encompassing the backbone particles in the final stage. The last step consists of brief, unambiguously dissolvable atomic elements run in Cartesian space (i.e., carried out using the default TIP3P water model) [96]. Despite not being improved for membrane proteins, HADDOCK has great performance (92% overall success rate when taking into account each of the 400 final-stage models) when the astute observation of interaction information is available. In its ab initio docking mode, nonetheless, the performance dropped to 35% for last-stage models [96].

ClusPro, a web-based server that allows two interacting proteins to dock directly with one another. ClustPro performs the following steps: (1) rigid body docking involves sampling billions of conformations, (2) root-mean-square deviation (RMSD)-based clustering of the 1000 lowest energy structures produced, (3) refinement of chosen structures using energy minimization, and (4) the identification of the largest clusters that will represent the most likely models of the complex [176]. Additionally, PIPER16 is a docking software based on the fast Fourier transform (FFT) correlation technique, that is used for the rigid body docking stage. Katchalski-Katzir and colleagues developed the FFT method, which significantly improved rigid body protein-protein docking [111, 177]. This approach places the first protein, which is referred to as the receptor, in the center of the coordinate system on a fixed grid, the second protein, which is referred to as the ligand, on a moveable grid, and records the interaction energy as a correlation function (or as a sum of a few correlation functions). The numerical efficiency of the approach is a result of the capacity to effectively calculate such energy functions using fast Fourier transforms, which allows for the ability to thoroughly sample billions of conformations of the two interacting proteins and evaluate the energies at each grid point. As a result, proteins may be docked using the FFT-based technique without any prior knowledge of the complex's structure. Despite the server being frequently utilized, runs typically end in 4 h or less [111, 177].

DOCK/PIERR algorithm uses the tertiary structures of the two separate proteins to predict, in atomic precision, the structure of the complex generated by those two proteins. Coarse scoring and stiff docking are both part of the initial step. First, using fast Fourier transforms (FFT) on a grid with a spacing of 1, a comprehensive collection of rigid transformations of one protein with respect to the other is sampled and scored. Combining a van der Waals term for shape complementarity with a residue potential PIE based on interface residue-residue interactions yields the score. A rigorous theory and large learning set were used to optimize the scoring function's parameter set (640 protein complexes). The highest scoring conformations were saved for predictions and then filtered for interface conflicts and clustering. The collection of conformations is specifically grouped in rigid body space, and then models with more than 45 atomic conflicts were filtered away [178].

On the basis of interface RMSD, the remaining models are reclustered. By eliminating models that are too similar to other models, clustering enables more varied ranking. The emphasis of the second stage calculations is on re-ranking using an atomic model. Only the top 1000 models determined by clustering, as stated above, are taken into consideration. The interface side chains for each model are redesigned, and a quick energy reduction is carried out. The structures are then rescored using a combination of the residue potential (PIE) and PISA, an atomic potential created specifically for protein contact surfaces. However, second stage is not intended to be a polishing stage. The structural changes are quite minor (RMSD 0.1) and are intended to raise the atomic score rather than the overall docking position. The system then outputs predictions for the top 10 models in the complex. With regard to the homodimer ensembles simulated in POPC bilayer and DPC micelle, where configurational entropy and surface curvature effects skew the handedness and topology of the homodimer ensemble, DOCK/PIERR reproduces key distinctions [178].

A generic membrane model offered by *RosettaMP* interacts with scoring, conformational sampling, and mutation methods that may be readily coupled to develop novel protocols. RosettaMP directly enlarges Rosetta3's architectural design. A pose to represent a biomolecule, a ScoreFunction to rate modeled structures and sequences, and Movers to sample new conformations of the pose, are the three fundamental items that are required by any Rosetta protocol for the modeling or design of activities. RosettaMP directly extends this architecture by modifying the original membrane ScoreFunction to rely on this membrane representation, adding an element to the pose representing the membrane bilayer, and implementing a new set of Movers to sample the conformational search space available in the membrane bilayer. RosettaMP significantly lowers the complexity barrier for the creation of new protocols to model and create MPs because of its interoperability with current Rosetta code and reusable representation of the lipid bilayer. This opens the door to many new, urgently required techniques [179].

The findings show that models of membrane protein complexes may be created using the aforementioned methods in a reasonably accurate manner. In order to compare the results of various algorithms, JabberDock was applied to an unbound benchmark of 20 transmembrane-helix proteins collected from three distinct benchmarks, however, as there is no common benchmark for transmembrane protein docking, the success rate was just 75%. These outcomes result in 22 individual comparisons from 20 cases, with seven cases correctly identified as compared to DOCK/PIPER's (two out of eight cases), eight cases correctly identified as compared to Memdock's (four out of eleven cases), and one case correctly identified as compared to HADDOCK's (one out of three cases) respectively.

When compared to the aforementioned methods, JabberDock has shown to be more effective at handling scenarios involving integral membrane protein dimers. This accomplishment is attributable to the molecular model (STID maps) that we use to dock proteins, which unifies electrostatics, dynamics, and protein structure into a single volumetric model. JabberDock should function effectively with transmembrane-solvent proteins as well, regardless of whether the ligand is extracellular, periplasmic, or cytoplasmic, given the positive results.

References

1. Li F, Egea PF, Vecchio AJ, Asial I, Gupta M, Paulino J, Bajaj R, Dickinson MS, Ferguson-Miller S, Monk BC (2021) Highlighting membrane protein structure and function: a celebration of the Protein Data Bank. J Biol Chem 296:100557
2. Wallin E, Heijne GV (1998) Genome-wide analysis of integral membrane proteins from eubacterial, archaean, and eukaryotic organisms. Protein Sci 7(4):1029–1038
3. Overington JP, Al-Lazikani B, Hopkins AL (2006) How many drug targets are there? Nat Rev Drug Discov 5(12):993–996
4. Bakheet TM, Doig AJ (2009) Properties and identification of human protein drug targets. Bioinformatics 25(4):451–457
5. Moraes I, Evans G, Sanchez-Weatherby J, Newstead S, Stewart PDS (2014) Membrane protein structure determination—The next generation. Biochimica et Biophysica Acta (BBA) – Biomembranes 1838(1, Part A):78–87. https://doi.org/10.1016/j.bbamem.2013.07.010
6. Jimonet P, Jäger R (2004) Strategies for designing GPCR-focused libraries and screening sets. Curr Opin Drug Discov Devel 7(3): 325–333
7. Schnur DM, Hermsmeier MA, Tebben AJ (2006) Are target-family-privileged substructures truly privileged? J Med Chem 49(6): 2000–2009
8. Carpenter EP, Beis K, Cameron AD, Iwata S (2008) Overcoming the challenges of membrane protein crystallography. Curr Opin Struct Biol 18(5):581–586
9. Moraes I, Quigley A (2021) Structural biology and structure–function relationships of membrane proteins. Biology 10(3):245
10. Moraes I, Evans G, Sanchez-Weatherby J, Newstead S, Stewart PDS (2014) Membrane protein structure determination—the next generation. Biochim Biophys Acta 1838(1): 78–87
11. Pan Y, Stocks BB, Brown L, Konermann L (2009) Structural characterization of an integral membrane protein in its natural lipid environment by oxidative methionine labeling and mass spectrometry. Anal Chem 81(1): 28–35
12. Zorman S, Botte M, Jiang Q, Collinson I, Schaffitzel C (2015) Advances and challenges of membrane–protein complex production. Curr Opin Struct Biol 32:123–130
13. Drew D, Fröderberg L, Baars L, de Gier J-WL (2003) Assembly and overexpression of membrane proteins in Escherichia coli. Biochim Biophys Acta 1610(1):3–10
14. Wagner S, Bader ML, Drew D, de Gier J-W (2006) Rationalizing membrane protein overexpression. Trends Biotechnol 24(8): 364–371
15. Kobilka BK (2007) G protein coupled receptor structure and activation. Biochim Biophys Acta 1768(4):794–807
16. Hardy D, Bill RM, Jawhari A, Rothnie AJ (2016) Overcoming bottlenecks in the membrane protein structural biology pipeline. Biochem Soc Trans 44(3):838–844
17. Schnell DJ, Hebert DN (2003) Protein translocons: multifunctional mediators of protein translocation across membranes. Cell 112(4): 491–505
18. Junge F, Schneider B, Reckel S, Schwarz D, Dötsch V, Bernhard F (2008) Large-scale production of functional membrane proteins. Cell Mol Life Sci 65(11):1729–1755
19. Privé GG (2007) Detergents for the stabilization and crystallization of membrane proteins. Methods 41(4):388–397
20. Katona G, Andreasson U, Landau EM, Andreasson L-E, Neutze R (2003) Lipidic cubic phase crystal structure of the photosynthetic reaction centre from Rhodobacter sphaeroides at 2.35 Å resolution. J Mol Biol 331(3):681–692
21. Cherezov V, Rosenbaum DM, Hanson MA, Rasmussen SG, Thian FS, Kobilka TS, Choi H-J, Kuhn P, Weis WI, Kobilka BK (2007) High-resolution crystal structure of an engineered human β2-adrenergic G protein–coupled receptor. Science 318(5854):1258–1265
22. Cherezov V, Peddi A, Muthusubramaniam L, Zheng YF, Caffrey M (2004) A robotic system for crystallizing membrane and soluble proteins in lipidic mesophases. Acta Crystallogr D Biol Crystallogr 60(10):1795–1807
23. Segelke B (2005) Macromolecular crystallization with microfluidic free-interface diffusion. Expert Rev Proteomics 2(2):165–172

24. Li L, Mustafi D, Fu Q, Tereshko V, Chen DL, Tice JD, Ismagilov RF (2006) Nanoliter microfluidic hybrid method for simultaneous screening and optimization validated with crystallization of membrane proteins. Proc Natl Acad Sci 103(51):19243–19248
25. Jiang Y, Lee A, Chen J, Ruta V, Cadene M, Chait BT, MacKinnon R (2003) X-ray structure of a voltage-dependent K+ channel. Nature 423(6935):33–41
26. Rasmussen SG, Choi H-J, Rosenbaum DM, Kobilka TS, Thian FS, Edwards PC, Burghammer M, Ratnala VR, Sanishvili R, Fischetti RF (2007) Crystal structure of the human β2 adrenergic G-protein-coupled receptor. Nature 450(7168):383–387
27. Manjasetty BA, Turnbull AP, Panjikar S, Büssow K, Chance MR (2008) Automated technologies and novel techniques to accelerate protein crystallography for structural genomics. Proteomics 8(4):612–625
28. Riekel C, Burghammer M, Schertler G (2005) Protein crystallography microdiffraction. Curr Opin Struct Biol 15(5):556–562
29. Wagner S, Baars L, Ytterberg AJ, Klussmeier A, Wagner CS, Nord O, Nygren P-A, van Wijk KJ, de Gier J-W (2007) Consequences of membrane protein overexpression in Escherichia coli. Mol Cell Proteomics 6(9):1527–1550
30. Rosenbusch J, Lustig A, Grabo M, Zulauf M, Regenass M (2001) Approaches to determining membrane protein structures to high resolution: do selections of subpopulations occur? Micron 32(1):75–90
31. Almeida JG, Preto AJ, Koukos PI, Bonvin AM, Moreira IS (2017) Membrane proteins structures: a review on computational modeling tools. Biochim Biophys Acta 1859(10): 2021–2039
32. Wagner S, Klepsch MM, Schlegel S, Appel A, Draheim R, Tarry M, Högbom M, Van Wijk KJ, Slotboom DJ, Persson JO (2008) Tuning Escherichia coli for membrane protein overexpression. Proc Natl Acad Sci 105(38): 14371–14376
33. Schlegel S, Löfblom J, Lee C, Hjelm A, Klepsch M, Strous M, Drew D, Slotboom DJ, de Gier J-W (2012) Optimizing membrane protein overexpression in the Escherichia coli strain Lemo21 (DE3). J Mol Biol 423(4):648–659
34. Hays FA, Roe-Zurz Z, Stroud RM (2010) Overexpression and purification of integral membrane proteins in yeast. In: Methods in enzymology, vol 470. Elsevier, Amsterdam, pp 695–707
35. Serebryany E, Zhu GA, Yan EC (2012) Artificial membrane-like environments for in vitro studies of purified G-protein coupled receptors. Biochim Biophys Acta 1818(2): 225–233
36. Chae PS, Rasmussen SG, Rana RR, Gotfryd K, Chandra R, Goren MA, Kruse AC, Nurva S, Loland CJ, Pierre Y (2010) Maltose–neopentyl glycol (MNG) amphiphiles for solubilization, stabilization and crystallization of membrane proteins. Nat Methods 7(12):1003–1008
37. Drew D, Newstead S, Sonoda Y, Kim H, Von Heijne G, Iwata S (2008) GFP-based optimization scheme for the overexpression and purification of eukaryotic membrane proteins in Saccharomyces cerevisiae. Nat Protoc 3(5): 784–798
38. Chun E, Thompson AA, Liu W, Roth CB, Griffith MT, Katritch V, Kunken J, Xu F, Cherezov V, Hanson MA (2012) Fusion partner toolchest for the stabilization and crystallization of G protein-coupled receptors. Structure 20(6):967–976
39. Tate CG, Schertler GF (2009) Engineering G protein-coupled receptors to facilitate their structure determination. Curr Opin Struct Biol 19(4):386–395
40. Steyaert J, Kobilka BK (2011) Nanobody stabilization of G protein-coupled receptor conformational states. Curr Opin Struct Biol 21(4):567–572
41. Serrano-Vega MJ, Tate CG (2009) Transferability of thermostabilizing mutations between β-adrenergic receptors. Mol Membr Biol 26(8):385–396
42. Stevens RC, Yokoyama S, Wilson IA (2001) Global efforts in structural genomics. Science 294(5540):89–92
43. Duke E, Johnson L (2010) Macromolecular crystallography at synchrotron radiation sources: current status and future developments. Proc R Soc A: Math Phys Eng Sci 466(2124):3421–3452
44. Venko K, Choudhury AR, Novič M (2017) Computational approaches for revealing the structure of membrane transporters: case study on bilitranslocase. Comput Struct Biotechnol J 15:232–242
45. Schlessinger A (2014) Characterizing the structure, function, and evolution of human solute carrier (SLC) transporters using computational approaches. In: Membrane transport mechanism. Springer, Berlin, Heidelberg, pp 23–57
46. Simons KT, Bonneau R, Ruczinski I, Baker D (1999) Ab initio protein structure prediction

of CASP III targets using ROSETTA. Proteins: Struct Funct Bioinf 37(S3):171–176

47. Hung L-H, Samudrala R (2003) PROTINFO: secondary and tertiary protein structure prediction. Nucleic Acids Res 31(13): 3296–3299
48. Pellegrini-Calace M, Carotti A, Jones D (2003) Folding in lipid membranes (FILM): a novel method for the prediction of small membrane protein 3D structures. Proteins: Struct Funct Bioinf 50(4):537–545
49. Jones DT (2001) Predicting novel protein folds by using FRAGFOLD. Proteins: Struct Funct Bioinf 45(S5):127–132
50. Yarov-Yarovoy V (2015) Rosetta structural modeling. In: Zheng J, Trudeau M (eds) Handbook of ion channels. CRC Press, Boca Raton
51. Kelm S, Shi J, Deane CM (2010) MEDELLER: homology-based coordinate generation for membrane proteins. Bioinformatics 26(22):2833–2840
52. Ebejer J-P, Hill JR, Kelm S, Shi J, Deane CM (2013) Memoir: template-based structure prediction for membrane proteins. Nucleic Acids Res 41(W1):W379–W383
53. Xu Y, Xu D (2000) Protein threading using PROSPECT: design and evaluation. Proteins: Struct Funct Bioinf 40(3):343–354
54. Xu J, Li M, Kim D, Xu Y (2003) RAPTOR: optimal protein threading by linear programming. J Bioinforma Comput Biol 1(01): 95–117
55. Kozma D, Tusnády GE (2015) TMFoldWeb: a web server for predicting transmembrane protein fold class. Biol Direct 10(1):1–8
56. Lobley A, Sadowski MI, Jones DT (2009) pGenTHREADER and pDomTHREADER: new methods for improved protein fold recognition and superfamily discrimination. Bioinformatics 25(14):1761–1767
57. Zhou H, Skolnick J (2009) Protein structure prediction by pro-Sp3-TASSER. Biophys J 96(6):2119–2127
58. Roy A, Kucukural A, Zhang Y (2010) I-TASSER: a unified platform for automated protein structure and function prediction. Nat Protoc 5(4):725–738
59. Kelm S, Shi J, Deane CM (2009) iMembrane: homology-based membrane-insertion of proteins. Bioinformatics 25(8):1086–1088
60. Daga PR, Patel RY, Doerksen RJ (2010) Template-based protein modeling: recent methodological advances. Curr Top Med Chem 10(1):84–94
61. Michino M, Chen J, Stevens RC, Brooks CL III (2010) FoldGPCR: structure prediction protocol for the transmembrane domain of G protein-coupled receptors from class A. Proteins: Struct Funct Bioinf 78(10): 2189–2201
62. Flock T, Venkatakrishnan A, Vinothkumar K, Babu MM (2012) Deciphering membrane protein structures from protein sequences. Genome Biol 13(6):1–4
63. Karawajczyk A, Drgan V, Medic N, Oboh G, Passamonti S, Novič M (2007) Properties of flavonoids influencing the binding to bilitranslocase investigated by neural network modelling. Biochem Pharmacol 73(2):308–320
64. Martinčič R, Venko K, Župerl Š, Novič M (2014) Chemometrics approach for the prediction of structure–activity relationship for membrane transporter bilitranslocase. SAR QSAR Environ Res 25(11):853–872
65. Brooks BR, Brooks CL III, Mackerell AD Jr, Nilsson L, Petrella RJ, Roux B, Won Y, Archontis G, Bartels C, Boresch S (2009) CHARMM: the biomolecular simulation program. J Comput Chem 30(10):1545–1614
66. Pearlman DA, Case DA, Caldwell JW, Ross WS, Cheatham TE III, DeBolt S, Ferguson D, Seibel G, Kollman P (1995) AMBER, a package of computer programs for applying molecular mechanics, normal mode analysis, molecular dynamics and free energy calculations to simulate the structural and energetic properties of molecules. Comput Phys Commun 91(1–3):1–41
67. Andersen OS, Koeppe RE (2007) Bilayer thickness and membrane protein function: an energetic perspective. Annu Rev Biophys Biomol Struct 36:107–130
68. Sugita Y, Okamoto Y (1999) Replica-exchange molecular dynamics method for protein folding. Chem Phys Lett 314(1–2): 141–151
69. Nymeyer H, Woolf TB, Garcia AE (2005) Folding is not required for bilayer insertion: replica exchange simulations of an α-helical peptide with an explicit lipid bilayer. Proteins: Struct Funct Bioinf 59(4):783–790
70. Gullingsrud J, Schulten K (2003) Gating of MscL studied by steered molecular dynamics. Biophys J 85(4):2087–2099
71. Bernardi RC, Melo MC, Schulten K (2015) Enhanced sampling techniques in molecular dynamics simulations of biological systems. Biochim Biophys Acta 1850(5):872–877
72. Furini S, Domene C (2016) Computational studies of transport in ion channels using

metadynamics. Biochim Biophys Acta 1858(7):1733–1740

73. Marrink SJ, Risselada HJ, Yefimov S, Tieleman DP, De Vries AH (2007) The MARTINI force field: coarse grained model for biomolecular simulations. J Phys Chem B 111(27): 7812–7824
74. Oren I, Fleishman SJ, Kessel A, Ben-Tal N (2004) Free diffusion of steroid hormones across biomembranes: a simplex search with implicit solvent model calculations. Biophys J 87(2):768–779
75. Baker NA (2005) Improving implicit solvent simulations: a Poisson-centric view. Curr Opin Struct Biol 15(2):137–143
76. Feig M, Onufriev A, Lee MS, Im W, Case DA, Brooks CL III (2004) Performance comparison of generalized born and Poisson methods in the calculation of electrostatic solvation energies for protein structures. J Comput Chem 25(2):265–284
77. Im W, Lee MS, Brooks CL III (2003) Generalized born model with a simple smoothing function. J Comput Chem 24(14):1691–1702
78. Tanizaki S, Feig M (2006) Molecular dynamics simulations of large integral membrane proteins with an implicit membrane model. J Phys Chem B 110(1):548–556
79. Pazos F, Helmer-Citterich M, Ausiello G, Valencia A (1997) Correlated mutations contain information about protein-protein interaction. J Mol Biol 271(4):511–523
80. Xue LC, Dobbs D, Honavar V (2011) HomPPI: a class of sequence homology based protein-protein interface prediction methods. BMC Bioinformatics 12(1):1–24
81. Ahmad S, Mizuguchi K (2011) Partner-aware prediction of interacting residues in protein-protein complexes from sequence data. PLoS One 6(12):e29104
82. Murakami Y, Mizuguchi K (2010) Applying the Naïve Bayes classifier with kernel density estimation to the prediction of protein–protein interaction sites. Bioinformatics 26(15): 1841–1848
83. Neuvirth H, Raz R, Schreiber G (2004) ProMate: a structure based prediction program to identify the location of protein–protein binding sites. J Mol Biol 338(1):181–199
84. Porollo A, Meller J (2007) Prediction-based fingerprints of protein–protein interactions. Proteins: Struct Funct Bioinf 66(3):630–645
85. de Vries SJ, Bonvin AM (2011) CPORT: a consensus interface predictor and its performance in prediction-driven docking with HADDOCK. PLoS One 6(3):e17695
86. Afsar Minhas FA, Geiss BJ, Ben-Hur A (2014) PAIRpred: partner-specific prediction of interacting residues from sequence and structure. Proteins: Struct Funct Bioinf 82(7): 1142–1155
87. Xue LC, Dobbs D, Bonvin AM, Honavar V (2015) Computational prediction of protein interfaces: a review of data driven methods. FEBS Lett 589(23):3516–3526
88. Shen H, Chou JJ (2008) MemBrain: improving the accuracy of predicting transmembrane helices. PLoS One 3(6):e2399
89. Naveed H, Jackups R Jr, Liang J (2009) Predicting weakly stable regions, oligomerization state, and protein–protein interfaces in transmembrane domains of outer membrane proteins. Proc Natl Acad Sci 106(31): 12735–12740
90. Hayat S, Walter P, Park Y, Helms V (2011) Prediction of the exposure status of transmembrane beta barrel residues from protein sequence. J Bioinforma Comput Biol 9(01): 43–65
91. Rudden LS, Degiacomi MT (2021) Transmembrane protein docking with jabberdock. J Chem Inf Model 61(3):1493–1499
92. Moreira IS, Fernandes PA, Ramos MJ (2010) Protein–protein docking dealing with the unknown. J Comput Chem 31(2):317–342
93. Lesk VI, Sternberg MJ (2008) 3D-Garden: a system for modelling protein–protein complexes based on conformational refinement of ensembles generated with the marching cubes algorithm. Bioinformatics 24(9): 1137–1144
94. Geng C, Xue LC, Roel-Touris J, Bonvin AM (2019) Finding the ΔΔG spot: are predictors of binding affinity changes upon mutations in protein–protein interactions ready for it? Wiley Interdiscip Rev Comput Mol Sci 9(5): e1410
95. Tovchigrechko A, Vakser IA (2005) Development and testing of an automated approach to protein docking. Proteins: Struct Funct Bioinf 60(2):296–301
96. Dominguez C, Boelens R, Bonvin AM (2003) HADDOCK: a protein–protein docking approach based on biochemical or biophysical information. J Am Chem Soc 125(7): 1731–1737
97. Casciari D, Seeber M, Fanelli F (2006) Quaternary structure predictions of transmembrane proteins starting from the monomer: a docking-based approach. BMC Bioinformatics 7(1):1–16
98. Alford RF, Samanta R, Gray JJ (2021) Diverse scientific benchmarks for implicit membrane

energy functions. J Chem Theory Comput 17(8):5248–5261
99. Vajda S (2005) Classification of protein complexes based on docking difficulty. Proteins: Struct Funct Bioinf 60(2):176–180
100. Goñi FM (2014) The basic structure and dynamics of cell membranes: an update of the Singer–Nicolson model. Biochim Biophys Acta 1838(6):1467–1476
101. Boes DM, Godoy-Hernandez A, McMillan DG (2021) Peripheral membrane proteins: promising therapeutic targets across domains of life. Membranes 11(5):346
102. Chatzigoulas A, Cournia Z (2022) Predicting protein–membrane interfaces of peripheral membrane proteins using ensemble machine learning. Brief Bioinform 23(2):bbab518
103. Taylor MJ, Husain K, Gartner ZJ, Mayor S, Vale RD (2017) A DNA-based T cell receptor reveals a role for receptor clustering in ligand discrimination. Cell 169(1):108–119.e120
104. Banjade S, Rosen MK (2014) Phase transitions of multivalent proteins can promote clustering of membrane receptors. elife 3: e04123
105. Christie S, Shi X, Smith AW (2020) Resolving membrane protein–protein interactions in live cells with pulsed interleaved excitation fluorescence cross-correlation spectroscopy. Acc Chem Res 53(4):792–799
106. Domene C, Bond PJ, Deol SS, Sansom MS (2003) Lipid/protein interactions and the membrane/water interfacial region. J Am Chem Soc 125(49):14966–14967
107. Killian JA, von Heijne G (2000) How proteins adapt to a membrane–water interface. Trends Biochem Sci 25(9):429–434
108. White SH, Wimley WC (1998) Hydrophobic interactions of peptides with membrane interfaces. Biochim Biophys Acta 1376(3): 339–352
109. Kotthoff I, Kundrotas PJ, Vakser IA (2022) DOCKGROUND membrane protein-protein set. PLoS One 17(5):e0267531
110. Janin J, Henrick K, Moult J, Eyck LT, Sternberg MJ, Vajda S, Vakser I, Wodak SJ (2003) CAPRI: a critical assessment of predicted interactions. Proteins: Struct Funct Bioinf 52(1):2–9
111. Koukos PI, Faro I, van Noort CW, Bonvin AM (2018) A membrane protein complex docking benchmark. J Mol Biol 430(24): 5246–5256
112. Vreven T, Moal IH, Vangone A, Pierce BG, Kastritis PL, Torchala M, Chaleil R, Jiménez-García B, Bates PA, Fernandez-Recio J (2015) Updates to the integrated protein–protein interaction benchmarks: docking benchmark version 5 and affinity benchmark version 2. J Mol Biol 427(19):3031–3041
113. Van Dijk M, Bonvin AM (2008) A protein–DNA docking benchmark. Nucleic Acids Res 36(14):e88
114. Hartshorn MJ, Verdonk ML, Chessari G, Brewerton SC, Mooij WT, Mortenson PN, Murray CW (2007) Diverse, high-quality test set for the validation of protein–ligand docking performance. J Med Chem 50(4): 726–741
115. Trellet M, Melquiond AS, Bonvin AM (2013) A unified conformational selection and induced fit approach to protein-peptide docking. PLoS One 8(3):e58769
116. Janin J (2005) Assessing predictions of protein–protein interaction: the CAPRI experiment. Protein Sci 14(2):278–283
117. Méndez R, Leplae R, Lensink MF, Wodak SJ (2005) Assessment of CAPRI predictions in rounds 3–5 shows progress in docking procedures. Proteins: Struct Funct Bioinf 60(2): 150–169
118. Lensink MF, Wodak SJ (2010) Docking and scoring protein interactions: CAPRI 2009. Proteins: Struct Funct Bioinf 78(15): 3073–3084
119. Qin S, Zhou HX (2010) Selection of near-native poses in CAPRI rounds 13–19. Proteins: Struct Funct Bioinf 78(15):3166–3173
120. Janin J (2013) The targets of CAPRI rounds 20–27. Proteins: Struct Funct Bioinf 81(12): 2075–2081
121. Lensink MF, Wodak SJ (2013) Docking, scoring, and affinity prediction in CAPRI. Proteins: Struct Funct Bioinf 81(12):2082–2095
122. Lensink MF, Méndez R, Wodak SJ (2007) Docking and scoring protein complexes: CAPRI 3rd Edition. Proteins: Struct Funct Bioinf 69(4):704–718
123. Lensink MF, Velankar S, Wodak SJ (2017) Modeling protein–protein and protein–peptide complexes: CAPRI 6th edition. Proteins: Struct Funct Bioinf 85(3):359–377
124. Lensink MF, Velankar S, Kryshtafovych A, Huang SY, Schneidman-Duhovny D, Sali A, Segura J, Fernandez-Fuentes N, Viswanath S, Elber R (2016) Prediction of homoprotein and heteroprotein complexes by protein docking and template-based modeling: a CASP-CAPRI experiment. Proteins: Struct Funct Bioinf 84:323–348
125. Lensink MF, Velankar S, Baek M, Heo L, Seok C, Wodak SJ (2018) The challenge of modeling protein assemblies: the CASP12-

CAPRI experiment. Proteins: Struct Funct Bioinf 86:257–273

126. Moonens K, Hamway Y, Neddermann M, Reschke M, Tegtmeyer N, Kruse T, Kammerer R, Mejías-Luque R, Singer BB, Backert S (2018) Helicobacter pylori adhesin HopQ disrupts trans dimerization in human CEACAM s. EMBO J 37(13):e98665
127. Lensink MF, Nadzirin N, Velankar S, Wodak SJ (2020) Modeling protein-protein, protein-peptide, and protein-oligosaccharide complexes: CAPRI 7th edition. Proteins: Struct Funct Bioinf 88(8):916–938
128. Lensink MF, Brysbaert G, Nadzirin N, Velankar S, Chaleil RA, Gerguri T, Bates PA, Laine E, Carbone A, Grudinin S (2019) Blind prediction of homo-and hetero-protein complexes: the CASP13-CAPRI experiment. Proteins: Struct Funct Bioinf 87(12):1200–1221
129. Tiefenauer L, Demarche S (2012) Challenges in the development of functional assays of membrane proteins. Materials 5(11): 2205–2242
130. Ezkurdia I, Bartoli L, Fariselli P, Casadio R, Valencia A, Tress ML (2009) Progress and challenges in predicting protein–protein interaction sites. Brief Bioinform 10(3): 233–246
131. Bordner AJ (2009) Predicting protein-protein binding sites in membrane proteins. BMC Bioinformatics 10(1):1–10
132. Hunke S, Müller VS (2012) Approaches to analyze protein–protein interactions of membrane proteins. InTech, Rijeka, pp 327–348
133. Liang J, Naveed H, Jimenez-Morales D, Adamian L, Lin M (2012) Computational studies of membrane proteins: models and predictions for biological understanding. Biochim Biophys Acta 1818(4):927–941
134. Tompa P, Fuxreiter M (2008) Fuzzy complexes: polymorphism and structural disorder in protein protein interactions. Trends Bio chem Sci 33(1):2–8
135. Koehler Leman J, Ulmschneider MB, Gray JJ (2015) Computational modeling of membrane proteins. Proteins 83(1):1–24. https://doi.org/10.1002/prot.24703
136. Joseph AP, Polles G, Alber F, Topf M (2017) Integrative modelling of cellular assemblies. Curr Opin Struct Biol 46:102–109. https://doi.org/10.1016/j.sbi.2017.07.001
137. Koehler J, Meiler J (2011) Expanding the utility of NMR restraints with paramagnetic compounds: background and practical aspects. Prog Nucl Magn Reson Spectrosc 59(4):360
138. Yu L, Sun C, Song D, Shen J, Xu N, Gunasekera A, Hajduk PJ, Olejniczak ET (2005) Nuclear magnetic resonance structural studies of a potassium channel– charybdotoxin complex. Biochemistry 44(48): 15834–15841
139. Berardi MJ, Shih WM, Harrison SC, Chou JJ (2011) Mitochondrial uncoupling protein 2 structure determined by NMR molecular fragment searching. Nature 476(7358): 109–113
140. Bayrhuber M, Meins T, Habeck M, Becker S, Giller K, Villinger S, Vonrhein C, Griesinger C, Zweckstetter M, Zeth K (2008) Structure of the human voltage-dependent anion channel. Proc Natl Acad Sci 105(40):15370–15375
141. Shahid SA, Bardiaux B, Franks WT, Krabben L, Habeck M, van Rossum B-J, Linke D (2012) Membrane-protein structure determination by solid-state NMR spectroscopy of microcrystals. Nat Methods 9(12): 1212–1217
142. Wang S, Munro RA, Shi L, Kawamura I, Okitsu T, Wada A, Kim S-Y, Jung K-H, Brown LS, Ladizhansky V (2013) Solid-state NMR spectroscopy structure determination of a lipid-embedded heptahelical membrane protein. Nat Methods 10(10):1007–1012
143. Liao M, Cao E, Julius D, Cheng Y (2013) Structure of the TRPV1 ion channel determined by electron cryo-microscopy. Nature 504(7478):107–112
144. Lazaridis T, Karplus M (1999) Effective energy function for proteins in solution. Proteins: Struct Funct Bioinf 35(2):133–152
145. Yarov-Yarovoy V, Schonbrun J, Baker D (2006) Multipass membrane protein structure prediction using Rosetta. Proteins: Struct Funct Bioinf 62(4):1010–1025
146. Spassov VZ, Yan L, Szalma S (2002) Introducing an implicit membrane in generalized Born/solvent accessibility continuum solvent models. J Phys Chem B 106(34):8726–8738
147. Ulmschneider JP, Ulmschneider MB (2007) Folding simulations of the transmembrane helix of virus protein U in an implicit membrane model. J Chem Theory Comput 3(6): 2335–2346
148. Monticelli L, Kandasamy SK, Periole X, Larson RG, Tieleman DP, Marrink S-J (2008) The MARTINI coarse-grained force field: extension to proteins. J Chem Theory Comput 4(5):819–834
149. Wassenaar TA, Ingólfsson HI, Prieß M, Marrink SJ, Schäfer LV (2013) Mixing

MARTINI: electrostatic coupling in hybrid atomistic–coarse-grained biomolecular simulations. J Phys Chem B 117(13):3516–3530
150. Adamian L, Liang J (2001) Helix-helix packing and interfacial pairwise interactions of residues in membrane proteins. J Mol Biol 311(4):891–907
151. Fleishman SJ, Ben-Tal N (2002) A novel scoring function for predicting the conformations of tightly packed pairs of transmembrane α-helices. J Mol Biol 321(2):363–378
152. Park Y, Elsner M, Staritzbichler R, Helms V (2004) Novel scoring function for modeling structures of oligomers of transmembrane α-helices. Proteins: Struct Funct Bioinf 57(3):577–585
153. Illergård K, Callegari S, Elofsson A (2010) MPRAP: an accessibility predictor for α-helical transmem-brane proteins that performs well inside and outside the membrane. BMC Bioinformatics 11(1):1–11
154. Adamian L, Nanda V, DeGrado WF, Liang J (2005) Empirical lipid propensities of amino acid residues in multispan alpha helical membrane proteins. Proteins: Struct Funct Bioinf 59(3):496–509
155. Adamian L, Liang J (2006b) Prediction of transmembrane helix orientation in polytopic membrane proteins. BMC Struct Biol 6:13
156. Mihajlovic M, Lazaridis T (2006) Calculations of pH-dependent binding of proteins to biological membranes. J Phys Chem B 110(7):3375–3384
157. Yuzlenko O, Lazaridis T (2011) Interactions between ionizable amino acid side chains at a lipid bilayer–water interface. J Phys Chem B 115(46):13674–13684
158. Forrest LR, Woolf TB (2003) Discrimination of native loop conformations in membrane proteins: decoy library design and evaluation of effective energy scoring functions. Proteins: Struct Funct Bioinf 52(4):492–509
159. Gao C, Stern HA (2007) Scoring function accuracy for membrane protein structure prediction. Proteins: Struct Funct Bioinf 68(1): 67–75
160. Koehler Leman J, Ulmschneider MB, Gray JJ (2015) Computational modeling of membrane proteins. Proteins: Struct Funct Bioinf 83(1):1–24
161. Ray A, Lindahl E, Wallner B (2010) Model quality assessment for membrane proteins. Bioinformatics 26(24):3067–3074
162. Heim AJ, Li Z (2012) Developing a high-quality scoring function for membrane protein structures based on specific inter-residue interactions. J Comput Aided Mol Des 26(3): 301–309
163. Hurwitz N, Wolfson HJ (2021) Memdock: an α-helical membrane protein docking algorithm. In: Computational design of membrane proteins. Springer, New York, pp 111–117
164. Lacapere J-J, Pebay-Peyroula E, Neumann J-M, Etchebest C (2007) Determining membrane protein structures: still a challenge! Trends Biochem Sci 32(6):259–270
165. Schneidman-Duhovny D, Inbar Y, Nussinov R, Wolfson HJ (2005) PatchDock and SymmDock: servers for rigid and symmetric docking. Nucleic Acids Res 33 (suppl_2):W363–W367
166. Mashiach E, Nussinov R, Wolfson HJ (2010) FiberDock: flexible induced-fit backbone refinement in molecular docking. Proteins: Struct Funct Bioinf 78(6):1503–1519
167. Schott-Verdugo S, Gohlke H (2019) PACKMOL-memgen: a simple-to-use, generalized workflow for membrane-protein–lipid-bilayer system building. J Chem Inf Model 59(6):2522–2528
168. Berendsen HJ, van der Spoel D, van Drunen R (1995) GROMACS: a message-passing parallel molecular dynamics implementation. Comput Phys Commun 91(1–3):43–56
169. Maier JA, Martinez C, Kasavajhala K, Wickstrom L, Hauser KE, Simmerling C (2015) ff14SB: improving the accuracy of protein side chain and backbone parameters from ff99SB. J Chem Theory Comput 11(8): 3696–3713
170. Jämbeck JP, Lyubartsev AP (2012) Derivation and systematic validation of a refined all-atom force field for phosphatidylcholine lipids. J Phys Chem B 116(10):3164–3179
171. Degiacomi MT, Dal Peraro M (2013) Macromolecular symmetric assembly prediction using swarm intelligence dynamic modeling. Structure 21(7):1097–1106
172. Jiménez-García B, Roel-Touris J, Romero-Durana M, Vidal M, Jiménez-González D, Fernández-Recio J (2018) LightDock: a new multi-scale approach to protein–protein docking. Bioinformatics 34(1):49–55
173. Krishnanand K, Ghose D (2009) Glowworm swarm optimization for simultaneous capture of multiple local optima of multimodal functions. Swarm Intell 3(2):87–124
174. Atilgan AR, Durell S, Jernigan RL, Demirel MC, Keskin O, Bahar I (2001) Anisotropy of fluctuation dynamics of proteins with an elastic network model. Biophys J 80(1):505–515

175. Zhou H, Zhou Y (2002) Distance-scaled, finite ideal-gas reference state improves structure-derived potentials of mean force for structure selection and stability prediction. Protein Sci 11(11):2714–2726
176. Kozakov D, Hall DR, Xia B, Porter KA, Padhorny D, Yueh C, Beglov D, Vajda S (2017) The ClusPro web server for protein–protein docking. Nat Protoc 12(2):255–278
177. Katchalski-Katzir E, Shariv I, Eisenstein M, Friesem AA, Aflalo C, Vakser IA (1992) Molecular surface recognition: determination of geometric fit between proteins and their ligands by correlation techniques. Proc Natl Acad Sci 89(6):2195–2199
178. Li B, Kihara D (2012) Protein docking prediction using predicted protein-protein interface. BMC Bioinformatics 13(1):1–17
179. van Noort CW, Honorato RV, Bonvin AM (2021) Information-driven modeling of biomolecular complexes. Curr Opin Struct Biol 70:70–77
180. Ritchie DW (2003) Evaluation of protein docking predictions using Hex 3.1 in CAPRI rounds 1 and 2. Proteins: Struct Funct Bioinf 52(1):98–106
181. Kundrotas PJ, Anishchenko I, Badal VD, Das M, Dauzhenka T, Vakser IA (2018) Modeling CAPRI targets 110–120 by template-based and free docking using contact potential and combined scoring function. Proteins: Struct Funct Bioinf 86:302–310

Chapter 13

Assessing Protein-Protein Docking Protocols: Case Studies of G-Protein-Coupled Receptor Interactions

Archana Sonawani, Amit Naglekar, Shalmali Kharche, and Durba Sengupta

Abstract

The interactions of G-protein-coupled receptors (GPCRs) with other proteins are critical in several cellular processes but resolving their structural dynamics remains challenging. An increasing number of GPCR complexes have been experimentally resolved but others including receptor variants are yet to be characterized, necessitating computational predictions of their interactions. Although integrative approaches with multi-scale simulations would provide rigorous estimates of their conformational dynamics, protein-protein docking remains a first tool of choice of many researchers due to the availability of open-source programs and easy to use web servers with reasonable predictive power. Protein-protein docking algorithms have limited ability to consider protein flexibility, environment effects, and entropy contributions and are usually a first step towards more integrative approaches. The two critical steps of docking: the sampling and scoring algorithms have improved considerably and their performance has been validated against experimental data. In this chapter, we provide an overview and generalized protocol of a few docking protocols using GPCRs as test cases. In particular, we demonstrate the interactions of GPCRs with extracellular protein ligands and an intracellular protein effectors (G-protein) predicted from docking approaches and test their limitations. The current chapter will help researchers critically assess docking protocols and predict experimentally testable structures of GPCR complexes.

Key words Protein-protein interactions, GPCRs, Chemokine, Ligand, G-protein, Effector

1 Introduction

Protein-protein interactions drive a multitude of biological processes such as in cell signaling pathways. During cell signaling, the first interactions occur in the membrane or at the membrane interface and involve one or more membrane protein partners. The biological importance of membrane proteins is highlighted by the increase in the number of therapeutics targeting these proteins, over the last few decades [1]. Although a large number of membrane protein structures have been resolved [2], determining their associations with the corresponding ligands or effector molecules

Agnieszka A. Kaczor (ed.), *Protein-Protein Docking: Methods and Protocols*, Methods in Molecular Biology, vol. 2780, https://doi.org/10.1007/978-1-0716-3985-6_13,
© The Author(s), under exclusive license to Springer Science+Business Media, LLC, part of Springer Nature 2024

remains challenging, in part due to the sensitivity of such protein-protein complexes to the membrane environment. The inherent dynamics [3] and specific and non-specific lipid effects [4, 5] of these complexes render identifying membrane protein-protein interactions even more difficult.

G-protein-coupled receptors (GPCRs) constitute the largest family of membrane proteins in humans. These receptors are implicated in several human diseases and therefore are promising drug targets [6, 7]. GPCRs are heptahelical transmembrane receptors connected by three extracellular loops (ECL) and three intracellular loops (ICL), an extracellular N-terminal domain, and a C-terminal domain at the intracellular site. GPCRs recognize a diverse range of molecules (ligands) such as small molecules, peptides, lipids, and proteins that bind either towards the extracellular site or the lumen of transmembrane domain [8]. Interactions of GPCRs with small molecule ligands are well characterized, but the molecular details of interactions with protein ligands such as chemokines are still emerging. At the intracellular site, GPCRs transduce signals through downstream signaling proteins/effectors such as G-proteins. Both these protein-protein interactions, i.e., GPCR-ligand and GPCR-effector interactions occur at the membrane interface and include both the receptor transmembrane helices and loop regions. Multi-scale computational approaches have been critical in analyzing different facets of GPCR function [9]. In this chapter, we focus on computational predictions of GPCR interactions with protein ligands (chemokines) and downstream effectors (G-proteins).

Chemokines are a large family of small secreted cytokine proteins that signal by binding to chemokine receptors [10]. To date, approximately 50 chemokines and 20 chemokine receptors have been identified but only a few complexes have been resolved. The chemokine receptors exhibit promiscuity for chemokines (one chemokine receptor can bind several different chemokines and vice versa) [11]. However, chemokines bind to their corresponding receptor subfamilies with high affinity and the diversity of chemokine-receptor pairs represents a test case for protein-protein interactions at the extracellular side.

The interaction between GPCRs and G-proteins is one of the most important cellular signaling events. In G-protein-mediated signaling, the binding of signaling molecule leads to the stabilization of the active state of the receptor that is capable of coupling to intracellular G-protein which further transduces the signal to downstream effector proteins [12, 13]. The heterotrimeric G-protein consists of three subunits (Gα, Gβ, and Gγ). Recent structural data shows the direct coupling between GPCRs and Gα [14]. The Gα proteins are classified into four families: $G\alpha_s$, $G\alpha_i$, $G\alpha_q$, and $G\alpha_{12}$ with receptor-specific coupling. During activation, the C-terminus of the G-protein α subunit binds in the pocket formed by the active state of GPCR [15, 16]. Although the number of G-proteins is limited, their interactions with GPCRs are still poorly understood.

1.1 Experimental Approaches

To characterize protein-protein interactions and understand the molecular function of the two interacting proteins, it is important to determine the three-dimensional (3D) structure of the protein-protein complex. Recent advancements in structural proteomics using x-ray crystallography, cryo-electron microscopy (cryo-EM), and nuclear magnetic resonance (NMR) spectroscopy techniques have resulted in a large-scale determination of protein complexes at atomic resolution. In particular, the 3D structures of chemokine receptors with chemokines CXCR4-vMIP-II [17], CXCR2-CXCL8 [18], CCR5-CCL5 [19], CCR5-5P7-CCL5 [20], CCR6-CCL20 [21] and small molecules (CXCR4-IT1t [22], CCR2-BMS681/MK-0812 [23], and CCR5-maraviroc [24] have been deposited in Protein Data Bank (PDB) and are consistent with the previous mutational data. Furthermore, structures of GPCRs in complex with G-proteins (β2AR-Gs [25], GLP-1R-Gs [26], calcitonin receptor-Gs [27], μOR-Gi [28], A1AR-Gi [29], rhodopsin-Gi [30], and 5-HT1B-Go [31] have been resolved till date. However, experimental resolution of protein complexes are inherently limited due to the technical difficulties and high cost. As a result, compared to the number of resolved protein structures, a significantly lower proportion of protein complex structures have been resolved and deposited in the PDB. In addition, variant proteins [32, 33] and post-translational modifications [34] have been shown to be functionally important and their effects on GPCR complexes remain unclear. Consequently, there is a need for computational approaches to predict protein-protein complexes.

1.2 Current Computational Approaches: Docking as a First Step

Computational methods are being increasingly applied to bridge the gap and predict structural models of protein-protein complexes with the aim to complement experimental approaches. Although integrative approaches to elucidate protein-protein interfaces would be more exhaustive, molecular docking can be considered as the first step to predict structural models of protein complexes [3, 35]. One of the first examples of protein-protein docking protocols was the prediction of the BPTI-trypsin protein-protein association [36]. Since then protein-protein docking techniques have greatly advanced and several docking algorithms have been developed to predict the three-dimensional structure of a complex [37–43].

The performance of docking methods is tested in the blind prediction challenge known as the Critical Assessment of Predicted Interactions (CAPRI; [44]) to monitor the improvements of docking and scoring algorithms as well as promote their development.

Current docking algorithms implement a range of efficient search and scoring strategies to improve both the reliability and accuracy of docking results. At the latest seventh meeting of the CAPRI, the predictions were evaluated for eight protein-protein complexes. Among the best five performing servers, ClusPro maintained its high ranking performance followed by HDOCK, MDOCKPP, LZERD, and HADDOCK servers [45].

Current docking protocols are usually fast as they either consider proteins as rigid bodies or sample only a limited flexibility of the receptor/ligand. The high computational efficiency of docking approaches allows them to provide a first step toward a more detailed analysis. Usually, docking predictions are combined with subsequent molecular dynamics (MD) simulations that samples the conformational dynamics of both receptor and ligand to provide more exhaustive results [46–49]. Alternatively, simulations can provide structural ensembles of protein conformations and each of the conformers can be docked to the partner protein to test for convergence of the structural ensemble of the complex. Similarly, short MD simulations can be used to improve docking predictions and to calculate more detailed interaction energies. In our recent study on CXCR4-CXCL12 interactions, we have performed ensemble docking with CXCL12 for the different N-terminal conformations of CXCR4 obtained from MD simulations. Further, short simulations of these docked complexes were carried out to identify the near-native bound complex [34]. For chemokine receptors, docking might be challenging due to a membrane-dependent conformational dynamics of binding interface lined by N-terminus of the receptors [50, 51]. Moreover, a detailed sampling of the conformational space is needed to explore possible binding modes and their dynamics [51].

The current developments in docking algorithms have improved the predictive power for protein-protein complexes substantially [52]. However, testing of different software to benchmark protein-protein complex prediction relevant to a biological system of interest is highly recommended. Experimentally resolved structures can be treated as positive controls against the test system. In fact, the choice of proper controls, both positive and negative controls, can help test and validate the docking approach used. Apart from the structures of components, additional information can be provided in the form of interacting residues or distance-based restraints, to reduce uncertainty and/or off-site binding. Further, to validate the docking predictions, one can analyze the fraction of native contacts reported in structural or biochemical experiments. Depending on the availability of the structure of (similar) co-complexes, RMSD of ligand from the reference can be quantified. In this chapter, we have discussed current docking approaches for protein-protein interactions and presented two test cases of GPCR-protein docking.

2 Methods

2.1 A General Approach to Protein-Protein Docking

We outline here a protocol that describes the different steps involved in a typical protein-protein docking framework (*see* Fig. 1). Protein-protein docking involves two individual proteins and usually the larger protein is referred to as the receptor while the smaller one is known as the ligand. In exceptional cases, such as GPCR-effector interaction, the effector molecules are treated as ligands. Several protein-protein docking programs are available for both academic and commercial use. A list of available docking programs is given in Table 1.

1. Generally, modeling a protein-protein complex involves experimentally determined structures of both the interacting/binding partners. The high resolution experimentally resolved structures can be obtained from Protein Data Bank (PDB; www.rcsb.org), a database for 3D structural data. If no structural information is available, protein structure can be predicted using homology modeling (based on sequence comparison). Thus, obtaining accurate results from docking of homology-modeled structures depend on the quality of the structural templates used for modeling.
2. To assess the quality of the docked complex generated by the docking algorithm, the first step would be to compare it with a known complex. Therefore, if the structure of the bound complex is available, both the receptor and ligand are extracted from the complex, and docking is performed to reconstruct the complex. For the test cases, the individual unbound structures of receptor and ligand are used for docking.
3. The first step of docking involves searching different ligand conformations and orientations to generate several receptor-ligand binding modes (poses) using search algorithms. The search algorithms are further classified as "rigid-body" docking and "flexible docking" based on the flexibility of either the receptor or ligand or both.

Rigid-Body Docking The algorithm considers both the receptor and ligand as rigid bodies. A global search for the ligand-protein is carried out in the six-dimensional space (3D translation and 3D rotation) relative to the static receptor protein. Rigid-body methods use either a global search method such as fast Fourier transform (FFT; [53]) or a local shape matching such as geometric hashing algorithm [54]. Protein conformational change is implicitly achieved by allowing certain degree of overlap between proteins.

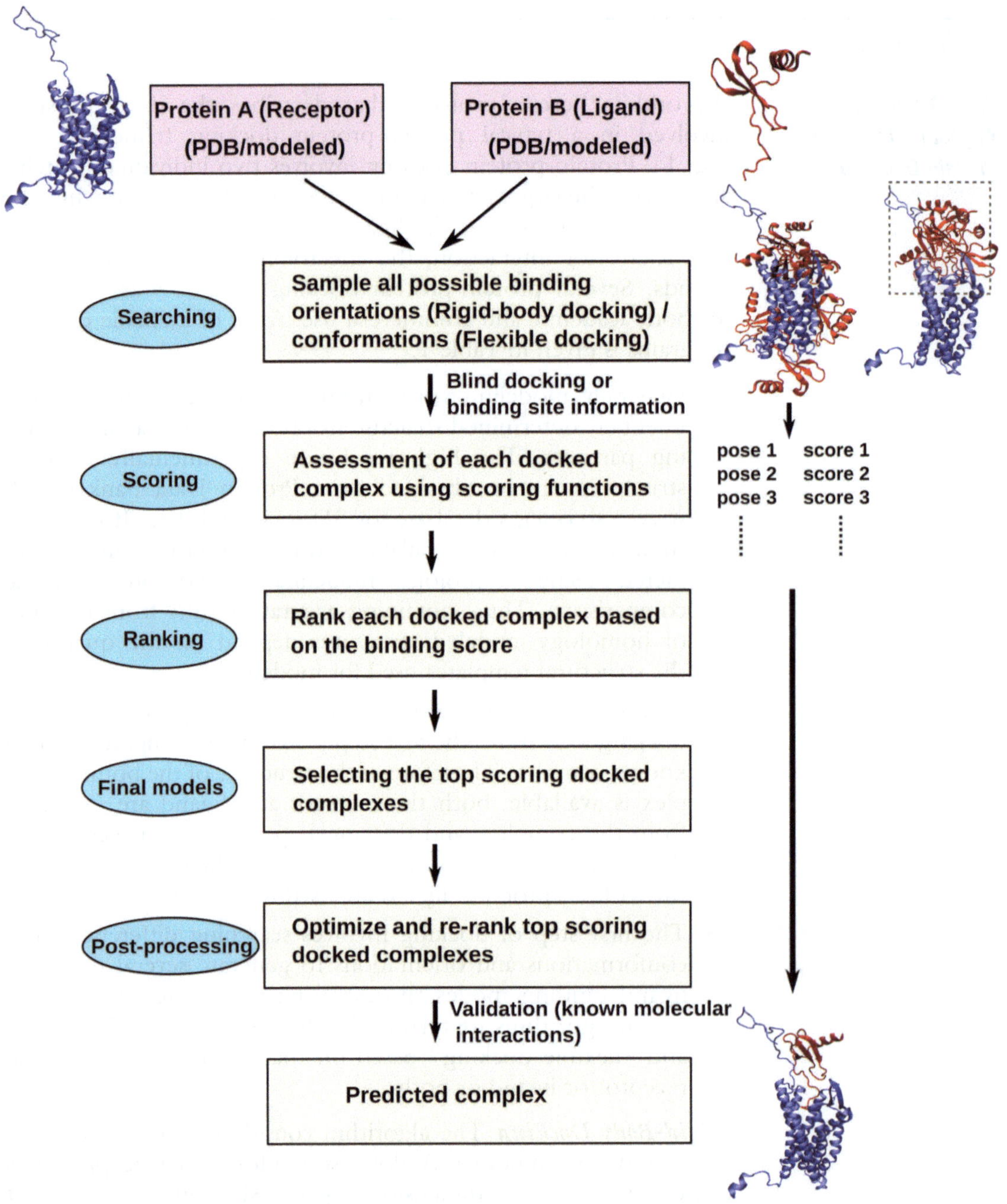

Fig. 1 Workflow of protein-protein docking protocol. The first step of the docking protocol is to obtain 3D structures of both proteins. Docking is carried out using two main steps: searching and scoring. All the poses are ranked using docked scores and the top-scoring poses are selected. Post-processing of these top poses aid in identifying near-native docked complex

Table 1
List of available protein-protein docking programs

	Search algorithm	Web/ Desktop	Website	Rigid/ Flexible	Free/ commercial
Docking programs					
ClusPro	FFT-based	Web	https://cluspro.org/	Rigid	Open access
PatchDock	Geometric hashing	Web/ Desktop	https://bioinfo3d.cs.tau.ac.il/PatchDock/	Rigid	Open source
MolFit	FFT-based	Desktop	http://www.weizmann.ac.il/Chemical_Research_Support//molfit/home.html	Rigid	Open access
GRAMM	FFT-based	Web/ Desktop	http://gramm.compbio.ku.edu/gramm	Rigid	Open access
HEX	SFT-based	Web/ Desktop	http://hexserver.loria.fr/	Rigid	Open access
RosettaDock	Randomized search	Web	http://rosettadock.graylab.jhu.edu	Flexible	Open access
DOT	FFT-based	Desktop	https://www.sdsc.edu/CCMS/DOT/	Rigid	Modified BSD License
PyDock	FFT-based	Desktop	https://life.bsc.es/pid/pydock	Rigid	Open source
PyDockWEB	FFT-based	Web	https://life.bsc.es/pid/pydockWEB	Rigid	Open access
HADDOCK	Randomized search	Web	https://wenmr.science.uu.nl/haddock	Flexible	Open access
ZDOCK	FFT-based	Web/ Desktop	https://zlab.umassmed.edu/zdock	Rigid	Open access and commercial (Accelrys License Pack)
LZerD	Geometric hashing	Web/ Desktop	https://lzerd.kiharalab.org/	Rigid	Open source
Multi-LZerd	Geometric hashing	Web/ Desktop	https://lzerd.kiharalab.org/	Rigid	Open source
SwarmDock	Randomized search	Web	http://bmm.cancerresearchuk.org/%7ESwarmDock/	Flexible	Open access
ATTRACT	Randomized search	Desktop	www.attract.ph.tum.de/services/ATTRACT/	Flexible	Open source under GPL3 License
PIPER	FFT-based	Desktop	https://www.schrodinger.com/piper	Rigid	Commercial

(continued)

Table 1 (continued)

	Search algorithm	Web/ Desktop	Website	Rigid/ Flexible	Free/ commercial
SymmDock	Geometric hashing	Web/ Desktop	http://bioinfo3d.cs.tau.ac.il/SymmDock/	Rigid	Open access
LightDock	Randomized search	Desktop	https://life.bsc.es/pid/lightdock	Flexible	Open source under GPL3 License
ICM-Pro	Randomized search	Desktop	https://www.molsoft.com/protein_protein_docking.html	Flexible	Commercial
NovaDock	Randomized search	Desktop	https://www.dnastar.com/software/nova-protein-modeling/novadock/	Flexible	Commercial
FRODOCK	SFT-based	Web	https://frodock.iqfr.csic.es/	Rigid	Open access
Post-docking programs					
FireDock	–	Web	http://bioinfo3d.cs.tau.ac.il/FireDock/	Flexible	Open source
FiberDock	–	Web	http://bioinfo3d.cs.tau.ac.il/FiberDock/php.php	Flexible	Open access
DOCK-PIE	–	Web	https://clsbweb.oden.utexas.edu/dock.html	Flexible	Open access
RDOCK	–	Desktop	https://www.3ds.com/products-services/biovia/	Flexible	Commercial
ZRANK	–	Desktop	https://www.3ds.com/products-services/biovia/	Flexible	Commercial
PyDockWEB	–	Web	https://life.bsc.es/pid/pydock/home	Rigid	Open access

Note: *FFT* fast Fourier transform, *SFT* spherical Fourier transform

- Advantages: Easy to perform, global exploration of the search space, high computational speed and less time-consuming approach. A high success rate in binding mode prediction for further flexible refinement.
- Limitations: The implicit approach is not able to consider large conformational changes such as side chain and backbone flexibility.

Flexible Docking Here, both the receptor and ligand are considered as flexible, although the extent of flexibility is limited or simplified. For a few benchmark systems, rigid docking and flexible docking have been shown to have similar performance [55]. The protein flexibility is explicitly considered by allowing either backbone or side-chain rearrangements, or both [38]. Flexible docking methods use randomized search where the initial ligand-binding orientations are refined using multi-scale modeling approach such as genetic algorithm [56] or Monte Carlo methods [57, 58].

- Advantages: Explicit representation of flexibility using variety of scoring functions.
- Limitations: Computational cost of conformational search is high, time-consuming approach, increasing the degree of freedom of protein flexibility increases the number of false-positives and poor ranking ability using current scoring functions.

Several docking algorithms provide option to incorporate biological information such as mutagenesis data, NMR restraints, SAXS profile, binding site residues at the interaction interface or blocking residues from being in the binding site for both receptor and ligand during the search process. These options are usually used to restrict the search space around the known set of residues that improves the computational efficiency of search process and increases the probability of finding optimal binding modes as the number of possible binding orientations would be reduced. Thus, the reliability of the docking results can be further improved by incorporating experimental information.

4. The second step of docking involves scoring, where each of the docked conformation is scored using a scoring function. The scoring functions estimate the binding score of the docked complexes and aim to distinguish the binding conformation which most accurately represents the native complex from the sampled conformations. Several scoring functions are available and further categorized as Physics-based, shape complementarity-based [59], knowledge-based [60], energy-based, and machine learning-based [61].
5. The docked conformations are then ranked according to their binding scores to identify the top binding modes that can be selected as final docked complexes.
6. In some cases, the initial docking steps may not be able to generate any near-native binding complex structures. As a result, the user can further optimize and re-rank a certain number of top binding poses obtained from initial docking steps using the post-docking approach [61]. During

optimization and refinement, protein flexibility is explicitly considered. Several post-docking algorithms are available (Table 1) which use more sophisticated scoring functions.

2.2 Computational Requirements

1. A computer with Microsoft Windows or Linux operating system, with at least RAM (4 GB) and hard disk space (16 GB), internet access and a web browser.
2. Most of the docking programs are accessible to academic users as easy-to-use web servers. Other downloadable programs are often open access or open source with a license. Several software are licensed differently for academic and commercial use.
3. Coordinate file for receptor and ligand in PDB format.
4. Software for visualization of docked complexes such as UCSF Chimera [62], PyMOL [63], and BIOVIA Discovery studio visualizer [64] is recommended.

3 Case Study 1: Chemokine Receptor CXCR4 and Chemokine Interaction

To highlight the concepts described in the protocol, we take as an example the study of chemokine receptor type 4 (CXCR4). CXCR4, a member of the class A GPCR family [65] is an important therapeutic target for several disease conditions including HIV-1 and cancer [66, 67]. CXCR4 shows specific binding to cognate ligand CXCL12 [68, 69]. The N-terminus of CXCR4 plays an important role in ligand affinity and selectivity by binding to the CXCL12 core whereas the N-terminus of the CXCL12 inserts into the large orthosteric binding pocket within the extracellular part of CXCR4 according to the two-step/two site model of chemokine binding to receptors [70–72]. The residues from the distal end of the N-terminus of CXCR4 and other receptors have not been resolved in the crystal structures [73] as it is intrinsically disordered and highly flexible as revealed from NMR studies [74, 75].

3.1 Protocol

3.1.1 System Setup

Here, we have used three docking software: ClusPro 2.0 [76], PatchDock [77], and ZDOCK [78] that is commercially available within Discovery Studio suite version 3.5 [64].

Receptor Protein The crystal structure of CXCR4 was taken from PDB (PDB entry: 3ODU; [22]). The co-complexed It1T (small molecule), T4 lysozyme and water were not considered and the corresponding atoms/residues were removed. Since the N-terminal domain has been shown to be important for ligand binding, the missing region must be reconstructed in this case. We have used homology modeling (Modeller v9.16; [79]) to build the missing N-terminal domain (residues 1–26) and the residues of the ICL3 using two templates: the NMR structure of the CXCR4 N-terminal domain (PDB: 2K04; residues 1–38) and truncated CXCR4 structure (residues 27–318). As the N-terminal domain of CXCR4 is highly flexible and GPCRs are membrane embedded dynamic molecules, the modeled conformation may not be in a near-native state. We inserted the modeled CXCR4 structure into the POPC (1-palmitoyl-2-oleoyl-sn-glycero-3-phosphocholine) bilayer and performed a short molecular dynamics simulation using GROMACS ver 5.1.0 software [80] and CHARMM36 force field [81, 82] to relax the modeled structure. The representative structure was chosen as a starting structure for docking.

Ligand Protein The crystal structure of CXCL12 was retrieved from PDB (PDB entry: 2J7Z). We selected chain B from the dimer for this study.

3.1.2 Defining Controls for Validation of Docking Protocols

The first step in any docking process is the validation of the docking programs. The docking program should be able to distinguish between binders and non-binders and rank them accordingly. Therefore, besides the test ligand CXCL12, we considered the viral chemokine vMIP-II (co-crystallized with CXCR4; [17]) (PDB: 4RWS) as positive control and non-cognate chemokines such as CXCL4, CXCL5, CXCL7, CXCL8, and CXCL10 (PDB: 1RHP, 2MGS, 1NAP, 1ILQ, and 1LV9, respectively) as negative controls.

3.1.3 Preparation of Proteins and Molecular Docking

Blind docking (without specifying blocked residues or filtering binding sites; refer to **step 3** of the protocol) was performed to find all possible binding modes between chemokines and CXCR4.

ClusPro The receptor and ligand coordinates in PDB format were uploaded on the ClusPro web server.

PatchDock The receptor and ligand coordinates in PDB format were uploaded on the PatchDock web server. The default values for clustering RMSD (4 Å) and complex type (default) were used.

ZDOCK Both the receptor and ligand structures were prepared using the "*Prepare Protein*" module for the addition of hydrogen atoms, missing atoms or residues, adding protonation, etc. The prepared protein structures were saved in PDB format. The proteins were then energy minimized using 2000 steps of the steepest descent algorithm to remove steric clashes. To dock the two proteins, in the *Protocols Explorer* → *Protein modeling* → *Dock Proteins (ZDOCK)* protocol was selected. This opened the parameters in the Parameters Explorer. The prepared and minimized receptor and ligand proteins were selected from the dropdown list. ZDOCK provides two sampling sizes 6 degrees or 15 degrees. An angular step size of 6 was selected that generates overall 54,000 poses. The 15 degrees generates lesser number of conformations and the search finishes faster, but here sampling size is more important than the speed of the protocol. The maximum number of clusters was set to 10 and cluster size (number of poses) was set to 100.

3.1.4 Refinement of Docked Complexes

The top docked complexes obtained from initial docking using ClusPro and PatchDock servers were further optimized/re-ranked using the post-docking program FireDock [83]. The model's file containing all the top docked complexes in PDB format was uploaded to the FireDock web server.

The clusters obtained from ZDOCK were further refined and re-ranked using the post-docking program RDOCK [84]. In the Discovery studio client, the ZDOCK Results 3D window was opened and the docked complexes were typed using CHARMm Polar H force field which is an essential step. In the *Protocols Explorer → Protein modeling → Refine Docked Proteins* (*RDOCK*) protocol was selected. This opened the parameters in the Parameters Explorer. The typed receptor and ligand proteins were selected from the dropdown list. The selected clusters were given as input to the *Input Poses* parameter.

3.1.5 Results

The final docked complex for each CXCR4-chemokine (Table 2 and Fig. 2) was selected based on these criteria: (1) the chemokines should dock at the known interaction site (similar to positive control vMIP-II and co-crystal complexes of related chemokine receptor-chemokines), (2) maximum known interactions as reported in literature, and (3) ranked based on the binding energies.

Table 2
Binding energies reported for different CXCR4-chemokine docked complexes

Docked complexes	PDB entry (chemokines)	ClusPro FireDock energy	PatchDock FireDock energy	ZDOCK RDOCK energy (kcal/mol)
CXCR4-vMIP-II	4RWS	−104.38	−25.15	−42.84
CXCR4-CXCL12	2J7Z	−68.08	−12.53	−56.84
CXCR4-CXCL4	1RHP	No expected pose	No expected pose	−19.58
CXCR4-CXCL5	2MGS	No expected pose	No expected pose	No expected pose
CXCR4-CXCL7	1NAP	No expected pose	−17.5	−28.82
CXCR4-CXCL8	1ILQ	No expected pose	No expected pose	No expected pose
CXCR4-CXCL10	1LV9	−128.36	13.14	No expected pose

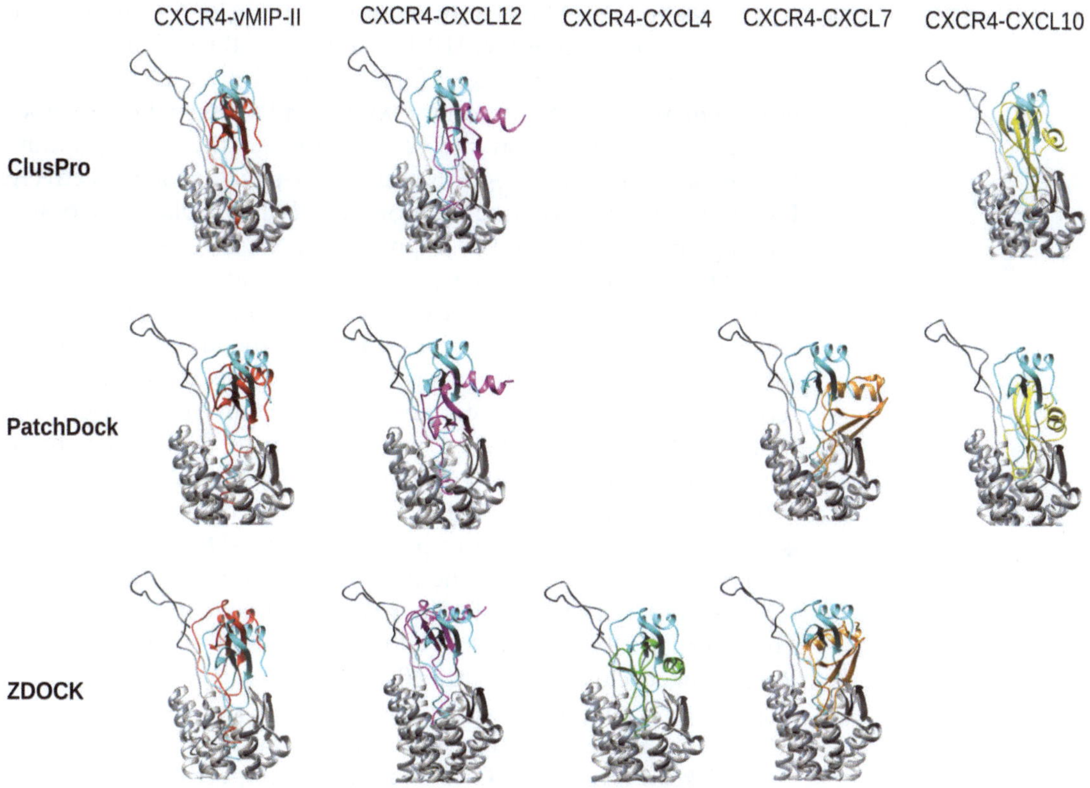

Fig. 2 The CXCR4-chemokine docked complexes predicted using ClusPro, PatchDock, and ZDOCK. The docked complexes of CXCR4 (dark grey) with positive control (vMIP-II: red), test ligand (CXCL12: magenta), and negative controls (CXCL4: green; CXCL7: orange and CXCL10: yellow) are shown superimposed to the co-crystal complex of CXCR4 (light grey) and vMIP-II (cyan). The complexes are represented in the cartoon format

The low energy structures at the centers of the 10–30 most populated clusters obtained from ClusPro web server and the docking results (top 100 solutions) obtained from PatchDock web server were downloaded and analyzed. The docking solutions in which chemokines docked at the known interaction sites were selected and further optimized and re-ranked using FireDock web server. All the input poses were then ranked according to binding energies. The poses were then scored for maximum interactions and lowest binding energy.

In case of ClusPro, the Firedock energies for CXCR4-vMIP-II and CXCR4-CXCL12 docked complexes were −104.38 and −68.08 kcal/mol respectively. The only non-cognate chemokine that could dock CXCR4 was CXCL10 and the corresponding FireDock energy was −128.36 kcal/mol. The binding energy of non-cognate chemokine CXCL10 was favorable compared to the cognate ligand CXCL12.

In case of PatchDock, the FireDock energies for the docked complexes of CXCR4-vMIP-II and CXCR4-CXCL12 were −25.15 and −12.53 kcal/mol respectively. Of the five non-cognate chemokines, only CXCL7 and CXCL10 could dock at the known interaction site of CXCR4 and the corresponding FireDock energies were −17.5 and −13.14 kcal/mol respectively. However, the non-cognate chemokine CXCL7 displayed favorable binding energy compared to cognate ligand CXCL12.

ZDOCK generated ten clusters containing several docked poses for each of the CXCR4-chemokine complex. The cluster or clusters in which the chemokines docked at the known interaction site were selected. All the poses from these clusters were further scored based on RDOCK energy and maximum interactions. The best pose for CXCR4-vMIP-II was from cluster 1 (with 61 members) and corresponded to a RDOCK energy of −42.84 kcal/mol. Whereas for CXCR4-CXCL12, the best pose belonged to cluster 4 (28 members) and corresponded to a RDOCK energy of −56.84 kcal/mol. Among the five non-cognate chemokines, only CXCL4 and CXCL7 could dock at the known interaction site. However, the best pose for CXCL4 and CXCL7 belonged to lower-ranked clusters 5 (with 20 members) and 8 (19 members) respectively. Compared to the positive control and test ligand, their RDOCK energies were significantly lower; −19.58 and −28.82 kcal/mol for CXCL4 and CXCL7 respectively.

To summarize, the ClusPro and PatchDock web servers were not able to distinguish between the binders and non-binders. The negative controls using both these algorithms displayed favorable binding energy compared to the test ligand and positive control. However, RDOCK energies were able to differentiate between the binders and non-binders. Thus, in this case study, the predictive power and performance of ZDOCK was better than the other protocols tested.

4 Case Study 2: Glucagon-Like Peptide-1 Receptor (GLP-1R) and Gα-Protein Interaction

The second test case discussed here is the Glucagon-like peptide-1 receptor (GLP-1R), a class B1 GPCR involved in glucose regulation and insulin secretion [85]. It is an important drug target for many metabolic diseases like obesity and type-2 diabetes [86]. The GLP-1R mediates the effects of GLP-1 peptide hormone by predominantly coupling to stimulatory G-protein ($G\alpha_s$) [87]. However, to date there are no reported interactions between GLP-R and inhibitory G-protein ($G\alpha_i$), although transient interactions with non-cognate effectors have been reported in GPCRs.

4.1 Protocol

4.1.1 System Setup

We have performed protein-protein docking of GLP-1R and Gα_s using three software viz. ClusPro [76], GRAMM [88], and PyDockWeb [89] and ZDOCK [78].

Receptor Protein The cryo-EM structure of GLP-1R was obtained from PDB (PDB entry: 6B3J; [26]) as a reference structure. The residues of the GLP-1 ligand, T4 lysozyme, and the crystal water were removed. The residues 127–136 connecting the extracellular domain (ECD) and transmembrane region (TM), ICL3 (residues 337–343) and the C-terminal region (residues 421–443) were missing from the cryo-EM structure. Therefore, the complete structure of GLP-1R (residues 24–443) was modeled using homology modeling. Robetta server (https://robetta.bakerlab.org/; [90, 91]) was used to build the missing region using reference structure as a template. The five models were obtained and the best model was selected based on the model confidence score = 0.85. Further, we performed energy minimization of the model in vacuum using 10,000 steps of the steepest descent algorithm (in GROMACS version 2018.1 [80] with CHARMM36 force field [81, 82] to remove steric clashes.

Effector Protein (Ligand) The cryo-EM structure of Gs protein was retrieved from PDB (PDB entry: 6B3J). As only the alpha subunit of the Gs protein interacts with the GLP-1R, we modeled the alpha subunit (chain A) using homology modeling. ModWeb server [92] was used to model the missing region (residues 9–394) using the crystal structure of Gs protein (PDB entry: 3SN6; chain A) as a template. The server generated 97 models and the reliable model was selected based on the lower z-DOPE score = −0.79 and RMSD = 0.71 Å with the reference structure.

4.1.2 Defining Controls for Validation of Docking Protocols

Similar to case study 1, we have defined positive and negative controls to validate the docking software. The cryo-EM structure of the Gα_s (complexed with GLP-1R; PDB entry: 6B3J) was used as the positive control whereas the modeled structure of the Gα_i was used as the negative control.

The missing region (residues 5–181 and 233–240) from the structure for the Gα_i (PDB entry: 6DDE) was modeled using (PDB entry: 1GOT; chain A) as a template in the SWISS-MODEL server [93]. Out of four models obtained from the server, the best model was selected based on the QMEANscore = −1.25.

4.1.3 Preparation of Proteins and Molecular Docking

We initially performed blind docking to sample all the possible binding orientations between GLP-1R and the effector protein. However, no expected pose was obtained using any of the four software (*see* Subheading 4.1.5) and, therefore, we carried out site-specific docking where the information of binding site residues (Table 3) was incorporated into the docking protocols.

Table 3
List of binding site residues used for site-specific docking

$G\alpha_s$ (Effector protein) PDB entry: 6B3J	$G\alpha_i$ (Effector protein) PDB entry: 6DDE	GLP-1R (Receptor) PDB entry: 6B3J
R380, D381, I383, Q384, R385, H397, L388, Q390, Y391, E392, L393, L394	L5, E28, E43, C192, D193, E207, N241, K270, E275, K279, K280, I285, C305, E308, T327, D337	R176, H180, Y250, L251, L254, L255, F257, S258, V327, V331, K334, S352, L356

ClusPro The coordinate files in PDB format for both receptor and ligand were uploaded on ClusPro web server. For site-specific docking, the binding site residues for both the receptor and effector (ligand) were provided separately in the *Advanced options* → *Attraction and Repulsion* → residues were provided in the *Attraction* field only.

GRAMM The coordinate files in PDB format for both receptor and ligand were uploaded on GRAMM web server. The number of top docking predictions was specified as 100. In the case of site-specific docking, in the *Advanced options* → *Interface residue constraints*, the binding site residues were provided for both the receptor and effector (ligand) separately.

PyDockWeb The receptor and ligand coordinate files in PDB format were uploaded on the PyDockWeb web server. For site-specific docking, the chains for receptor and effector molecules were selected → binding site residues for respective chains were provided.

ZDOCK The coordinates for the receptor and effector (ligand) in PDB format were uploaded on the ZDOCK web server.

4.1.4 Refinement of Docked Complexes

The top docked complexes obtained from initial site-specific docking using ClusPro, GRAMM and pyDockWEB servers were further optimized/re-ranked using the post-docking program FireDock [83]. The model's file containing all docked complexes in PDB format was uploaded to the FireDock web server.

4.1.5 Results

The final docked complex for each GLP-1R-Gα is shown in Fig. 3. The conformers were validated against the reported interaction site of the Gα protein from the cryo-EM co-complex. The low energy structures at the centers of the 30 most populated clusters obtained from ClusPro web server and top 100 docked solutions obtained from GRAMM web server were downloaded and analyzed separately for blind and site-specific docking (*see* Subheading 4.1.3). In the case of blind docking, none of the docking algorithms could

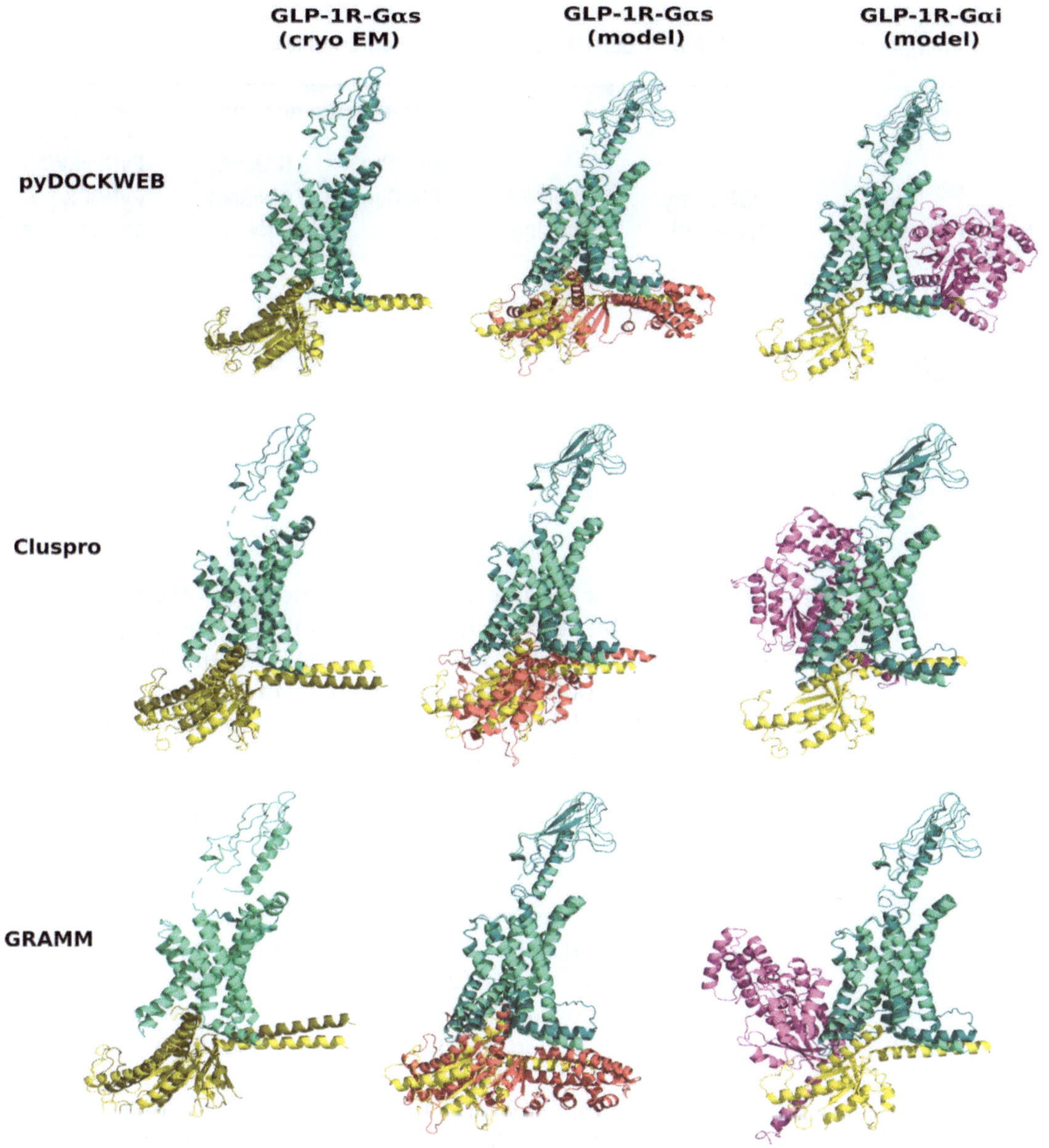

Fig. 3 The site-specific GLP-1R-Gα docked complexes predicted using PyDockWeb, ClusPro, and GRAMM. The docked complexes of GLP-1R (deep teal) with positive control (cryo $G\alpha_S$: deep olive), test ligand (model $G\alpha_S$: deep salmon), and negative control (model $G\alpha_i$: violet) are shown superimposed to the cryo-EM complex of GLP-1R (aquamarine) and $G\alpha_S$ (yellow). The complexes are represented in the cartoon format

dock the G-protein (ligand) at the expected site and, therefore, post-docking analysis was not performed. For site-specific docking, the solutions obtained from ClusPro and GRAMM in which Gα docked at known interaction sites were selected and further optimized and re-ranked using the Firedock web server. All the input poses were then ranked according to binding energies. The poses

Table 4
Binding energies reported for different GLP-1R-G-protein docked complexes

Docked complexes	PDB entry (effector)	Blind docking	Site-specific docking: ClusPro FireDock energy	Site-specific docking: GRAMM Firedock energy	Site-specific docking: PyDockWeb PyDOCK energy
GLP-1R and $G\alpha_s$ (both cryo-EM)	6B3J	No expected pose[a]	–	–	−92.34
GLP-1R and $G\alpha_s$ (both modeled)	6B3J		–	–	−57.54
GLP-1R and $G\alpha_i$ (both modeled)	6DDE		No expected pose	No expected pose	No expected pose

[a]No expected pose was obtained using all the three web servers

with the lowest binding energy were selected as the final docked complex (Table 4). Upon superimposing the docked poses with GLP-1R-$G\alpha_s$, we observed that PyDockWEB provided the most accurate prediction for the reproducibility of the cryo-EM complex with the lowest RMSD of 0.46 nm and also for the test molecule, modeled $G\alpha_s$ 0.75 nm. In pyDockWeb, the PyDock energies for GLP-1R-$G\alpha_s$ (cryo-EM) and GLP-1R-$G\alpha_s$ (modeled) docked complexes were −92.34 and −57.54 kcal/mol respectively. As expected, the $G\alpha_i$ did not dock at the known interaction site of GLP-1R.

In summary, we have performed blind and site-specific docking for predicting GLP-1R-G-protein interactions. None of the four docking software was able to reproduce the cryo-EM complex using a blind docking approach. Similarly, the test and negative controls (modeled $G\alpha_s$ and $G\alpha_i$ respectively) also were not docked at the expected site of GLP-1R. However, in site-specific docking, all three software could reproduce the cryo-EM complex. PyDOCKWEB and GRAMM outperformed ClusPro software in predicting the interactions between modeled GLP-1R and $G\alpha_s$ whereas the non-cognate $G\alpha_i$ didn't dock at the known interaction site of GLP-1R in all the protocols.

5 To Dock or Not to Dock

In recent years, with the rapid development of hardware and computational algorithms, much progress has been made in the field of protein-protein docking. Despite development of several docking algorithms, success in the identification of native-like complexes based on scoring functions and search methods is limited. One of the simplifications in docking is the estimation of binding energies in terms of scoring functions, largely based on the enthalpic parameters and empirical nature of the entropic term. Furthermore, there are no scoring functions developed for incorporating the effect of post-translational modification (PTM) of proteins which regulate the protein-protein interactions by modifying the sites of contact. Therefore, there is a need for development of PTM parameters and their implementation in docking approaches. Further, docking of transmembrane proteins is limited by the search space imposed by the lipid bilayer and there is still room for improvement to the scoring functions accounting for the membrane environment accurately. Importantly, protein-lipid interactions have been shown to modulate the conformational dynamics of protein complexes by both specific and non-specific effects [94–96] and are naturally not incorporated in current docking protocols. The limitations of docking with respect to rigid protein conformations and approximate scoring functions often increase the false-positive rates while identifying specific protein-protein interactions. Further, docking of homology-modeled structures is more challenging compared to docking of experimental structures. The major factor in selection of appropriate templates for homology modeling is sequence similarity. Moreover, incorporating environmental effects exerted by receptor-lipid interactions [4, 97] remains a challenge. Further, although water has been known to contribute to the thermodynamics of ligand binding, the surrounding water is taken into account as a solvent accessible surface or an implicit factor in most of the docking methods. Some of these factors may actually lead to the improved prediction from a particular protocol, for a given system of interest. One of the case studies shown here, the chemokine-receptor system, demanded extensive testing to come up with a reproducible protocol and appropriate controls, although chemokines are small proteins (approximately 70 residues). Currently, docking-based virtual screening tools are often used in computational approaches for identifying small molecule hits in the drug discovery process. However, protein-protein interactions are difficult targets for drug discovery as the interface is large, flat, and interface residues are continuous or discontinuous resulting in high affinity binding. Thus, there remains a need to improve the predictive power of protein-protein docking protocols and critically test each protocol for the system of choice.

6 Conclusion

Molecular docking has been established as the most widely used computational technique for the prediction of protein-protein interactions. Docking has been an active field of development for software algorithms as well as identifying the molecular basis of protein interactions. The CAPRI challenge critically assesses the accuracy and robustness of the docking algorithms in periodic intervals. One of the major challenges faced by these software is defining functions to calculate simplified interaction energies that enable fast calculations for assessing molecule libraries while minimizing the computational cost.

Docking can often be used as a black box; however, it is of utmost importance to assess the performance and accuracy of the algorithms to choose a method suitable for a given system. For small molecule docking, there are benchmarking databases available with known interactions between targets and ligands. However, protein-protein docking mostly lacks the availability of such structural information. Thus, one has to rely on the available three-dimensional structures of co-complexes relevant to a given system and also incorporate experimental data that would be useful in maximizing the chances of getting the most accurate binding poses. We hope the current chapter would help future researchers optimize their docking protocols and critically assess them in the future.

References

1. Rask-Andersen M, Masuram S, Schioth HB (2014) The druggable genome: evaluation of drug targets in clinical trials suggests major shifts in molecular class and indication. Annu Rev Pharmacol Toxicol 54:9–26
2. Hilger D, Masureel M, Kobilka BK (2018) Structure and dynamics of GPCR signaling complexes. Nat Struct Mol Biol 25:4–12
3. Kharche S, Sengupta D (2020) Dynamic protein interfaces and conformational landscapes of membrane protein complexes. Curr Opin Struct Biol 61:191–197
4. Pawar AB, Sengupta D (2018) Effect of membrane composition on receptor association: implications of cancer lipidomics on ErbB receptors. J Membr Biol 251:359–368
5. Sengupta D, Kumar GA, Chattopadhyay A (2017) Interaction of membrane cholesterol with GPCRs: implications in receptor oligomerization. In: Herrick-Davis K, Milligan G, Di Giovanni G (eds) G-protein-coupled receptor dimers, The receptors, vol 33. Humana Press, Cham, p 415
6. Hauser AS, Attwood MM, Rask-Andersen M et al (2017) Trends in GPCR drug discovery: new agents, targets and indications. Nat Rev Drug Discov 16:829–842
7. Yang D, Zhou Q, Labroska V et al (2021) G protein-coupled receptors: structure-and function-based drug discovery. Signal Transduct Target Ther 6:1–27
8. Latorraca NR, Venkatakrishnan AJ, Dror RO (2017) GPCR dynamics: structures in motion. Chem Rev 117:139–155
9. Sengupta D, Joshi M, Athale CA et al (2016) What can simulations tell us about GPCRs: integrating the scales. Methods Cell Biol 132: 429–452
10. Hughes CE, Nibbs RJB (2018) A guide to chemokines and their receptors. FEBS J 285: 2944–2971

11. Zlotnik A (2006) Involvement of chemokine receptors in organ-specific metastasis. Contrib Microbiol 13:191–199
12. Tuteja N (2009) Signaling through G protein coupled receptors. Plant Signal Behav 4:942–947
13. Gurevich VV, Gurevich EV (2017) Molecular mechanisms of GPCR signaling: a structural perspective. Int J Mol Sci 18:2519–2535
14. Moreira IS (2014) Structural features of the G-protein/GPCR interactions. Biochim Biophys Acta 1840:16–33
15. García-Nafría J, Tate CG (2019) Cryo-EM structures of GPCRs coupled to G_s, G_i and G_o. Mol Cell Endocrinol 488:1–13
16. Rosenbaum DM, Rasmussen SGF, Kobilka BK (2009) The structure and function of G-protein-coupled receptors. Nature 459:356–363
17. Qin L, Kufareva I, Holden LG et al (2015) Crystal structure of the chemokine receptor CXCR4 in complex with a viral chemokine. Science 347:1117–1122
18. Liu K, Wu L, Yuan S et al (2020) Structural basis of CXC chemokine receptor 2 activation and signalling. Nature 585:135–140
19. Zhang H, Chen K, Tan Q et al (2021) Structural basis for chemokine recognition and receptor activation of chemokine receptor CCR5. Nat Commun 12:1–12
20. Zheng Y, Han GW, Abagyan R et al (2017) Structure of CC chemokine receptor 5 with a potent chemokine antagonist reveals mechanisms of chemokine recognition and molecular mimicry by HIV. Immunity 46:1005–1017
21. Wasilko DJ, Johnson ZL, Ammirati M et al (2020) Structural basis for chemokine receptor CCR6 activation by the endogenous protein ligand CCL20. Nat Commun 11:1–9
22. Wu B, Chien EYT, Mol CD et al (2010) Structures of the CXCR4 chemokine GPCR with small-molecule and cyclic peptide antagonists. Science 330:1066–1071
23. Apel AK, Cheng RKY, Tautermann CS et al (2019) Crystal structure of CC chemokine receptor 2A in complex with an orthosteric antagonist provides insights for the design of selective antagonists. Structure 27:427–438
24. Tan Q, Zhu Y, Li J et al (2013) Structure of the CCR5 chemokine receptor-HIV entry inhibitor Maraviroc complex. Science 341:1387–1390
25. Rasmussen S, DeVree B, Zou Y et al (2011) Crystal structure of the β2 adrenergic receptor-Gs protein complex. Nature 477:549–555
26. Liang YL, Khoshouei M, Glukhova A et al (2018) Phase-plate cryo-EM structure of a biased agonist-bound human GLP-1 receptor-Gs complex. Nature 555:121–125
27. Liang YL, Khoshouei M, Radjainia M et al (2017) Phase-plate cryo-EM structure of a class B GPCR-G protein complex. Nature 546:118–123
28. Koehl A, Hu H, Maeda S et al (2018) Structure of the μ opioid receptor-Gi protein complex. Nature 558:547–552
29. Draper-Joyce CJ, Khoshouei M, Thal DM et al (2018) Structure of the adenosine-bound human adenosine A_1 receptor-Gi complex. Nature 558:559–563
30. Kang Y, Kuybeda O, Waal P et al (2018) Cryo-EM structure of human rhodopsin bound to an inhibitory G protein. Nature 558:553–558
31. García-Nafría J, Nehmé R, Edwards PC et al (2018) Cryo-EM structure of the serotonin 5-HT_{1B} receptor coupled to heterotrimeric G_o. Nature 558:620–623
32. Nikte SV, Sonar K, Tandale A et al (2021) Loss of a water-mediated network results in reduced agonist affinity in a β2-adrenergic receptor clinical variant. Biochimica et Biophysica Acta-Proteins Proteomics 1869:140605
33. Tandale A, Joshi M, Sengupta D (2016) Structural insights and functional implications of inter-individual variability in β2-adrenergic receptor. Sci Rep 6:1–11
34. Sonawani A, Kharche S, Dasgupta D et al (2022) Allosteric modulation of the chemokine receptor-chemokine CXCR4-CXCL12 complex by tyrosine sulfation. Int J Biol Macromol 206:812–822
35. Chavent M, Duncan AL, Sansom MS (2016) Molecular dynamics simulations of membrane proteins and their interactions: from nanoscale to mesoscale. Curr Opin Struct Biol 40:8–16
36. Wodak SJ, Janin J (1978) Computer analysis of protein-protein interaction. J Mol Biol 124:323–342
37. Moal IH, Moretti R, Baker D et al (2013) Scoring functions for protein-protein interactions. Curr Opin Struct Biol 23:862–867
38. Andrusier N, Mashiach E, Nussinov R et al (2008) Principles of flexible protein-protein docking. Proteins 73:271–289
39. Zacharias M (2010) Accounting for conformational changes during protein-protein docking. Curr Opin Struct Biol 20:180–186
40. Smith GR, Sternberg MJE (2002) Prediction of protein-protein interactions by docking methods. Curr Opin Struct Biol 12:28–35

41. Ritchie D (2008) Recent progress and future directions in protein-protein docking. Curr Protein Pept Sci 9:1–15
42. Bonvin AM (2006) Flexible protein-protein docking. Curr Opin Struct Biol 16:194–200
43. Vajda S, Kozakov D (2009) Convergence and combination of methods in protein-protein docking. Curr Opin Struct Biol 19:164–170
44. Janin J, Henrick K, Moult J (2003) CAPRI: a critical assessment of PRedicted interactions. Proteins Struct Funct Bioinform 52:2–9
45. Lensink MF, Nadzirin N, Velankar S et al (2019) Modeling protein-protein, protein-peptide, and protein-oligosaccharide complexes: CAPRI 7th edition. Proteins Struct Funct Bioinform 88:916–938
46. Ziarek JJ, Kleist AB, London N et al (2017) Structural basis for chemokine recognition by a G protein-coupled receptor and implications for receptor activation. Sci Signal 10:1–14
47. Xu L, Li Y, Sun H et al (2013) Structural basis of the interactions between CXCR4 and CXCL12/SDF-1 revealed by theoretical approaches. Mol BioSyst 9:2107–2117
48. Tamamis P, Floudas CA (2014) Elucidating a key component of cancer metastasis: CXCL12 (SDF-1 α) binding to CXCR4. J Chem Inf Model 54:1174–1188
49. Liou JW, Chang FT, Chung Y et al (2014) In silico analysis reveals sequential interactions and protein conformational changes during the binding of chemokine CXCL-8 to its receptor CXCR1. PLoS One 9:e94178
50. Kharche S, Joshi M, Sengupta D et al (2018) Membrane-induced organization and dynamics of the N-terminal domain of chemokine receptor CXCR1:insights from atomistic simulations. Chem Phys Lipids 210:142–148
51. Kharche S, Joshi M, Chattopadhyay A et al (2021) Conformational plasticity and dynamic interactions of the N-terminal domain of the chemokine receptor CXCR1. PLoS Comp Biol 17:e1008593
52. Desta IT, Porter KA, Xia B et al (2020) Performance and its limits in rigid body protein-protein docking. Structure 28:1071–1081
53. Katchalski-Katzir E, Shariv I, Eisenstein M et al (1992) Molecular surface recognition: determination of geometric fit between proteins and their ligands by correlation techniques. Proc Natl Acad Sci USA 89:2195–2199
54. Lamdan Y, Wolfson HJ (1988) Geometric hashing: a general and efficient model-based recognition scheme. In: Second international conference on computer vision, Tampa, FL, USA, 5–8 December 1988
55. Alogheli H, Olanders G, Schaal W, Brandt P, Karlén A (2017) Docking of macrocycles: comparing rigid and flexible docking in glide. J Chem Inf Model 57(2):190–202
56. Gardiner EJ, Willett P, Artymiuk PJ (2001) Protein docking using a genetic algorithm. Proteins Struct Funct Genet 44:44–56
57. Lorenzen S, Zhang Y (2007) Monte Carlo refinement of rigid-body protein docking structures with backbone displacement and side-chain optimization. Protein Sci 16:2716–2725
58. Zhang Z, Schindler CEM, Lange OF et al (2015) Application of enhanced sampling Monte Carlo methods for high-resolution protein-protein docking in Rosetta. PLoS One 10:1–17
59. Geppert T, Proschak E, Scheider G (2010) Protein-protein docking by shape-complementarity and property matching. J Comput Chem 31:1919–1928
60. Huang SY, Zou X (2008) An iterative knowledge-based scoring function for protein-protein recognition. Proteins 72:557–579
61. Bordner AJ, Gorin AA (2007) Protein docking using surface matching and supervised machine learning. Proteins 68:488–502
62. Pettersen EF, Goddard TD, Huang CC et al (2004) UCSF Chimera—a visualization system for exploratory research and analysis. J Comput Chem 25:1605–1612
63. The PyMOL molecular graphics system, Version 1.2r3pre, Schrodinger, LLC
64. Biovia, Dassault systems, Discovery Studio3.5, Dassault Systèmes, San Diego, 2013
65. Loetscher M, Geiser T, O'Reilly T et al (1994) Cloning of a human seven-transmembrane domain receptor, LESTR, that is highly expressed in leukocytes. J Biol Chem 269:232–237
66. Murakami T, Yamamoto N (2000) Roles of chemokines and chemokine receptors in HIV-1 infection. Int J Hematol 72:412–417
67. Balkwill F (2004) The significance of cancer cell expression of the chemokine receptor CXCR4. Semin Cancer Biol 14:171–179
68. Oberlin E, Amara A, Bachelerie F et al (1996) The CXC chemokine SDF-1 is the ligand for LESTR/fusin and prevents infection by T-cell-line-adapted HIV-1. Nature 382:833–835
69. Busillo JM, Benovic JL (2007) Regulation of CXCR4 signaling. Biochim Biophys Acta 1768:952–963
70. Monteclaro FS, Charo IF (1996) The amino-terminal extracellular domain of the

MCP-1receptor, but not the RANTES/MIP-1 α receptor, confers chemokine selectivity. Evidence for a two-step mechanism for MCP-1 receptor activation. J Biol Chem 271:19084–19092

71. Crump MP, Gong JH, Loetscher P et al (1997) Solution structure and basis for functional activity of stromal cell-derived factor-1 dissociation of CXCR4 activation from binding and inhibition of HIV-1. EMBO J 16:6996–7007
72. Kofuku Y, Yoshiura C, Ueda T et al (2009) Structural basis of the interaction between chemokine stromal cell-derived factor-1/CXCL12 and its G-protein-coupled receptor CXCR4. J Biol Chem 284:35240–35250
73. Sonawani A, Kharche S, Dasgupta D et al (2022) Insights into the dynamic interactions at chemokine-receptor interfaces and mechanistic models of chemokine binding. J Struct Biol 214:107877
74. Ziarek JJ, Kleist AB, London N et al (2017) Structural basis for chemokine recognition by a G protein-coupled receptor and implications for receptor activation. Sci Signal 10:eaah5756
75. Veldkamp CT, Seibert C, Peterson FC et al (2008) Structural basis of CXCR4 sulfotyrosine recognition by the chemokine SDF-1/CXCL12. Sci Signal 16:1–9
76. Kozakov D, Hall DR, Xia B et al (2017) The ClusPro web server for protein-protein docking. Nat Protoc 12:255–278
77. Duhovny D, Nussinov R, Wolfson HJ (2002) Efficient unbound docking of rigid molecules. In: Guigo R, Gusfield D (eds) Algorithms in bioinformatics, Lecture notes in computer science, vol 2452. Springer, Berlin/Heidelberg, p 185
78. Chen R, Li L, Weng Z (2003) ZDOCK: an initial-stage protein-docking algorithm. Proteins 52:80–87
79. Fiser A, Šali A (2003) MODELLER: generation and refinement of homology based protein structure models. Methods Enzymol 374:461–491
80. Abraham MJ, Murtola T, Schulz R et al (2015) Gromacs: high performance molecular simulations through multi-level parallelism from laptops to supercomputers. SoftwareX 1–2:19–25
81. Best RB, Zhu X, Shim J et al (2012) Optimization of the additive CHARMM all-atom protein force field targeting improved sampling of the backbone φ, ψ and side-chain χ1 and χ2 Dihedral Angles. J Chem Theory Comput 8:3257–3273
82. Huang J, Mackerell AD Jr (2013) CHARMM36 all-atom additive protein force field: validation based on comparison to NMR data. J Comput Chem 34:2135–2145
83. Mashiach E, Schneidman-Duhovny D, Andrusier N et al (2008) FireDock: a web server for fast interaction refinement in molecular docking. Nucleic Acids Res 36:W229–W232
84. Li L, Chen R, Weng Z (2003) RDOCK: refinement of rigid-body protein docking predictions. Proteins Struct Funct Genet 53:693–707
85. Graaf C, Donnelly D, Wootten D et al (2016) Glucagon-like peptide-1 and its class B G protein-coupled receptors: a long march to therapeutic successes. Pharmacol Rev 68:954–1013
86. Drucker DJ, Habener JF, Holst JJ (2017) Discovery, characterization, and clinical development of the glucagon-like peptides. J Clin Invest 127:4217–4227
87. Thorens B (1992) Expression cloning of the pancreatic beta cell receptor for the gluco-incretin hormone glucagon-like peptide 1. Proc Natl Acad Sci USA 89:8641–8645
88. Vakser IA (1996) Long-distance potentials: an approach to the multiple-minima problem in ligand-receptor interaction. Protein Eng 9:37–41
89. Jimenez-Garcia B, Pons C, Fernandez-Recio J (2013) pyDockWEB: a web server for rigid-body protein-protein docking using electrostatics and desolvation scoring. Bioinformatics 29:1698–1699
90. Kim DE, Chivian D, Baker D (2004) Protein structure prediction and analysis using the Robetta server. Nucleic Acids Res 32:W526–W531
91. Baek M, DiMaio F, Anishchenko I, Dauparas J, Ovchinnikov S, Lee GR et al (2021) Accurate prediction of protein structures and interactions using a three-track neural network. Science 373(6557):871–876
92. Eswar N, John B, Mirkovic N et al (2003) Tools for comparative protein structure modeling and analysis. Nucleic Acids Res 31:3375–3380
93. Waterhouse A, Bertoni M, Bienert S et al (2018) SWISS-MODEL: homology modelling of protein structures and complexes. Nucleic Acids Res 46:W296–W303
94. Pawar AB, Sengupta D (2021) Role of cholesterol in transmembrane dimerization of the

ErbB2 growth factor receptor. J Membr Biol 254:301–310

95. Prasanna X, Chattopadhyay A, Sengupta D (2015) Role of lipid-mediated effects in β_2-adrenergic receptor dimerization. Adv Exp Med Biol 842:247–261
96. Gopal SM, Pawar AB, Wassenaar TA et al (2020) Lipid-dependent conformational landscape of the ErbB2 growth factor receptor dimers. Chem Phys Lipids 230:104911
97. Sengupta D, Prasanna X, Mohole M et al (2018) Exploring GPCR-lipid interactions by molecular dynamics simulations: excitements, challenges, and the way forward. J Phys Chem B 122:5727–5737

Chapter 14

Protein-Protein Docking Approach to GPCR Oligomerization

Dmytro Khylyuk

Abstract

G-protein-coupled receptors (GPCRs), the largest family of human membrane proteins, play a crucial role in cellular control and are the target of approximately one-third of all drugs on the market. Targeting these complexes with selectivity or formulating small molecules capable of modulating receptor-receptor interactions could potentially offer novel avenues for drug discovery, fostering the development of more refined and safer pharmacotherapies. Due to the lack of experimentally derived X-ray crystallography spectra of GPCR oligomers, there is growing evidence supporting the development of new in silico approaches for predicting GPCR self-assembling structures. The significance of GPCR oligomerization, the challenges in modeling these structures, and the potential of protein-protein docking algorithms to address these challenges are discussed. The study also underscores the use of various software solutions for modeling GPCR oligomeric structures and presents practical cases where these techniques have been successfully applied.

Key words G-protein-coupled receptors (GPCRs), Oligomerization, Protein-protein docking, Docking protocols

1 Introduction

G-protein-coupled receptors (GPCRs) represent the largest family of human membrane proteins with nearly 800 members, which act as the "control panel" of the cell. GPCR families are divided into five groups, based on their sequence and structural similarity: rhodopsin (family A), secretin (family B), glutamate (family C), adhesion, and frizzled/taste. The rhodopsin family is the largest and most diverse of all families. GPCRs can detect the presence of a speaking diverse collection of molecules outside the cell and initiate a variety of intracellular signaling cascades in response. Drugs can induce, block, or modulate these effects, essentially "switching" on and off just like in the control panel. Indeed, GPCRs represent the largest class of drug targets, with roughly one-third of all drugs on the market acting by binding to a GPCR and modifying its intracellular signaling profile [1]. In 2017, 475 drugs (~34% of all drugs

Agnieszka A. Kaczor (ed.), *Protein-Protein Docking: Methods and Protocols*, Methods in Molecular Biology, vol. 2780, https://doi.org/10.1007/978-1-0716-3985-6_14,
© The Author(s), under exclusive license to Springer Science+Business Media, LLC, part of Springer Nature 2024

approved by the US Food and Drug Administration (FDA)) expressed their pharmaceutical properties through influence on 108 unique GPCRs. Approximately 321 agents have been in clinical trials, of which ~20% target 66 potentially novel G protein-coupled receptor (GPCR) targets that do not have any approved drugs in clinical use [2].

GPCR oligomerization has been an area of interest and controversy for many years. Recently there has been increasing evidence that both homodimers and heterodimers play an indispensable role in GPCR signaling. There is growing evidence that GPCRs exist as dimers (or oligomers) and that these dimers may be important for G-protein activation for at least some GPCR families. Dimerization is clearly an important mechanism of receptor activation for the glutamate family of PCRs, where ligand-induced changes in the dimer interface of the amino-terminal ligand-binding domain have been demonstrated by crystallography [3].

Targeting these complexes selectively or designing small molecules that affect receptor-receptor interactions might provide new opportunities for novel drug discovery, in case of the development of more selective and safer drugs. However, nowadays we can postulate, that there is a precipice between the amount of X-ray GPCR oligomer spectrums, compared to all enzyme's structures, which are available in the PBD portal (https://www.rcsb.org/). Molecular modeling approaches have to solve the problem of the lack of experimental data for straight drug design, targeted to GPCR dimer and oligomer conjugates. Nevertheless, there are some pitfalls and limitations to using these techniques. For this reason, the aim of this chapter is to provide useful information about the possible prediction techniques of the GPCR dimerization (oligomerization) using protein-protein docking algorithms, some cases, and to discuss the recent milestones in the in silico predictions of the GPCR self-assembling process.

2 Software Solutions

Protein-protein docking algorithms are widely used tools for modeling GPCR oligomeric structures, which consider two monomers (either identical or different) as interacting proteins as rigid bodies and produce a number of plausible solutions. The method does not need powerful computational tools and can provide possible dimer configurations, which can then be either used to design experiments to test the possible dimer configurations.

The docking procedure consists generally of three stages: (i) the searching stage which involves the generation of GPCR dimers, (ii) sampling stage which discards decoy solutions, and (iii) the refinement (optional) and ranking stage (scoring, filtering). Thereby obtained GPCR dimeric complexes are scored according to various criteria, which can include, on one hand, biophysical parameters such as geometric and chemical complementarity, electrostatics, H-bonding interactions, and solvation [4].

In case of protein-protein docking as applied to GPCR dimers, rigid-body docking is one of the main disadvantages in view of the highly flexible GPCR regions (intracellular and extracellular loops) which may take part, at least to certain extent in GPCR dimer formation [5]. Moreover, conformational changes in transmembrane and loop regions upon dimer formation cannot be satisfactorily considered by the rigid approach.

Casciari et al. proposed a computational protocol for effective predictions of the supramolecular organization of integral transmembrane proteins, starting from the monomer systems. Predictions were performed by ZDOCK 2.1 for homo-oligomeric membrane proteins with known quaternary structures. Obtained structures possessed Cα-RMSDs lower than 2.5 Å from the native oligomer, regardless of the resolution of the structural models [6].

One of the most significant research, dedicated to GPCR oligomerizations, was the complex benchmark of widely used protein-protein docking software. Kaczor et al. estimated ZDOCK [7], ClusPro [8], HEX [9], GRAMM-X [10], PatchDock [11], SymmDock [12], HADDOCK [13], and 3D-GARDEN [14] for reconstructing native complexes of transmembrane proteins of known three-dimensional structures. GRAMM-X [10] software solution demonstrated the best results with RMSD lower than 0.7 Å in 8 out of 12 cases. The problem was that in four cases involving seven transmembrane receptors, the RMSD was at least 9 Å. This was considered a total failure in detecting true protein-protein interactions (PPI).. The results show that the docking success depends on the applied docking algorithm and scoring functions, but also on the inherent structural features of the transmembrane proteins [15]. This failure may be attributed to the inherent structural features of GPCR dimers, particularly the relatively small interface compared to other transmembrane proteins for which prediction was successful. Furthermore, protein-protein docking software is preferably designed for water-soluble proteins, and as no membrane filter is applied, appropriate solutions are hidden among hundreds of solutions incompatible with membrane integration [16]. Hurwitz et al. described the novel algorithm, named Memdock [17], for docking α-helical membrane proteins which takes into consideration the lipid bilayer environment for docking as well as for refining and ranking the docking candidates. It was shown that the proposed algorithm improves docking accuracy and candidate ranking compared to the existing protein-protein docking algorithm. This improvement was achieved by introducing sampling and scoring specific for membrane proteins. Membrane consistent orientation was enforced in both docking and refinement steps as an integral part of the algorithms rather than a post-processing step. In addition, the refined docking candidates undergo re-ranking using a scoring function formulated from a combination of knowledge-based potentials derived from statistical analyses of known 3D structures of α-helical membrane proteins.

Memdock protocol is freely available for academic users (http://bioinfo3d.cs.tau.ac.il/Memdock/index.html).

Transmembrane proteins are often modeled using RosettaDock, which uses real-space Monte Carlo minimization (MCM) on both rigid-body and side-chain degrees of freedom to identify the lowest free energy docked arrangement of two protein structures [18]. This algorithm makes use of the biophysical theory of an encounter complex formation followed by a transition to a bound state [19]. Also, it's possible to use ROSIE (Rosetta Online Server that Includes Everyone) web server for protein-protein docking, which is also a single server framework that can accommodate a range of Rosetta protocols [20].

It is clear that docking techniques for GPCR oligomerization are far from their milestones. For the development and benchmarking of the docking protocols, it was proposed new high-quality sets of membrane protein complexes. The new dataset consists of 456 non-redundant alpha-helical binary interfaces, which is significantly larger and more representative than the previously developed sets. The authors suggest that this set will become the basis for the development of docking and scoring in the future [21].

3 Practical Cases

Because of the paradigm shift in GPCR-oriented drug discovery, the scientific community has gained interest in the development of novel types of molecules including homo- and hetero-bivalent ligands. Bivalent ligands have long been thought to act solely as connectors of the two pharmacophore groups. On the other hand, experimental studies have shown that binding affinity and signaling potency are increased by bivalent ligands compared to their monovalent counterparts. Consequently, this suggests that these ligands are capable of modulating the dynamics of GPCR oligomers. There are some examples of the possible practical usage in the modern drug design.

Jörg et al. have developed and tested a multicomponent protocol to model the inactive conformation of the dopamine D2 receptor dimer [22]. This model was used to study interactions of the D2 receptor homo-bivalent antagonists based on ropinirole and clozapine, 1,4-disubstituted aromatic piperidines/piperazines, and arylamidoalkyl-substituted phenylpiperazine pharmacophores. The most potent homo-bivalent ligands based on ropinitole scaffold with 22-, 26-, and 30-atom spacers demonstrated approximately from 20- to 80-fold greater potency ($EC_{50} = 3.9$, 6.2, and

14 nM, respectively) compared to initial ropinirole (304 nM) in a GTPgS functional assay, which confirms the availability of the predicted oligomer structures for constructing more biologically effective chemical entities [18].

Increasing dopamine 2 receptor (D_2R) homodimer formation has been correlated with the pathophysiology of schizophrenia [23]. Therefore, targeting these D_2R dimers might offer new information about the pathophysiology of diseases related to this GPCR dimer, potentially opening new therapeutic avenues. With this aim to explain the ability of some D_2R receptor antagonists to change the D_2R receptor homodimerization, computational modeling of D_2R receptors has been performed using the Rosie web server [20]. Molecular modeling revealed that D_2R antagonist spiperone demonstrated specific interactions with Tyr199 and Phe390, which may be the reason for decreasing the level of $D_{2Long}R$ dimer formation significantly by 40–60% in real-time and after long-term ($\geq$16 h) incubations [24]. A protein-protein docking technique with ROSETTA without a membrane topology filter was applied to build the heterodimer of the mGluR2-5HT2A receptors [25]. However, in this case it was not possible to indicate the correct model using scoring results, thus the output was evaluated according to visual inspection based on experimental evidence resulting in the TM4,5-TM4,5 heterodimer model [26].

Protein-protein docking was used to construct a model of the A_{2A} Adenosine-D_2 Dopamine receptor heterodimer with a TM-IV/V interface using the program HADDOCK2.1 [27]. Mansoor et al. have constructed a bivalent-ligand-bound Adenosine 2A receptor (A2AR) dimer within the context of a modeled tetramer, which includes A2AR and dopamine 2 receptor homodimers along with their cognate G-proteins. Additionally, they have proposed the structures of homo-bivalent ligands, which may influence the interactions between pharmacophore groups and ligand-binding residues. This, in turn, could modulate the allosteric communication network and the water channel formed within the receptor. However, this research is more theoretical because of the absence of practical biological assays either in vitro or in vivo, which may confirm the possible applications in the treatments [28].

Phenomena of the heterodimer formation, which consist of two different types of GPCR allow the development of hetero-bivalent ligands, where one part of the molecule possesses the agonistic activity to the part of the heterodimers, and another part of the structure demonstrates antagonistic influence to another part. An example of such a chimera is the formation of Mu opioid receptor (MOR) and chemokine receptor (CXCR4) heterodimers, which have garnered a great deal of interest in recent years. Both MOR and CXCR4 are known to form homodimers and heterodimers with other G protein-coupled receptors. Such

MOR-CXCR4 heterodimers have been constructed using HEX 8.0.0. In addition, MOR agonist/CXCR4 antagonist bivalent ligands have been proposed as a result of docking simulation. The characterization of their pharmacological properties both in vitro and in vivo has been performed. One of the constructed bivalent ligands showed promising antinociceptive effect by targeting putative MOR-CXCR4 heterodimers, which may serve as a novel chemical probe for further developing more potent bivalent ligands with potential application in analgesic therapies for chronic pain management [29].

Docking experiments to create GPCR dimers (both homo- and heterodimers) has not been reported often in recent years, as the phenomenon of GPRC dimerization has gained attention only during the last few years. In 2013, PubMed search with keywords "GPRC dimerization docking" yielded only eight publications [30]. Even now, every publication with GPCR oligomer modeling is a significant occurrence. The development of machine-learning techniques [31, 32] and the appearance of new experimental data may push the quality and practical usability of predicted models to the new levels.

References

1. Latorraca NR, Venkatakrishnan AJ, Dror RO (2017) GPCR dynamics: structures in motion. Chem Rev 117(1):139–155
2. Hauser AS, Attwood MM, Rask-Andersen M, Schiöth HB, Gloriam DE (2017) Trends in GPCR drug discovery: new agents, targets and indications. Nat Rev Drug Discov 16(12):829–842
3. Kobilka BK (2007) G protein coupled receptor structure and activation. Biochim Biophys Acta 1768(4):794–807
4. Selent J, Kaczor AA (2011) Oligomerization of G protein-coupled receptors: computational methods. Curr Med Chem 18(30):4588–4605
5. Łukasiewicz S, Polit A, Kędracka-Krok S, Wędzony K et al (2010) Hetero-dimerization of serotonin 5-HT2A and dopamine D2 receptors. Biochim Biophys Acta 1803(12): 1347–1358
6. Casciari D, Seeber M, Fanelli F (2006) Quaternary structure predictions of transmembrane proteins starting from the monomer: a docking-based approach. BMC Bioinformatics 7(1):1–16
7. Chen R, Li L, Weng Z (2003) ZDOCK: an initial-stage protein-docking algorithm. Proteins 52(1):80–87
8. Comeau SR, Gatchell DW, Vajda S, Camacho CJ (2004) ClusPro: an automated docking and discrimination method for the prediction of protein complexes. Bioinformatics 20(1): 45–50
9. Macindoe G, Mavridis L, Venkatraman V, Devignes MD et al (2010) HexServer: an FFT-based protein docking server powered by graphics processors. Nucleic Acids Res 38 (suppl_2):W445–W449
10. Tovchigrechko A, Vakser IA (2006) GRAMM-X public web server for protein–protein docking. Nucleic Acids Res 34(suppl_2):W310–W314
11. Schneidman-Duhovny D, Inbar Y, Nussinov R (2005) PatchDock and SymmDock: servers for rigid and symmetric docking. Nucleic Acids Res 33(suppl_2):W363–W367
12. Schneidman-Duhovny D, Inbar Y, Nussinov R (2005) Geometry-based flexible and symmetric protein docking. Proteins 60(2):224–231
13. Dominguez C, Boelens R, Bonvin AM (2003) HADDOCK: a protein−protein docking approach based on biochemical or biophysical information. J Am Chem Soc 125(7): 1731–1737
14. Lesk VI, Sternberg MJ (2008) 3D-Garden: a system for modelling protein–protein complexes based on conformational refinement of ensembles generated with the marching cubes algorithm. Bioinformatics 24(9):1137–1144

15. Kaczor AA, Selent J, Sanz F et al (2013) Modeling complexes of transmembrane proteins: systematic analysis of protein- protein docking tools. Mol Inform 32(8):717–733
16. Kaczor AA, Guixà-González R, Carrio P et al (2015) Multi-component protein–protein docking based protocol with external scoring for modeling dimers of G protein-coupled receptors. Mol Inform 34(4):246–255
17. Hurwitz N, Schneidman-Duhovny D, Wolfson HJ (2016) Memdock: an α-helical membrane protein docking algorithm. Bioinformatics 32(16):2444–2450
18. Kaczor AA, Jörg M, Capuano B (2016) The dopamine D2 receptor dimer and its interaction with homobivalent antagonists: homology modeling, docking and molecular dynamics. J Mol Model 22(9):1–14
19. Lyskov S, Gray JJ (2008) The RosettaDock server for local protein–protein docking. Nucleic Acids Res 36(suppl_2):W233–W238
20. Moretti R, Lyskov S, Das R, Meiler J (2018) Web-accessible molecular modeling with Rosetta: the Rosetta online server that includes everyone (ROSIE). Protein Sci 27(1):259–268
21. Kotthoff I, Kundrotas PJ, Vakser IA (2022) DOCKGROUND membrane protein-protein set. PLoS One 17(5):e0267531
22. Jörg M, Kaczor AA, Mak FS, Lee KC et al (2014) Investigation of novel ropinirole analogues: synthesis, pharmacological evaluation and computational analysis of dopamine D 2 receptor functionalized congeners and homobivalent ligands. MedChemComm 5(7): 891–898
23. Wang M, Pei L, Fletcher PJ et al (2010) Schizophrenia, amphetamine-induced sensitized state and acute amphetamine exposure all show a common alteration: increased dopamine D2 receptor dimerization. Mol Brain 3(1):1–9
24. Wouters E, Marín AR, Dalton JA et al (2019) Distinct dopamine D2 receptor antagonists differentially impact D2 receptor oligomerization. Int J Mol Sci 20(7):1686
25. Bruno A, Guadix AE, Costantino G (2009) Molecular dynamics simulation of the heterodimeric mGluR2/5HT2A complex. An atomistic resolution study of a potential new target in psychiatric conditions. J Chem Inf Model 49(6):1602–1616
26. Bartuzi D, Kaczor AA, Targowska-Duda KM (2017) Recent advances and applications of molecular docking to G protein-coupled receptors. Molecules 22(2):340
27. Borroto-Escuela DO, Rodriguez D, Romero-Fernandez W et al (2018) Mapping the interface of a GPCR dimer: a structural model of the A2A adenosine and D2 dopamine receptor heteromer. Front Pharmacol 30(9):829
28. Mansoor S, Kayık G, Durdagi S et al (2022) Mechanistic insight into the impact of a bivalent ligand on the structure and dynamics of a GPCR oligomer. Comput Struct Biotechnol J 20:925–936
29. Ma H, Li M, Pagare PP et al (2022) Novel bivalent ligands carrying potential antinociceptive effects by targeting putative mu opioid receptor and chemokine receptor CXCR4 heterodimers. Bioorg Chem 120:105641
30. Kaczor AA, Selent J, Poso A (2013) Structure-based molecular modeling approaches to GPCR oligomerization. In: Methods in cell biology, vol 117. Academic Press, pp 91–104
31. Khamis MA, Gomaa W, Ahmed WF (2015) Machine learning in computational docking. Artif Intell Med 63(3):135–152
32. Zhang Y, Wang Y, Zhou W et al (2019) A combined drug discovery strategy based on machine learning and molecular docking. Chem Biol Drug Des 93(5):685–699

Chapter 15

Refinement of Docked Protein-Protein Complexes Using Repulsive Scaling Replica Exchange Simulations

Till Siebenmorgen, Yasmin Saremi Nanji, and Martin Zacharias

Abstract

Accurate prediction and evaluation of protein-protein complex structures is of major importance to understand the cellular interactome. Predicted complex structures based on deep learning approaches or traditional docking methods require often structural refinement and rescoring for realistic evaluation. Standard molecular dynamics (MD) simulations are time-consuming and often do not structurally improve docking solutions. Better refinement can be achieved with our recently developed replica-exchange-based scheme employing different levels of repulsive biasing between proteins in each replica simulation (RS-REMD). The bias acts specifically on the intermolecular interactions based on an increase in effective pairwise van der Waals radii without changing interactions within each protein or with the solvent. It allows for an improvement of the predicted protein-protein complex structure and simultaneous realistic free energy scoring of protein-protein complexes. The setup of RS-REMD simulations is described in detail including the application on two examples (all necessary scripts and input files can be obtained from https://gitlab.com/TillCyrill/mmib).

Key words Molecular dynamics of protein complexes, Molecular dynamics refinement, Blind protein-protein docking, Protein-protein complex formation, Flexible docking refinement, Molecular dynamics-based docking

1 Introduction

Basically all biological processes within and between cells involve proteins and protein-protein interactions. Most proteins do not function individually as protein monomers but rather as subunits of large, often transiently formed, assemblies or molecular machines. The qualitative and quantitative characterization of these protein-protein interaction networks is an ongoing experimental effort [1, 2]. It is desirable to assign structural models to the associated protein-protein interactions and indeed there are efforts underway to use docking methods or more recently deep learning-based methods to solve this task [3]. Deep learning-based approaches such as AlphaFold2 [4] and RoseTTAfold [5] show

Agnieszka A. Kaczor (ed.), *Protein-Protein Docking: Methods and Protocols*, Methods in Molecular Biology, vol. 2780, https://doi.org/10.1007/978-1-0716-3985-6_15,
© The Author(s), under exclusive license to Springer Science+Business Media, LLC, part of Springer Nature 2024

great promise to provide a structurally fully resolved picture of protein interactomes [6]. However, docking methods and to some degree also deep learning methods can frequently produce unrealistic (incorrect) complexes or complexes of very limited accuracy that require further refinement before becoming useful for biochemists and structural biologists [7]. In principle, atomistic molecular dynamics (MD) simulation including full flexibility of proteins with explicit water should be ideally suited to predict realistic interaction geometries. Indeed, in recent years near-native protein-protein complex structures have been recovered from multiple long MD simulations [8]. The application of such approaches is, however, still limited due to the large computational demand. Promising strategies based on advanced sampling have been designed to identify native binding protein-protein complexes [8, 9]. For example, tempered binding to explore several binding events of protein-protein cases on the special-purpose computer ANTON has been very successful [8]. In tempered binding, the interaction strength between the protein partners is scaled in regular steps to enhance the simulation of rare events. Other more efficient approaches use perturbed distance restraints to allow reversible protein-protein association during MD simulations [10] or use replica exchange, in which several copies of the system (replicas) are simulated in parallel, performing exchange attempts between the different replicas in regular time intervals [11]. The standard T-REMD method with different temperatures along the replicas, however, is strongly limited by the size of the simulation system. We have recently developed a special REMD approach, termed RS-REMD, that employs various levels of a repulsive biasing potential between proteins along the replica ladder [9, 12, 13]. The method largely avoids the kinetic trapping of proteins in non-native complex geometries and allowed rapid refinement of docked complexes. The method was further extended to perform simulations to also allow extracting a free energy score for binding. On benchmark tests in many cases, near-native solutions could be detected as those with the lowest free energy score. In the following, we first give a brief overview of the principles of the method followed by a detailed description of applications to protein-protein docking examples in Subheadings 2 and 3.

1.1 Generating Initial Docked Protein-Protein Complex Structures

The RS-REMD method can be applied to protein-protein complexes (or also protein-peptide or protein-drug complexes) obtained from arbitrary sources such as manually docked complexes, models obtained as output from deep learning-based methods such as AlphaFold2 [4] or any other docking engine. In the present examples, we used the ATTRACT docking program [14–16] for generating start structures.

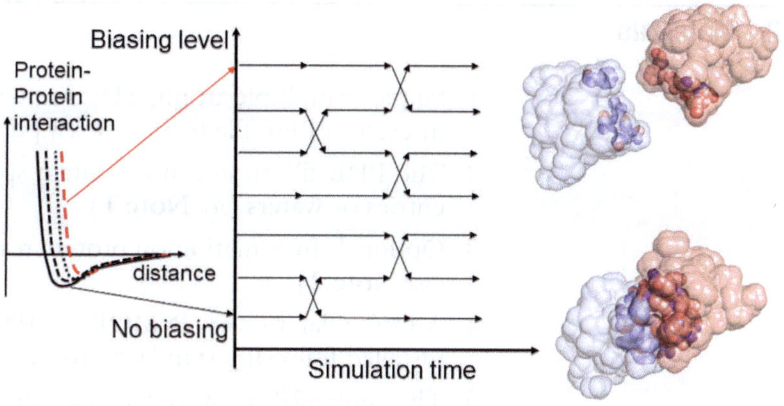

Fig. 1 Illustration of the RS-REMD method. Parallel MD simulations (replicas, middle panel) are performed on the same docked protein-protein complex (in explicit aqueous solvent). In each replica different Lennard-Jones (LJ) interactions between the protein and partners are used with increased van der Waals radii and reduced attraction along the parallel running replicas (left panel). It results in the dissociation of trapped complexes in the higher replicas (right panel) and rapid exploration of alternative complex geometries that can exchange eventually to lower replicas including the lowest replica under the control of the original force field

1.2 General Principle and Setup of RS-REMD Simulations

In principle, molecular dynamics (MD) simulations allow one to follow the association process under realistic conditions including full partner flexibility and surrounding solvent. However, due to the many local binding energy minima at the surface of protein partners, MD simulations are frequently trapped for long times in transient association states. We have designed a replica-exchange-based scheme employing different levels of repulsive biasing between partners in each replica simulation [9, 13]. The bias acts only on intermolecular interactions based on an increase in effective pairwise van der Waals radii (repulsive scaling (RS)-REMD) without affecting interactions within each protein or with the solvent (a schematic illustration of the approach is presented in Fig. 1).

We have extended the replica-exchange-based scheme to simultaneously refine and score protein-protein complexes [13]. The approach is computationally demanding compared to approaches that require single complex structures. However, it offers a route for refinement and a realistic ranking of predicted protein-protein docking geometries, including all energetic and entropic contributions to the binding, explicit solvation, and full flexibility of partner structures.

2 Materials

1. Single or multiple atomic 3D structures of protein molecules of interest in PDB file format (www.pdb.org).
2. The PDB file should not contain special residues, HETATM entries or waters (*see* **Note 1**).
3. Optional: Information on protein residues involved in binding (*see* **Note 2**).
4. A molecular viewer (PyMOL, VMD, Rasmol, etc.) should be installed following standard procedures.
5. The Amber20 software package should be installed and available (https://ambermd.org) including the Parmed module.
6. A compute cluster should be available preferably with GPU (Graphic Processing Unit) nodes or a PC with a high-end GPU card that can be used to run pmemd.cuda from the Amber20 package. In principle, the examples also work with the pmemd CPU code but at least an order of magnitude slower compared to the GPU code.
7. Optional: For generating docked complexes: ATTRACT software (available at www.t38.ph.tum.de). However, any other source to generate docked complexes can also be used.
8. All structures and input files to run the examples can be downloaded from the GitLab site: https://gitlab.com/TillCyrill/mmib. It includes the lj.py script for generating modified force field topology files. The lj.py script is the key program to set up the RS-REMD simulations.

3 Methods

3.1 System Preparation and Equilibration

The MD simulations are conducted with the pmemd.cuda module of the Amber20 software package [17]. It requires access or installation of the Amber20 software which can be obtained from (https://ambermd.org). The user needs an atomic 3D starting structure of the protein-protein complex of interest in PDB file format. This structure can be the result of a docking search or the output of a complex prediction by a deep learning approach or from a docking program. For all examples, the protein force field ff14SB [18] was used. For the setup of MD simulations, one can follow standard procedures briefly outlined below. In our case, the protein complexes were solvated using the TIP3P water model [19] in an octahedron box with periodic boundary conditions (minimum distance between protein and box edge was 15 Å). The charges were neutralized with Na^+ and Cl^- ions. First, a force field topology and coordinate start file of the complex are needed in Amber format. This can be achieved following standard procedures to set up simulations using the leap module of the Amber20 package. A typical input for the tleap program and a given start PDB file (myprotein.pdb) is given below. The tleap command is directly stated in the terminal.

```
tleap
>> source leaprc.protein.ff14SB
>> source leaprc.water.tip3p
>> mol = loadpdb myprotein.pdb
>> solvateoct mol TIP3PBOX 15.0
>> addions2 mol Na+ 0
>> addions2 mol Cl- 0
>> addions mol Na+ 16 Cl- 16
>> saveamberparm mol myprotein.prmtop myprotein.rst
>> quit
```

For the simulation we use the hydrogen mass repartition option [20] available in the parmed module of Amber20 allowing for a time step of 4 fs.

```
parmed myprotein.prmtop
>> Hmassrepartition
>> outparm myprotein_hrpt.prmtop
```

The starting protein-protein complex needs to be energy minimized first using a standard energy minimization input file (min.in) and the following command line:

```
pmemd.cuda -O -i min.in -c myprotein.rst -p myprotein_hrpt.
prmtop -r myprotein_min.rst -o myprotein_min.out
```

The energy-minimized structure (*myprotein_min.rst*) can be used for gradually heating up to 300 K in steps of 100 K and keeping the protein-protein complex close to the start structure using position restraints on non-hydrogen atoms (20 ps simulation for each step) at constant volume. At 300 K the positional restraints are removed over 40 ps of simulation time at a constant pressure. The Langevin thermostat is used for controlling the temperature.

```
pmemd.cuda -O -i heat.in -c myprotein_min.rst -p mypro-
tein_hrpt.prmtop -r myprotein_heat.rst -o myprotein_heat.out
-ref myprotein.rst -x myprotein_heat.nc
```

Table 1
Lennard-Jones scaling parameters for RS-REMD simulations in explicit solvent using 16 replicas

Replica number	*d* (in Å)[a]	*e* (scaling of ε)[a]
1	0.0	1.0
2	0.01	0.99
3	0.02	0.98
4	0.04	0.97
5	0.08	0.96
6	0.12	0.94
7	0.16	0.92
8	0.2	0.9
9	0.24	0.88
10	0.28	0.86
11	0.32	0.84
12	0.38	0.82
13	0.44	0.8
14	0.5	0.78
15	0.58	0.76
16	0.68	0.74

[a]The *d* and *e* parameters indicate the changes of pairwise Lennard-Jones parameters between atoms of the two partner proteins (parameters for interactions within each protein or with the solvent are not modified in the replica runs)

3.2 Run Repulsive Scaling Replica Exchange MD Simulation

Next, the repulsive scaling RS-REMD simulations are performed (typically for 5 ns per replica, exchange attempts every picosecond), i.e., H-REMD simulations with 16 replicas and increasing bias between the ligand and receptor by modifying intermolecular LJ-parameters (Table 1). The introduced bias is associated with lower (attractive) potential well depths ϵ and higher (repulsive) effective pairwise van der Waals radii *d*. Running the RS-REMD leads to the dissociation of protein-protein complexes in the higher replicas (with large *d* and reduced ϵ) and rapid exploration of the partner surface. Exchanges with lower replicas (and the reference replica without biasing) result in the accumulation of favorable complex geometries mostly but not always in the lowest replica under the control of the original force field.

The required modified force field topology files for each replica run are generated with the lj.py script (requires the parmed module of Amber20). A receptor and ligand mask indicates the residue number range for the first protein (e.g., receptor _mask = :1 − *x*, with x the number of residues in the first protein) and for the second protein (e.g., ligand_mask = :(x + 1)-last residue number in the complex). The lj.py script also requires the name of the complex topology file, the number of replicas [16], and an indicator for the type of scaling (4 is the standard setting). It also needs the 16 scaling parameters for the change in van der Waals radii (after *–d*, *see* also Table 1), the same for the variable *e* (16 scaling values after the *–e*) and finally the receptor and ligand residue masks and the name of the output topology files (*see* **Note 3**).

```
python lj.py myprotein_hrpt.prmtop 16 4 \
-d 0.00 0.01 0.02 0.04 0.08 0.12 0.16 0.20 0.24 0.28 0.32 0.38
0.44 0.50 0.58 0.68 \
-e 1.00 0.99 0.98 0.97 0.96 0.94 0.92 0.90 0.88 0.86 0.84 0.82
0.80 0.78 0.76 0.74 \
-r receptor_mask -l ligand_mask -o system.top | tee lj.log
```

This step generates 16 new topology files, each with a different scaling the LJ-parameters for the intermolecular interactions (*see* **Note 3**). The next step is to generate a remd.groupfile to start the different replica simulations simultaneously on a workstation equipped with a GPU (with the pmemd.cuda.MPI module of Amber20). The remd.groupfile requires the following line for each replica (16 lines as given below with the label J growing from 1 to 16).

```
-O -rem 3 -i remd.in -o remd_J.out -c myprotein_lastheat.rst -r
remd_J.rst  -x  remd_J.nc  -p  system_J.top  -inf  mdinfo.J  -l
logfile.J
```

The RS-REMD simulation is started as given below (see also Amber20 manual).

```
mpirun -np 16 pmemd.cuda_SPFP.MPI -O -ng 16 -groupfile remd.
groupfile
```

The free energy difference between the bound and unbound states of the complexes can also be estimated with a perturbation approach along the biasing (α) coordinate associated with a replica number (Table 1). The introduced bias in each replica is calculated using trajectory re-evaluation. The trajectory of replica $\alpha + 1$ is evaluated with the LJ parameter set of replica α, and the LJ parameter set of replica $\alpha + 1$. The associated free energy difference can be obtained using the BAR (Bennet-acceptance ratio) method. This approach can also be extended from only evaluating adjacent biases along the α path $\Delta U\alpha, \alpha + 1$ to evaluating all pairs of biases $\Delta U_{i,j}$, with $i, j = 1, \ldots, N$. The corresponding free energy difference is calculated using the multi-Bennet-acceptance ratio method

(MBAR). For calculating the binding free energies, only the second half of the simulation trajectories are typically used, and the first half was considered as equilibration time. The statistical errors can be calculated by splitting the simulations into five parts and calculating the mean and standard deviation.

3.3 MMGBSA Scoring and Analysis of Results

Individual frames of RS-REMD simulation runs can be scored by calculating the MMGBSA (Molecular Mechanics Generalized Born Surface Area) interaction energy. In this simple scoring method, the difference in total potential energy of the protein complex and the individual partner molecules is calculated after replacing the explicit solvent with an implicit (Generalized Born) model. Calculations can be performed using the ptraj module of the Amber20 package (an example code to run ptraj is given in **Note 4**).

The MMGBSA interaction energy term is fast to calculate but neglects several contributions (e.g., entropic contributions to binding, conformational changes upon binding). As described above (Subheading 3.2), it is also possible to estimate the free energy of binding using directly the sampling from the RS-REMD simulations.

3.4 Example 1: Refinement and Scoring of a Docked Complex Structure

In the first example we perform RS-REMD refinement simulations on a docking solution for the complex PDB 1syx obtained by the ATTRACT docking method and unbound partner structures. We start the protein-protein docking refinement by preparing the input files, generating the topology files and heating up the system as described above. For convenience, the steps for preparation, heating, and equilibration (described in Subheading 3.1) are included in python scripts (the scripts are provided at https://gitlab.com/TillCyrill/mmib).

```
pdb4amber -i ../docked/m1/bs.pdb -o m1/docked_4amb.pdb
python leap.py
python heat.py
source heat.run
```

As a next step we start a 5 ns production RS-REMD simulation with 16 replicas as described in Subheading 3. The necessary commands (described in Subheading 3) are all executed in the python script prep_RS.py

```
python prep_RS.py
source rs.run
```

3.4.1 Analysis of the RS-REMD Simulation

We start the analysis by stripping waters and ions from the trajectories. We first calculate the root-mean-square deviation ($Rmsd_{lig}$, the Rmsd of the non-hydrogen atoms of the smaller protein

Table 2
Analysis of RS-REMD refinement simulation

Replica	$Rmsd_{lig}$ first frame (Å)*	$Rmsd_{lig}$ last frame (Å**)	MMGBSA (kcal/mol)***
1	8.57	8.74	−35.53
2	8.46	5.47	−23.00
3	7.30	83.40	1.80
4	8.50	6.90	−29.65
5	8.50	**4.90**	**−39.47**
6	8.32	5.62	−30.83
7	7.95	6.19	−17.08
8	8.18	18.25	−1.29

*$Rmsd_{lig}$ of the first frame (after heating and equilibration), the last frame (**) and the MMGBSA score (***) of the last frame of the RS-REMD simulation. Results for the first eight replicas are shown. The replica with the lowest $Rmsd_{lig}$ also corresponds to the replica with the lowest MMGBSA interaction energy (in bold)

partner, ligand, after the best superposition of the receptor onto the native complex) for every replica. In the test case, this is possible because we know the experimental complex structure. The initial $Rmsd_{lig}$ of the docking solution is ~8 Å (Table 2, *see* **Note 5**). At the final stage of the simulation we find in replica 5 the refined structure with the lowest $Rmsd_{lig}$ (*see* **Note 6**). Without knowledge of the experimental structure, an appropriate score is required to identify the near-native solution. We calculate an MMGBSA score of the last frame for every replica. Replica 5 has the most favorable MMGBSA score of −39.47 kcal/mol (*see* Table 2). The analysis steps employ the ptraj library and are all included in 3 python scripts:

```
python stripped.py
python rmsd_distance.py
python score.py
```

3.5 Example 2: Application to an Ensemble Docking in Explicit Water

An advantage of the RS-REMD method is the possibility to use an ensemble of (up to 16) different start structures for refinement instead of refining single start structures as described in Example 1. The application is illustrated for the PDB 1qa9 protein-protein complex. In the ensemble mode, the RS-REMD simulations run with every replica starting from a different docking pose (Fig. 2). From a set of 50 top-scoring docking solutions (from the ATTRACT docking program) based on unbound partner structures, the 16 models with the lowest MMGBSA score were used as start structures for the RS-REMD simulations (Fig. 2). The results for the replicas with smallest and largest final deviation from the native complex, respectively, are summarized in Table 3. All steps

Fig. 2 (Left panel) Superposition of the ensemble of 16 docked starting structures for the complex 1qa9. The superposition is shown with respect to the receptor partner protein (green cartoon) and all 16 partner proteins (ligands) indicated by different colors (tints of blue). (Right panel) Placement of the ligand-protein with the best free energy score (violet) after RS-REMD refinement and scoring onto the native complex geometry (green cartoon)

Table 3
Analysis of RS-REMD simulations starting from an ensemble of docking solutions

Replica	$Rmsd_{lig}$ first frame (Å)	$Rmsd_{lig}$ last frame (Å)	MMGBSA of last frame (kcal/mol)	RS-REMD free energy score (kcal/mol)
2	5.3	2.9	−12.82	−14.75
4	4.1	45.6	−15.01	−3.79

are otherwise the same as described in Subheadings 3.1 and 3.2 and are included in the following python scripts (*see* https://gitlab.com/TillCyrill/mmib):

```
python score.py
python reshape_best_models.py
python drag.py
python prep_RS.py
source rs.run
```

3.5.1 Analysis and MMGBSA Scoring of the First Example

We first analyze the results by stripping the trajectories and calculating the $Rmsd_{lig}$ and MMGBSA interaction energy with the following commands:

```
python stripped.py -topDir Rep16 -trajDir Rep16 --autoimage T
python rmsd_distance.py
python score.py
```

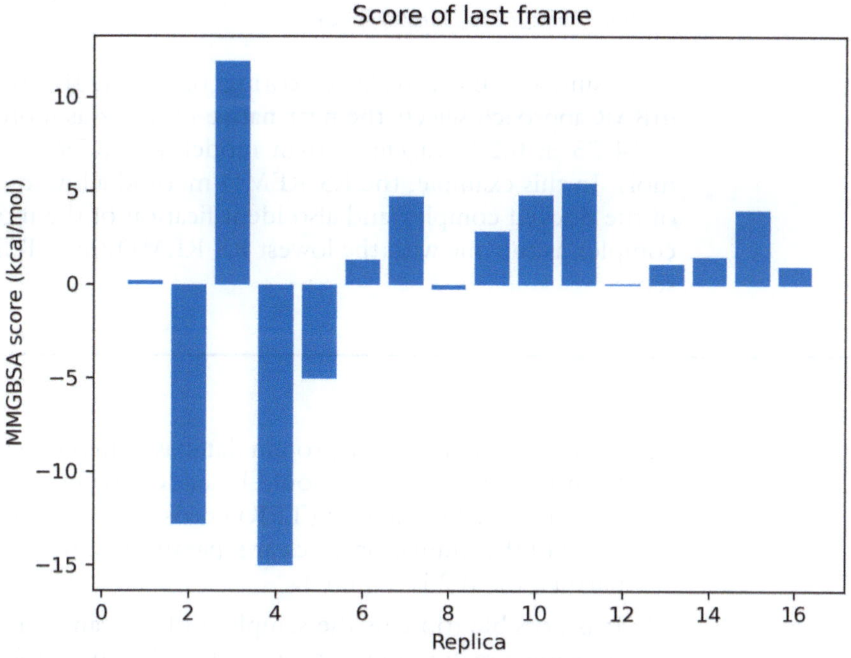

Fig. 3 Bar plot of the calculated MMGBSA scores of the final sampled structure for each replica. The structures with the lowest MMGBSA score are located in replica 2 and replica 4 and were used for further analysis

The resulting $Rmsd_{lig}$ values indicate a geometry with increased $Rmsd_{lig}$ in the reference replica (8 Å) but a drastic improvement for the second replica (2.9 Å). The MMGBSA interaction energy scores the structure in replica 4 best (−15 kcal/mol), followed by replica 2 (−12 kcal/mol) (*see* Fig. 3). However, the structure in replica 4 is a local energy minimum that is quite far away from the native binding site ($Rmsd_{lig}$ of 45 Å). Hence, the MMGBSA score fails, in this case, to identify a near-native solution as the best scoring solution (the structures in the other replicas have all more unfavorable MMGBSA scoring, not shown).

3.5.2 Binding Free Energy Estimation Using mBAR Combined with RS-REMD

We perform another RS-REMD run starting from the two best MMGBSA scored final structures (in replica 2 and 4, see the previous paragraph). The same parameter files as for the first run can be used. The RS-REMD simulations for both models take approximately 10 h on a workstation with a single GPU. The free energy change along the dissociation of the partner proteins is then calculated using MBAR (*see* Subheading 3).

```
python prep_RS_freeEnergy.py --models 2,4
source freeEnergy.run
python stripped.py -topDir Rep16 -trajDir m2,m4 --autoimage T
python calculate_replica_biases.py
```

```
python mbar_free_energies.py
```

Using the more realistic scoring based on the RS-REMD/mBAR approach selects the near-native model 2 as more favorable (−14.75 ± 0.28 kcal/mol) than model 4 (−3.79 ± 0.17 kcal/mol). In this example, the RS-REMD method allowed refinement of the docked complex and also identification of the most realistic complex as the one with the lowest RS-REMD/mBAR free energy score.

4 Notes

1. Protein structure in the protein database often misses residues. Such missing segments should be added by molecular modelling or a chain separation (TER) needs to be included. For the setup of the simulations, a clear separation between the protein partner 1 and 2 is required.
2. It is possible to focus the sampling of relevant complex states on known regions of receptors or ligands. This information can be included during the simulations in form of additional restraints. This option will not be covered in our present description.
3. The names of the generated force field topology files are system_1.top, system_2.top, ..., system_16.top. The "system_16.top" file represents the topology file with the strongest bias whereas the system_1.top represents the force field of the original parameter topology file.
4. Python code to be included for performing an MMGBSA trajectory analysis using the ptraj module of Amber20.

```
def calc_energy(topName, mask, traj):
  options = sander.gas_input(8)
  ene = pt.energy_decomposition(traj_com, top="radii_"+top
  Name, mm_options=options)
  return ene

def calc_energies(traj, receptor_mask, ligand_mask):
  complex_mask = "{}|{}".format(receptor_mask, ligand_mask)
  com_ene = calc_energy("complex.top", complex_mask, traj)
  lig_ene = calc_energy("ligand.top", ligand_mask, traj)
  rec_ene = calc_energy("receptor.top", receptor_mask, traj)

  score_vdw = com_ene["vdw"]-(lig_ene["vdw"]+rec_ene["vdw"])
  score_elec = com_ene["elec"]-(lig_ene["elec"]+rec_ene
  ["elec"])
  score_gb = com_ene["gb"]-(lig_ene["gb"]+rec_ene["gb"])

  return score_vdw+score_elec+score_gb
```

5. Although we start from one docking solution the initial $Rmsd_{lig}$ of structures in each replica can slightly differ because of a short equilibration prior to the RS-REMD data-gathering phase.
6. Due to the frequent exchanges between replicas, it is possible the most favorable structure at the end of an RS-REMD simulation is not found in the reference replica but by chance is located in other replicas in the finally sampled frame.

Acknowledgments

The authors are grateful for financial support from the Deutsche Forschungsgemeinschaft (SFB1035, Project B02).

References

1. Marsh JA, Teichmann SA (2015) Structure, dynamics, assembly, and evolution of protein complexes. Annu Rev Biochem 84:551–575
2. Cong Q, Anishchenko I, Ovchinnikov S, Baker D (2019) Protein interaction networks revealed by proteome coevolution. Science 65:185–189
3. Humphreys IR, Pei J, Baek M, Krishnakumar A, Anishchenko I, Ovchinnikov S et al (2021) Computed structures of core eukaryotic protein complexes. Science 374: eabm4805
4. Jumper J, Evans R, Pritzel A, Green T, Figurnov M, Ronneberger O et al (2021) Highly accurate protein structure prediction with AlphaFold. Nature 596:583–589
5. Baek M, DiMaio F, Anishchenko I, Dauparas J, Ovchinnikov S, Lee GR et al (2021) Accurate prediction of protein structures and interactions using a three-track neural network. Science 373:871–876
6. Bryant P, Pozzati G, Elofsson A (2022) Improved prediction of protein-protein interactions using AlphaFold2. Nat Commun 13: 1265
7. Zacharias M (2010) Accounting for conformational changes during protein-protein docking. Curr Opin Struct Biol 20:180–186
8. Pan AC, Jacobson D, Yatsenko K, Sritharan D, Weinreich TM, Shaw DE (2019) Atomic-level characterization of protein-protein association. Proc Natl Acad Sci 116:4244–4249
9. Siebenmorgen T, Engelhard M, Zacharias M (2020) Prediction of protein-protein complexes using replica exchange with repulsive scaling. J Comput Chem 41:1436–1447
10. Perthold JW, Oostenbrink C (2017) Simulation of reversible protein-protein binding and calculation of binding free energies using perturbed distance restraints. J Chem Theory Comput 13:5697–5708
11. Ostermeir K, Zacharias M (2017) Accelerated flexible protein-ligand docking using Hamiltonian replica exchange with a repulsive biasing potential. PLoS One 12:e0172072
12. Siebenmorgen T, Zacharias M (2019) Evaluation of predicted protein-protein complexes by binding free energy simulations. J Chem Theory Comput 15:2071–2086
13. Siebenmorgen T, Zacharias M (2020) Efficient refinement and free energy scoring of predicted protein-protein complexes using replica exchange with repulsive scaling. J Chem Inf Model 60:5552–5562
14. Glashagen G, de Vries S, Uciechowska-Kaczmarzyk U, Samsonov SA, Murail S, Tuffery P et al (2020) Coarse-grained and atomic resolution biomolecular docking with the ATTRACT approach. Proteins 88:1018–1028
15. de Vries SJ, Schindler CEM, Chauvot de Beauchêne I, Zacharias M (2015) A web interface for easy flexible protein-protein docking with ATTRACT. Biophys J 108:462–465
16. Zacharias M (2003) Protein-protein docking with a reduced protein model accounting for side-chain flexibility. Protein Sci 12:1271–1282
17. Case DA, Cheatham TE, Darden T, Gohlke H, Luo R, Merz KM et al (2005) The Amber biomolecular simulation programs. J Comput Chem 26:1668–1688

18. Maier JA, Martinez C, Kasavajhala K, Wickstrom L, Hauser KE, Simmerling C (2015) ff14SB: improving the accuracy of protein side chain and backbone parameters from ff99SB. J Chem Theory Comput 11:3696–3713
19. Jorgensen WL, Chandrasekhar J, Madura JD, Impey RW, Klein ML (1983) Comparison of simple potential functions for simulating liquid water. J Chem Phys 79:926–935
20. Hopkins CW, Le Grand S, Walker RC, Roitberg AE (2015) Long-time-step molecular dynamics through hydrogen mass repartitioning. J Chem Theory Comput 11:1864–1874

Chapter 16

Exploiting the Role of Features for Antigens-Antibodies Interaction Site Prediction

Michela Quadrini and Carlo Ferrari

Abstract

Antibodies are a class of proteins that recognize and neutralize pathogens by binding to their antigens. They are the most significant category of biopharmaceuticals for both diagnostic and therapeutic applications. Understanding how antibodies interact with their antigens plays a fundamental role in drug and vaccine design and helps to comprise the complex antigen binding mechanisms. Computational methods for predicting interaction sites of antibody-antigen are of great value due to the overall cost of experimental methods. Machine learning methods and deep learning techniques obtained promising results.

In this work, we predict antibody interaction interface sites by applying HSS-PPI, a hybrid method defined to predict the interface sites of general proteins. The approach abstracts the proteins in terms of hierarchical representation and uses a graph convolutional network to classify the amino acids between interface and non-interface. Moreover, we also equipped the amino acids with different sets of physico-chemical features together with structural ones to describe the residues. Analyzing the results, we observe that the structural features play a fundamental role in the amino acid descriptions. We compare the obtained performances, evaluated using standard metrics, with the ones obtained with SVM with 3D Zernike descriptors, Parapred, Paratome, and Antibody i-Patch.

Key words Interaction prediction, Graph convolutional neural networks, Feature selection, Machine learning methods

1 Introduction

Antibodies (Abs) are proteins that circulate the body through the blood and fluids and play a pivotal role in the immune system [1]. They recognize and eliminate pathogens by binding to their antigens with high affinity and specificity. Thanks to this capability, the antibodies are used for designing several drugs [2]. The antibodies represent the most important class of biopharmaceuticals [3]. One of the most notable applications is the development of antibody-drug conjugates (ADGs), i.e., monoclonal Abs covalently linked to small drug molecules [4]. ADGs exploit the ability of

Agnieszka A. Kaczor (ed.), *Protein-Protein Docking: Methods and Protocols*, Methods in Molecular Biology, vol. 2780,
https://doi.org/10.1007/978-1-0716-3985-6_16,
© The Author(s), under exclusive license to Springer Science+Business Media, LLC, part of Springer Nature 2024

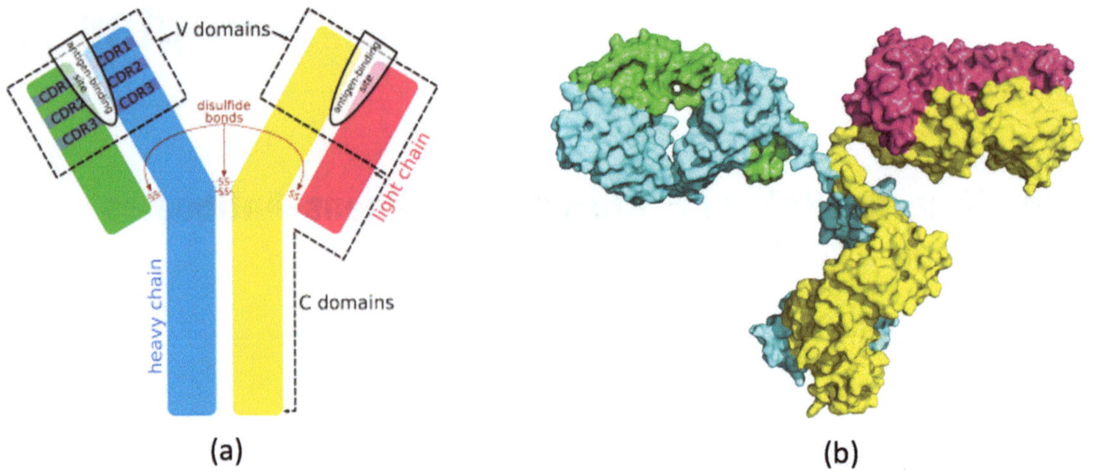

Fig. 1 Schematic representation of the structure of an antibody molecule in (**a**) and the molecular surface in (**b**)

antibodies to precisely bind the drug molecule on a specific target, such as cancer cells [5].

Antibodies are roughly Y-shaped molecules consisting of three equal-sized portions loosely connected by a flexible tether [6]. They consist of two heavy (H) and two light (L) chains linked together by disulfide bonds. Figure 1 shows a simplified schematic illustration of an antibody molecule and an example of molecular surface representation. The light and heavy chains are composed of constant (C) and variable (V) regions that determine the functional properties of the antibody and contribute to the antigen-binding site, respectively [7].

The two light chain types, kappa (κ) and lambda (λ), are found in antibodies and do not determine their functional differences. By contrast, the five heavy chains, called μ, δ, γ, α, and ϵ, define the class and thus the functional activities of antibodies. The corresponding five classes are immunoglobulin M (IgM), immunoglobulin D (IgD), immunoglobulin G (IgG), immunoglobulin A (IgA), and immunoglobulin E (IgE). IgG can be split into four subclasses, IgG1, IgG2, and IgG3, while IgA consists of IgA1 and IgA2 subclasses. Table 1 summarizes the antibody classes with the relative main functions.

Characterization of antibody-antigen interactions can identify immune recognition mechanisms and provide information on the development of the immune response. Experimental methods, such as X-ray crystallography, nuclear magnetic resonance spectroscopy and cryoelectron microscopy, represent the gold standard for determining structures and characterizing antibody-antigen binding sites [8]. Unfortunately, they are labor-intensive, time-consuming, and have high costs. Other experimental methods like alanine scanning [9] and H–D exchange mass spectrometry [10] are faster, cheaper, and of lower resolution. Moreover, they still require

Table 1
Classes of antibodies and the relative main functions

Classes	Subclass
IgG (subclasses: IgG1, IgG2, IgG3, IgG4)	Secreted during secondary response Major form of circulating antibodies
IgA (subclasses: IgA1, IgA2)	Major form of antibodies in external secretions
IgE	Triggers immediate allergic reactions
IgM	Secreted during primary response
IgD	Exact function unknown

remarkable effort. Higher-throughput methods such as multiplexed surface plasmon resonance can characterize many interactions simultaneously. However, these methods do not provide information on interaction localizations [11]. Computational methods thus play a fundamental role in predicting antibody-antigens interaction sites. They could help characterize the relationship between the shape of the antibody, its function, and the conservation of that function across viral strains giving information that can guide the development of efficacious vaccines. The prediction of antibody-antigen binding interfaces can be considered a special case of protein-protein interface site prediction. However, specific methods have been developed for antibody-antigen site prediction due to the significant importance of these interactions [12]. These methods can be roughly divided into sequence-based, and structure-based. The sequence-based ones take advantage of information derived from only amino acid sequences [13]. Regarding the sequence-based methods, the representative ones include ABCPred [14], proABC [15], and Parapred [16]. ABCPred takes sequence windows ranging from 10 to 20 amino acids as input having a target residue and uses recurrent artificial neural networks (RNNs) to classify the residues [14]. ProABC predicts the site-specific interactions in antibody-antigen complexes by taking advantage of random forest automatic learning techniques [15]. Parapred relies on a recurrent neural network approach to leverage local and global features of amino acids [16]. Regarding the methods based on structures, the notable ones are DiscoTope [17], Ellipro [18], and PEPITO [19]. DiscoTope, developed by Andersen et al., uses a combination of amino acid statistics, spatial context, and surface accessibility of amino acids [17]. Ellipro, developed by Ponomarenko et al., approximates a protein surface patch by an ellipsoid [18]. PEPITO, introduced by Sweredoski and Baldi, utilizes a combination of amino acid propensity scores and half sphere exposure values [19].

In this work, we apply HSS-PPI, a hybrid structural method developed for general protein-protein interaction sites and proposed in [20, 21], to predict the antibody-antigens interaction. HSS-PPI abstracts the proteins in terms of spatial and sequential neighboring among amino acid pairs by introducing the hierarchical representation. Such abstraction can capture the structural properties that promote the interactions and can be used to represent proteins with unknown structures by codifying only their sequential neighboring. The method takes advantage of graph convolutional networks, a deep learning framework, to classify the interaction of amino acids. Moreover, the amino acids of each protein/antibody are equipped with features. To understand the role of features in predicting antibody-antigens sites, we describe the residues with different sets of physicochemical and structural features. In particular, the two considered sets of physicochemical characteristics are selected from the AAIndex1 database [22] by a fuzzy clustering method [23] and by Randomized Logistic Regression used in [24]. Moreover, we also consider structural features, i.e., accessible surface area, relative solvent-accessible surface area, torsion angles PHI and PSI, and the number of residue contacts. Finally, we introduce the one-hot encoding vector to identify the amino acid type. Operatively, we evaluate HSS-PPI considering the two sets of physicochemical features equipped with structural features. Moreover, physicochemical features and structural ones are considered separately. We evaluate the method using the physicochemical and structural features separately. The performance of the approaches is evaluated in terms of the area under the receiver operating characteristic curve (AU-ROC) and precision-recall curve (PR-ROC), the accuracy, the precision, the recall, the F-measure, and the Matthews correlation coefficient. Such results show that spatial features play a fundamental role in antibody-antigens prediction. We also compare the obtained performances with the ones obtained with SVM with 3D Zernike descriptors, Parapred, and Antibody i-Patch by the receiver operating characteristic curve and precision-recall curve.

The chapter is organized as follows. Subheading 2 presents the methods and materials used to predict the antibody-antigens interaction. The method based on GCN and hierarchical representation is introduced in Subheading 2.1, while the description of the dataset and the metrics used to evaluate the methods are in Subheadings 2.2 and 2.3. The results relative to evaluation are discussed in Subheading 3. The chapter ends with some conclusions and discussion about future directions, in Subheading 4.

2 Materials

In this section, we introduce the methods based on GCN with hierarchical representation, that we use to predict the interaction sites. Moreover, we also describe the dataset and metrics used to test the approach and compare them with some competitors in the literature.

2.1 Hierarchical Representation and Graph Convolutional Network

The folding process of the amino acid sequence determines three-dimensional structures that show a hierarchy. To formalize such a hierarchy, we formalize two relationships among pairs of amino acids, *spatial* and *sequential neighboring*. Two amino acids are sequential neighboring if they are consecutive in the sequence, while they are sequential neighboring if their distance in the space is less than a fixed threshold. These relations are quantified as follows.

$$a_{(i,j)} = \begin{cases} 1, & \text{if } i \text{ and } j \text{ are sequential neighboring residues,} \\ \dfrac{1}{1+x}, & \text{if } i \text{ and } j \text{ are spatial neighboring residues,} \\ 0, & \text{otherwise.} \end{cases}$$

To formally represent the proteins, starting from the PDB file of each protein, we can compute the neighboring relations between each pair of amino acids and formalize their quantification into a matrix A, called *hierarchical representation matrix.* Formally, each element $a_{(i,j)}$ of the matrix A corresponds is 1 if the corresponding amino acids i and j are sequential residues in the chain or is equal to $\frac{1}{1+x}$, if x the distance between the two C_α atoms, is less than a predetermined threshold, and 0 otherwise. As a consequence, the elements $a_{(i,j)}$ are values between 0 and 1. This representation permits the interpretation of proteins as weighted undirected graphs whose nodes are amino acids. Moreover, it is useful to represent a protein whose 3D structure is unknown by codifying only the sequential neighboring. This feature is important, for example, in the engineering phase when the whole structure is still unknown.

2.1.1 Input Features

HSS-PPI is a hybrid structural method. Each amino acid is equipped with physicochemical and structural features and a one-hot encoding of the amino acid type. The considered structural features are the accessible surface area, relative solvent accessible surface area, torsion angles Φ and Ψ, and the number of residue contacts. The accessible surface area measures the protein surface area accessible to the surrounding solvent, while the relative accessible surface area is a degree of residue solvent exposure [25]. Contact number is the number of spatial neighboring residues, while the torsion angles Φ and Ψ describe the rotations of the polypeptide

Table 2
HQI8 Indices

Entry name	Description
BLAM930101	Alpha helix propensity of position 44 in T4 lysozyme
BIOV880101	Information value for accessibility; average fraction 35%
MAXF760101	Normalized frequency of alpha-helix
TSAJ990101	Volumes including the crystallographic waters using the ProtOr
NAKH920108	AA composition of MEM of multi-spanning proteins
CEDJ970104	Composition of amino acids in intracellular proteins (percent)
LIFS790101	Conformational preference for all beta-strands
MIYS990104	Optimized relative partition energies—method C

backbone around the bonds between $N - C_\alpha$ and $C_\alpha - C_\alpha$, respectively.

The structural features are associated with physicochemical ones. We consider two different sets of physicochemical properties selected from AAindex [22]. AAindex is a public database of numerical values that quantify the physicochemical properties of residues and their pairs. It contains 566 indices. The first set of features we use in HSSP consists of a selection of eight indexes, called HQI8, obtained using a consensus fuzzy clustering method on the database indices [23]. The list of these indices is reported in the Table 2 with the relative descriptions. Moreover, we also take into account 20 physicochemical indexes, reported in the Table 3 as proposed in [24].

These 20 physicochemical properties have been obtained from AAindex by Randomized Logistic Regression, a feature selection method in order to identify the most relevant feature [26]. Thanks to the scores assigned to features by the method, we are able to order the features according to their importance to discriminate between interacting and non-interacting sites. By setting the number of repetitions to 2000 and the minimum score threshold to 0.3, a set of 27 best-scoring indices is identified, shown in Fig. 2. This set is reduced to 20 indices by removing redundancy so that no two indices have a Pearson correlation coefficient of over 70% in absolute value.

Finally, a one-hot encoding describes the amino acid type ($d = 22$: 20 for the regular amino acids plus one for ASX, used when impossible to distinguish between Aspartate or Asparagine, and one for GLX, used when impossible to distinguish between Glutamate or Glutamine).

Table 3
The list of selected features ordered by decreasing importance

Entry name	Description
INM940104	Normalized flexibility parameters (B-values) for each residue surrounded by two rigid neighbors
MAXF760103	Normalized frequency of zeta *R*
KARP850103	Flexibility parameter for two rigid neighbors
QIAN880139	Weights for coil at the window position of 6
FASG760104	pK-N
OOBM770102	Short- and medium-range non-bonded energy per atom
RICJ880108	Relative preference value at N5
CHOP780214	Frequency of the third residue in turn
FODM020101	Propensity of amino acids within pi-helices
RACS820102	Average relative fractional occurrence in AR(i)
YUTK870102	Unfolding Gibbs energy in water, pH 9.0
OOBM850104	Optimized average non-bonded energy per atom
FASG760105	pK-C
WEBA780101	RF value in high salt chromatography
WERD780104	Free energy change of epsilon(i) to alpha(Rh)
SNEP660101	Principal component I
VASM830102	Relative population of conformational state C
KARP850101	Flexibility parameter for no rigid neighbors
PALJ810107	Normalized frequency of alpha-helix in all-alpha class
RACS820107	Average relative fractional occurrence in A0($i-1$)

2.1.2 Graph Convolutional Networks

Graph convolutional network is an approach for semi-supervised learning proposed in [27] by Kipf and Welling. Such an approach based on a neural network operates graphs-structured data by learning a function regarding the features on graph G to predict the labels on each node. The model takes the adjacency matrix A of G and the input feature matrix X as input.

Each adjacency matrix A can represent proteins represented as a weighted undirected graph $G(V, E, w)$, where $V = \{v, v_2, \ldots, v_n\}$ is the set of n nodes/residues, $E = \{e_1, e_2, \ldots, e_m\} \subseteq V \times V$ is the set of m edges/residue relationships, and $w : E \rightarrow [0, 1]$ are the edge weights that quantify the relationships between pairs of residues. Therefore, the adjacent matrix A presents the proteins as a weighted undirected graph by quantifying the neighboring relations between pairs of residues. Let m be the number of input

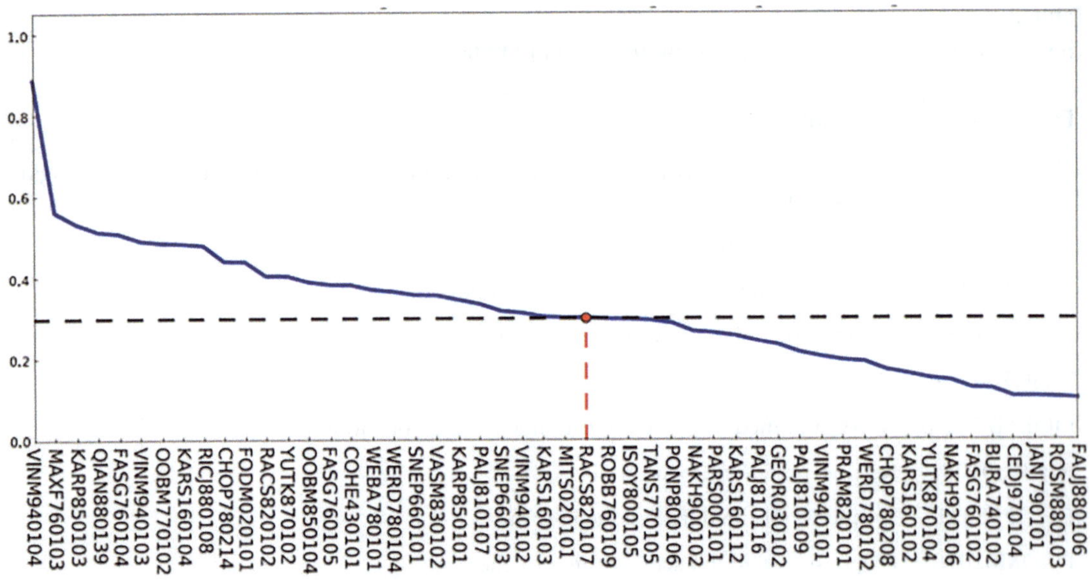

Fig. 2 List of the top important 50 indices

features for each residue of the considered protein. The input feature matrix is $X \in \mathbb{R}^m$, where each row represents the values of the m features of the corresponding node/residue. Finally, $L \in \{0,1\}^n$ denotes the vector of residue labels for the current protein. A non-linear function (propagation rule) formalizes a neural network layer

$$Z^{(h+1)} = f\left(Z^{(h)}, A\right), h = 1, \ldots, H,$$

where $Z^{(0)}$ is the input feature matrix, while $Z^{(h)}$ is a feature matrix, whose each row encodes the feature representation of the relative node. At each layer, these features determine the next layer's feature aggregations by a propagation rule. The propagation rule used in this approach is

$$f\left(Z^{(h)}, A\right) = \sigma\left(\widehat{D}^{-\frac{1}{2}}\widehat{A}\widehat{D}^{-\frac{1}{2}}Z^{(h)}W^{(h)}\right),$$

where I is the identity matrix, $\widehat{A} = A + I$, and $\widehat{D}$ is the diagonal node degree of $\widehat{A}$. Therefore, $\widehat{D}$ is a diagonal matrix, with element $d_{i,\,i}$ equal to the number of incident edges of node v_i incremented by one. $W^{(h)}$ is the weight matrix for layer h, and σ is a non-linear activation function. In this work, the implemented activation function is the Rectified Linear Unit (ReLU). The aggregation of features related to each node is computed in vector form using the following equation:

$$g_{v_i}^{(h+1)} = \sigma\left(\sum_j \frac{1}{c_{i,j}} g_{v_j}^{(l)} W^{(l)}\right)$$

Table 4
Number of complexes and in each class of the dataset and the relative rate of interacting and non-interacting residues

	Complexes	Positive (%)	Negative (%)
Train	213	8.73	91.27
Validation	106	8.87	91.13
Test	153	8.82	91.18

Positive examples are residue pairs that participate in the interface, negative examples are pairs that do not

where j indexes the neighboring nodes of v_i, and $c_{i,\ j}$ i s a normalization constant obtained from the adjacency matrix to account for the degree difference between v_i and v_j.

2.2 Dataset: Paratope Prediction

The used dataset, defined in [28], consists of 471 antibody-antigen complexes. Each sequence pair of complexes differ by more than 95%. Each complex contains structural information about the antibodies and their related antigens and has a resolution of more than 3 Å, as is common practice in the literature. We defined an antibody residue as part of the paratope if at least one of its heavy atoms (thus excluding hydrogen atoms) is at a distance of 4.5 Å from a heavy atom of the antigen. This threshold distance is the most common choice in the literature [29]. The entire dataset is split into three groups, training, validation, and test set. The list of complexes is reported in Appendix, while the rates of interacting and non-interacting residues of training, validation, and test set are in Table 4.

2.3 Evaluation

Predicting the sites of antibody-antigen complexes can be formulated as a binary classification problem where each residue can be either interfacial or non-interfacial.

We evaluate the performance of HSS-PPI considering different of features using six evaluation metrics: Accuracy (Acc), Precision (P), Recall (R), F-measure (F_1), the area under the receiver operator characteristic curve (AUC), and Matthews correlation coefficient (MCC).

$$\mathrm{Acc} = \frac{\mathrm{TP} + \mathrm{TN}}{\mathrm{TP} + \mathrm{TN} + \mathrm{TP} + \mathrm{FN}}$$

$$P = \frac{\mathrm{TP}}{\mathrm{TP} + \mathrm{FP}}$$

$$R = \frac{\mathrm{TP}}{\mathrm{TP} + \mathrm{FN}}$$

$$F_1 = 2\,\frac{P \times R}{P + R}$$

$$MCC = \frac{TP \times TN - FP \times FN}{\sqrt{(TP + FP)(TP + FN)(TN + FP)(TN + FN)}}$$

where TP is the true positive, i.e., the number of interaction amino acids classified correctly, FN is the false negative, i.e., the number of interaction residues identified incorrectly, FP represents false positive, i.e., the number of non-interaction sites identified incorrectly, TN is the true negative, i.e., the number of non-interaction sites identified correctly.

Moreover, we compare the performance of all considered approaches using the receiver operating characteristic (ROC) and the precision-recall (PR) curve plots and their area under the curve (AUC). A receiver operating characteristic curve (ROC) is a standard technique for evaluating the performance of the method over a range of trade-offs between true positive (TP) and false positive (FP) error rates. ROC curve is a plot of sensitivity, or the ability of the model to predict an event correctly, versus 1-specificity for the possible cut-off classification probability of all possible values p. Instead, the PR curve is obtained by plotting the precision values against the corresponding recall for all threshold values of p.

3 Results and Discussions

We apply HSS-PPI, a hybrid structural approach, initially developed for general protein-protein interaction sites, to predict the antibody-antigens interaction. We consider different sets of features, both physicochemical and structural, and study their influence on the method. We take into account two groups of physicochemical properties: the first one is HQI8, obtained by a selection of eight indexes from the index by a fuzzy method, while the second one consists of 20 indices selected from the index by Randomized Logistic Regression, a feature selection method to identify the most relevant feature. We test our approach using the set of structural features and the two groups of physicochemical ones. Operatively we first apply HSS-PPI on the two sets of physicochemical characteristics separately. Then, we add the five structural features. Finally, we also consider only the structural properties. In each experiment, the one-hot vector is also included one-hot encoding vector that describes the amino acid type to the feature set. All used features are described in Subheading 2.1. To perform the experiments, we used Stochastic Gradient Descent as an optimization algorithm, 5000 epochs with a learning rate and a dropout value equal to 0.001 and 0.5, respectively. Moreover, we set two hidden layers with 35 and 32 features. Each protein is abstracted into a hierarchical representation described in Subheading 2.1, and we consider all pairs of residues less the 8 Å. The framework is implemented in Python using Biopython and

Table 5
Measures of F1 score, classification accuracy, precision, recall, MCC, and ROC-AUC obtained on the test set of the antigens classes

HSS-PPI with	F1	Accuracy	Precision	Recall	MCC	AUC-ROC	PR-ROC
Structural features	0.629	0.888	0.547	0.786	0.590	0.940	0.632
HQI8	0.611	0.884	0.536	0.764	0.571	0.935	0.612
HQI8 and structural features	0.631	0.891	0.560	0.773	0.593	0.941	0.632
20	0.621	0.889	0.549	0.764	0.581	0.939	0.624
20 and structural features	0.636	0.892	0.561	0.782	0.598	0.943	0.640

TensorFlow 2.0. We use the PDBParser and DSSP modules of BioPython to abstract the shape into a graph and extract the structural features, respectively. The framework can be downloaded at https://gitlab.com/sebastiandaberdaku/hss-ppi. To train the model, we use HPC Server with eight 12-Core Intel Xeon Gold 5118 CPUs running at 2.30 GHz and using 1.5 TB RAM. We set 32 parallel threads under OS Fedora Linux 25. The performances of the approaches are reported in Table 5.

The performances of the HSS-PPI method computed with different metrics, F1, Accuracy, Precision, Recall, MCC, AUC-ROC, and PR-ROC, show how the structural features are able to improve the results. Moreover, HSS-PPI achieves remarkable results only with structural features. The results are consistent with the biological hypothesis, which states the protein structure or shape is strictly related to its functions and behaviors. The approach is compared with other predictors, Paratome, Antibody i-Patch, and Parapred. The Paratome web server takes as input the heavy (H) and light (L) chain sequences or the PDB structure and classifies residues in interface/non-interface. In our experiment, we consider the PDB structures of Abs. The Antibody i-Patch server takes the structures of Abs with its possible binding partner as input and predicts the residue's probability of interaction by a score. SVM with 3D Zernike descriptors.

The Parapred is a sequence-based method. It takes the variable domain sequences or the full H and L chain sequences as inputs, and it returns a score indicating the likelihood of belonging to the paratope of each residue. The methods have been trained and tested on the sets described in Subheading 2.2. The prediction performance comparison for each method in given terms of ROC and PR curves shown in Fig. 3 and was carried out on the complete test set. The results of the Paratome method are represented as a single point in the ROC and PR curve comparison plots since this approach provides a binary classification output differently from

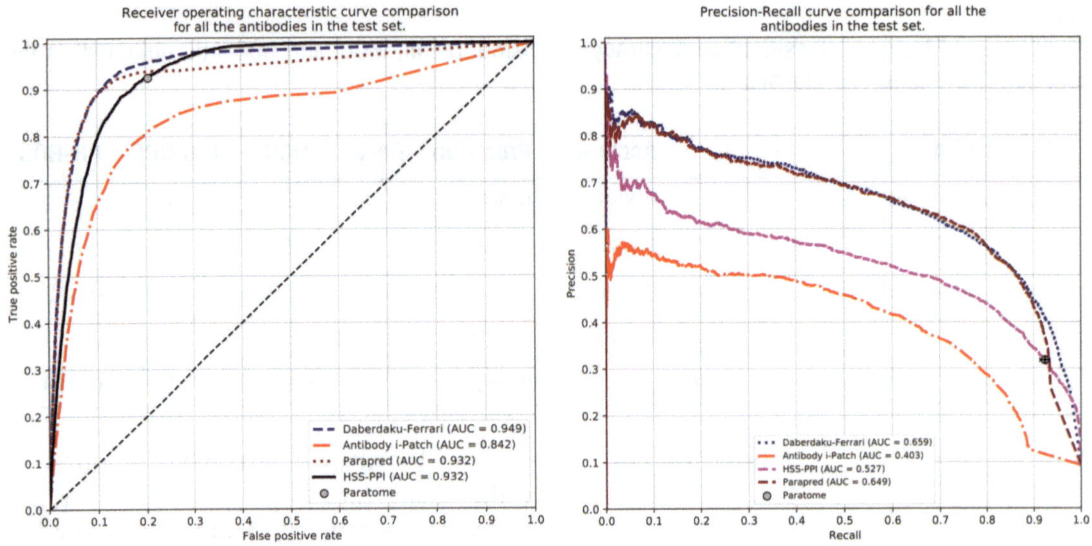

Fig. 3 Average ROC and PR curve comparison of the different paratope prediction methods on the complete test set

all the other methods that give a classification score which can be thresholded as output.

4 Conclusion and Future Work

In this chapter, we apply HSS-PPI, a hybrid structure-based method, to predict the interaction sites in antibodies. Prediction of antibody and antigens interfaces play a fundamental role in understanding the biological mechanisms and immunological research aimed at antibody design and manipulation. The method considers two kinds of relations among residues, spatial and sequential relations, to abstract the antibodies and uses a GCN to classify the residues as interacting or non-interacting.

This classification is obtained by assigning a score to each amino acid that indicates the interaction probability. The method is hybrid since it uses physicochemical and structural features. In this work, we have applied two sets of features equipped with the one-hot encoding vector that describes the amino acid type. We also evaluate the method by adding structural features (accessible surface area, relative solvent accessible surface area, torsion angles Φ and Ψ, and the number of residue contacts) to two sets of physicochemical ones. Finally, we also consider only structural features.

The results show that structural features play a fundamental role according to the biological hypothesis that links the shape or spatial configuration of the molecules with their functions and behaviors. This aspect suggests that the structural characteristics can describe residues to discriminate between interacting and non-interacting. We can note the spatial configurations of antibodies, determined by chemical and physical conditions during the folding, guide the interaction with other molecules, like antigens. Apparently, spatial features in predicting antibody-antigens interaction with deep learning methods play a fundamental role. We apply the approach to an antibody-antigens dataset proposed in [24] and compare it with SVMwith 3D Zernike descriptors, Parapred, Paratome, and Antibody i-Patch.

As immediate future work, we will consider other spatial features, including the free energy of dissociation, presence/absence of hydrogen bonds and salt bridges as proposed in [30]. We also intend to select a subset by Randomized Logistic Regression, a feature selection method able to assign a score to features according to their relevances. Another future direction is to determine local or sequential and global or spatial features related to the sequence and structure of antibodies to characterize the amino acids between interfaces and non-interfaces. Regarding the sequential features, we will investigate the role of the frequencies of particular kinds of amino acids into subsequences present in sliding windows of a length determined by entropy, like [21, 31]. Moreover, it is also interesting to capture and consider structural patterns to structures proposed for understanding the relationship between RNA sequences and secondary structures in [32, 33]. Motivated by the observations related to the obtained results and our previous results obtained from RNA secondary structures comparison [34–36], we believe that it is fundamental to define a dataset by considering the structural differences among proteins to represent the proteins in the dataset exhaustively. Although some structural classifications have been proposed in the literature, like SCOP [16], our approaches obtained as extensions of the ones defined to compare RNAs will work on polynomial time by neglecting the sequence of amino acids.

The prediction of the interactions among entities plays fundamental roles in various scenarios in the immune response, like determining metabolic pathways of the cells or signal transmission. In this scenario, other approaches have already been defined in the literature. Methods based on agents, even though a multiagent system is more suitable to reproduce the dynamics of a cell-cell interaction network than the geometry of specific site [37], are abstracted through the agents' perception capabilities. An idea successfully exploited in the simulation of metabolic pathways [38, 39]. However, from a multiscale perspective, the same approach can be applied to the individual units that make up the antibody interface sites and could provide representations complementary to those obtained with our machine-learning method [40, 41].

Acknowledgments

MQ is supported by the "GNCS-INdAM".

Appendix

The complete list of the structures employed in our study is given in the following Tables 6, 7, and 8.

Table 6
Training set

PDB ID	Heavy chain	Light chain	Antigen chain (s)	PDB ID	Heavy chain	Light chain	Antigen chain (s)
1A3R	L	H	P	1ACY	L	H	P
1A3R	L	H	P	1ACY	L	H	P
1AHW	D	E	F	1BGX	L	H	T
1BVK	D	E	F	1CE1	L	H	P
1CZ8	L	H	W	1DQJ	A	B	C
1E4W	L	H	P	1E6J	L	H	P
1EGJ	L	H	A	1F58	L	H	P
1FBI	P	Q	Y	1FE8	L	H	A
1FRG	L	H	P	1FSK	K	L	J
1H0D	A	B	C	1HYS	C	D	B
1IGC	L	H	A	1IQD	A	B	C
1JHL	L	H	A	1JRH	L	H	I
1KB5	L	H	A;B	1KC5	L	H	P
1KCS	L	H	P	1LK3	L	H	A
1MPA	L	H	P	1N0X	L	H	P
1N8Z	A	B	C	1NAK	L	H	P
1NCA	L	H	N	1NFD	G	H	D
1NMA	L	H	N	1NSN	L	H	S
1OB1	D	E	F	1ORS	A	B	C
1OSP	L	H	O	1P2C	D	E	F
1PKQ	F	G	J	1Q1J	L	H	P

(continued)

Table 6 (continued)

PDB ID	Heavy chain	Light chain	Antigen chain (s)	PDB ID	Heavy chain	Light chain	Antigen chain (s)
1RJL	A	B	C	1TET	L	H	P
1TJI	L	H	P	1TZH	L	H	V
1UWX	L	H	P	1VFB	A	B	C
1W72	L	H	A;C	1YJD	L	H	C
1YMH	A	B	E	1YQV	L	H	Y
1YY9	C	D	A	1ZTX	L	H	E
2AEP	L	H	A	2AP2	C	D	Q
2B0S	L	H	P	2BDN	L	H	A
2BRR	X	Y	P	2CK0	L	H	P
2CMR	L	H	A	2EH8	L	H	P
2GHW	b	B	A	2GSI	G	H	Z
2H1P	L	H	P	2HFG	L	H	R
2HH0	L	H	P	2HRP	L	H	P
2J4W	L	H	D	2J88	L	H	A
2NR6	C	D	A	2OQJ	G	H	I
2OR9	L	H	P	2OSL	L	H	P
2R56	L	H	A	2V17	L	H	A
2VXQ	L	H	A	2XQB	L	H	A
2XQY	K	J	E	2XTJ	B	D	A
2YPV	L	H	A	3B2U	X	W	V
3B9K	L	H	B	3BKY	L	H	P
3BN9	C	D	B	3CVH	L	H	A,C
3CXD	L	H	P	3GJF	L	H	A;C
3H3B	D	d	B	3HMX	L	H	A
3IDX	L	H	G	3L5W	L	H	I
3LD8	B	C	A	3LIZ	L	H	A
3LZF	L	H	A	3MA9	L	H	A
3MXW	L	H	A	3NH7	L	H	A
3NPS	C	B	A	3O0R	L	H	C;B
3O2D	L	H	A	3PNW	D	E	F
3Q1S	L	H	I	3Q3G	C	D	G;E

(continued)

Table 6
(continued)

PDB ID	Heavy chain	Light chain	Antigen chain (s)	PDB ID	Heavy chain	Light chain	Antigen chain (s)
3QNZ	A	B	C	3QWO	L	H	P
3RKD	L	H	A	3S35	L	H	X
3SDY	L	H	A;B	3SE8	L	H	G
3SO3	B	C	A	3T2N	M	I	B
3U4E	L	H	G	3UX9	D	D	C
3UZE	a	A	C	3V6O	E	C	A
3VI3	L	H	D	3WKM	L	H	A
3ZTN	L	H	A;B	4AG4	L	H	A
4ALA	L	H	C	4BZ1	L	H	A
4CNI	L	H	C	4D3C	L	H	A
4DKF	L	H	A	4DN4	L	H	M
4DQO	L	H	C	4DTG	L	H	K
4DW2	L	H	U	4ENE	D	C	A;B
4ETQ	L	H	C	4F3F	A	B	C
4G6J	L	H	A	4H88	L	H	A
4HC1	N	M	B	4HCR	N	M	B
4HKX	B	A	E	4HLZ	H	G	B
4HT1	L	H	T	4I2X	A	B	E
4I3S	L	H	G	4IJ3	B	C	A
4JPV	L	H	G	4JR9	L	H	A
4K2U	M	I	B	4K3J	L	H	B
4KI5	F	E	M	4KI5	D	C	M
4KUC	E	F	I	4KXZ	P	Q	D
4LIQ	L	H	E	4LMQ	L	H	D
4LQF	L	H	A	4LSP	L	H	G
4LSU	L	H	G	4LVH	C	B	A
4LVN	B	C	A	4NP4	L	H	A
4NZR	L	H	M	4NZT	L	H	M
4O4Y	L	H	A	4O58	L	H	A
4OGY	L	H	A	4OKV	B	A	F
4OT1	L	H	A	4R8W	L	H	B

(continued)

Table 6 (continued)

PDB ID	Heavy chain	Light chain	Antigen chain (s)	PDB ID	Heavy chain	Light chain	Antigen chain (s)
4RDQ	F	G	A;E	4RGN	G	F	S
4RGN	L	H	S	4RWY	L	H	A
4UU9	L	H	D	4XP4	L	H	A
4XTR	D	C	B	4Y5Y	E	D	F
4YDJ	L	H	G	4YDV	L	H	P
4YE4	L	H	G	4YWG	L	H	G
4ZS7	L	H	A	4ZTO	L	H	P
5ANM	C	D	E	5B3J	F	E	C
5BV7	L	H	A	5CZX	L	H	A
5D96	I	J	D	5DD0	L	H	P
5DFV	D	C	A	5DFW	H	H	A
5DHV	L	H	M	5DS8	L	H	P
5DSC	L	H	N	5E94	C	D	H
5EN2	B	A	C	5F96	L	H	G
5FB8	A	B	C	5GGS	D	C	Y
5GJS	L	H	A;B	5HI4	L	H	A;B
5IKC	L	H	N	5J13	B	C	A
5JQ6	L	H	A	5KW9	L	H	A
5L0Q	B	C	A	5LCV	L	H	A
5TE7	L	H	G	5TH9	L	H	A
5TL5	L	H	A	5U3M	L	H	A
5UKR	L	H	G	5V7J	E	D	B
5V7J	L	H	G	5VEB	B	A	X
5W6G	L	H	A	5WUX	L	H	E

Table 7
Validation set

PDB ID	Heavy chain	Light chain	Antigen chain (s)	PDB ID	Heavy chain	Light chain	Antigen chain (s)
1AR1	D	C	B	1BOG	A	B	C
1CU4	L	H	P	1EO8	L	H	A
1F90	L	H	E	1FJ1	C	D	E
1FNS	L	H	A	1FPT	L	H	P
1G9N	L	H	G	1GGI	L	H	P
1HIM	H	L	P	1I8K	A	B	C
1JPS	L	H	T	1KB9	K	J	E
1KCR	L	H	P	1MHP	L	H	A
1NL0	L	H	G	1PZ5	A	B	C
1QKZ	L	H	P	1SM3	L	H	P
1SY6	L	H	A	1TZI	A	B	V
1UJ3	A	B	C	1V7M	L	H	V
1WEJ	L	H	F	1XGY	L	H	P
1XIW	G	H	B;E	2ARJ	L	H	Q
2DD8	L	H	S	2FX7	L	H	P
2G5B	G	H	L	2HKF	L	H	P
2IGF	L	H	P	2IH3	B	A	C
2IPU	K	G	P	2J6E	L	H	A;B
2NYY	C	D	A	2OZ4	L	H	A
2Q8A	L	H	A	2QHR	L	H	P
2QQN	L	H	A	2QSC	L	H	P
2R29	L	H	A	2UZI	L	H	R
2VDK	L	H	A	2VXT	L	H	I
2W9E	L	H	A	2YC1	E	D	F
2ZCH	L	H	P	3BKJ	L	H	A
3D85	A	B	C	3EYF	C	D	F
3GBM	L	H	B	3GGW	C	D	F
3GI9	L	H	C	3H42	L	H	A;B
3HI6	L	H	A	3IU3	L	H	I
3JWD	L	H	A	3KR3	L	H	D
3KS0	L	H	B	3L5X	L	H	A

(continued)

Table 7 (continued)

PDB ID	Heavy chain	Light chain	Antigen chain (s)	PDB ID	Heavy chain	Light chain	Antigen chain (s)
3MJ9	L	H	A	3MLR	L	H	P
3UC0	L	H	A	3UJI	L	H	P
3W9E	B	A	C	3WIH	L	H	A
3ZKM	L	H	A	4CMH	C	B	A
4D9Q	D	E	B	4EDW	L	H	V
4ERS	L	H	A	4F2M	B	A	E
4F37	L	H	A	4FFV	L	H	A
4FFY	L	H	A	4FQJ	L	H	A
4G3Y	L	H	C	4G6F	L	H	P
4HF5	L	H	A	4I9W	D	E	A;B
4J4P	L	H	A;B	4J6R	L	H	G
4JZO	E	D	I	4LST	L	H	G
4PLJ	L	H	A;B	4UD3	B	B	J
4XMP	L	H	G	4XVJ	L	H	A
4XVS	L	H	G	4ZSO	C	D	F
5B71	A	B	E	5BV7	B	C	A
5DMG	D	C	Z	5E8E	A	B	H
5F3B	B	A	C	5FCU	L	H	G
5H35	G	F	E	5HBT	C	D	B
5MVZ	L	H	U				

Table 8
Test set

PDB ID	Heavy chain	Light chain	Antigen chain (s)	PDB ID	Heavy chain	Light chain	Antigen chain (s)
1CIC	A	B	D;C	1IAI	L	H	M;I
1IKF	L	H	C	2EIZ	A	B	C
2LTQ	E	F	D	2XRA	L	H	A
2XWT	B	A	C	2Y5T	B	A	G;E
2Y6S	L	H	Q	3BGF	L	H	S
3E8U	L	H	P	3EOA	L	H	I
3G04	A	B	C	3G5V	A	B	C
3G5Y	A	B	E	3GHE	L	H	P
3GO1	L	H	P	3H0T	A	B	C
3HI1	L	H	G	3HR5	L	H	R
3I50	L	H	E	3IET	C	D	Q
3IFO	L	H	P	3LEY	L	H	P
3MLW	L	H	P	3MLX	L	H	P
3MNZ	A	B	P	3NFP	L	H	I
3O41	L	H	P	3P0Y	L	H	A
3PP4	L	H	P	3QG6	L	H	C
3S37	L	H	X	3SKJ	L	H	E
3SQO	L	H	A	3THM	L	H	F
3U0T	C	D	E	3U9P	M	K	C
3UJJ	L	H	P	4BZ2	L	H	A
4EDX	L	H	W;V	4FP8	L	H	A
4G6A	L	H	A	4G6M	L	H	A
4GLR	I	H	A	4GMS	L	H	A
4H8W	L	H	G	4HHA	A	B	P
4HIX	L	H	A	4HJ0	Q	P	A
4HPO	L	H	P	4HPY	L	H	P
4HS6	L	H	Y	4HS8	L	H	A
4HZL	L	H	E	4I77	L	H	Z
4IRZ	L	H	A	4J8R	C	D	J
4JG0	L	H	P	4JO3	L	H	P
4JPW	L	H	G	4JZJ	L	H	C

(continued)

Table 8 (continued)

PDB ID	Heavy chain	Light chain	Antigen chain (s)	PDB ID	Heavy chain	Light chain	Antigen chain (s)
4K7P	L	H	X;Y	4L5F	L	H	E
4LKX	B	A	R	4M1G	L	H	A;B
4M5Z	L	H	A	4MWF	L	H	C
4N0Y	L	H	A	4N8C	L	H	X
4N9G	N	M	Z	4NRX	L	H	P
4O9H	L	H	A	4PY8	J	I	B
4QCI	L	H	C	4QHU	L	H	C
4QNP	L	H	B	4QTI	L	H	U
4QWW	C	D	B	4RAV	B	A	E;F
4RGM	L	H	S	4TSA	L	H	A
4TUJ	B	A	E	4U0R	C	B	A
4WEB	L	H	E	4XAK	L	H	A
4XH2	L	H	H	4XNM	L	H	C
4XXD	D	E	F	4Y5V	H	G	I
4YDK	L	H	G	4YDL	L	H	G
4YHP	L	H	P	4YK4	B	C	A
4YO0	B	A	E	4YQX	L	H	A
4YUE	L	H	C	4Z0X	L	A	B
4Z5R	P	Q	G	4ZFO	L	H	K
4ZPT	L	H	S	5A3I	H	G	E
5AUM	L	H	C	5C0N	L	H	A
5C7X	L	H	A;B	5CSZ	L	H	D
5D8J	L	H	A	5D93	B	C	A
5DLM	L	H	X	5DRZ	L	H	P

References

1. Frank SA (2020) Immunology and evolution of infectious disease. Princeton University Press, Princeton
2. Birrer MJ, Moore KN, Betella I, Bates RC (2019) Antibody-drug conjugate-based therapeutics: state of the science. J Natl Cancer Inst 111(6):538–549
3. Lu R-M, Hwang Y-C, Liu I-J, Lee C-C, Tsai H-Z, Li H-J, Wu H-C (2020) Development of therapeutic antibodies for the treatment of diseases. J Biomed Sci 27:1–30
4. Ye X, Ku Z, Zhang N, Fu T-M, An Z (2022) Recent progress in development of monoclonal antibodies against human cytomegalovirus. Curr Opin Virol 52:166–173
5. Khongorzul P, Ling CJ, Khan FU, Ihsan AU, Zhang J (2020) Antibody–drug conjugates: a comprehensive review. Mol Cancer Res 1:3–19
6. Norman PS (1995) Immunobiology: the immune system in health and disease. J Allergy Clin Immunol 96:274
7. Moser M, Leo O (2010) Key concepts in immunology. Vaccine 28:C2–C13
8. Bai X-C, McMullan G, Scheres SHW (2015) How cryo-EM is revolutionizing structural biology. Trends Biochem Sci 40:49–57
9. Weiss GA, Watanabe CK, Zhong A, Goddard A, Sidhu SS (2000) Rapid mapping of protein functional epitopes by combinatorial alanine scanning. Proc Natl Acad Sci 97:8950–8954
10. Gallagher ES, Hudgens JW (2016) Mapping protein–ligand interactions with proteolytic fragmentation, hydrogen/deuterium exchange-mass spectrometry. Methods Enzymol 566:357–404
11. Säfsten P (2009) Epitope mapping by surface plasmon resonance. Methods Mol Biol 524: 67–76
12. Esmaielbeiki R, Krawczyk K, Knapp B, Nebel J-C, Deane CM (2016) Progress and challenges in predicting protein interfaces. Brief Bioinform 17:117–131
13. Yao B, Zheng D, Liang S, Zhang C (2013) Conformational B-cell epitope prediction on antigen protein structures: a review of current algorithms and comparison with common binding site prediction methods. PLoS One 8: e62249
14. Saha S, Raghava GP, Singh. (2006) Prediction of continuous B-cell epitopes in an antigen using recurrent neural network. Proteins 65: 40–48
15. Olimpieri PP, Chailyan A, Tramontano A, Marcatili P (2013) Prediction of site-specific interactions in antibody-antigen complexes: the proABC method and server. Bioinformatics 29:2285–2291
16. Liberis E, Velickovic P, Sormanni P, Vendruscolo M, Liò P (2018) Parapred: antibody paratope prediction using convolutional and recurrent neural networks. Bioinformatics 34:2944–2950
17. Kringelum JV, Lundegaard C, Lund O, Nielsen M (2012) Reliable B cell epitope predictions: impacts of method development and improved benchmarking. PLoS Comput Biol 8:e1002829
18. Ponomarenko J, Bui H-H, Li W, Fusseder N, Bourne PE, Sette A, Peters B (2008) ElliPro: a new structure-based tool for the prediction of antibody epitopes. BMC Bioinformatics 9:1–8
19. Sweredoski MJ, Baldi P (2008) PEPITO: improved discontinuous B-cell epitope prediction using multiple distance thresholds and half sphere exposur. Bioinformatics 24:1459–1460
20. Quadrini M, Daberdaku S, Ferrari C (2022) Hierarchical representation for PPI sites prediction. BMC Bioinformatics 23:1–34
21. Quadrini M, Daberdaku S, Ferrari C (2020) Hierarchical representation and graph convolutional networks for the prediction of protein–protein interaction sites. In: Nicosia G et al (eds) Machine learning, optimization, and data science. LOD 2020. Lecture notes in computer science, vol 12566. Springer, Cham, pp 409–420
22. Kawashima S, Pokarowski P, Pokarowska M, Kolinski A, Katayama T, Kanehisa M (2007) AAindex: amino acid index database, progress report 2008. Nucleic Acids Res 36:D202–D205
23. Saha I, Maulik U, Bandyopadhyay S, Plewczynski D (2012) Fuzzy clustering of physicochemical and biochemical properties of amino acids. Amino Acids 43:583–594
24. Daberdaku S (2018) Structure-based antibody paratope prediction with 3D Zernike descriptors and SVM. In: International meeting on computational intelligence methods for bioinformatics and biostatistics. Springer, Cham, pp 27–49
25. Rose GD, Geselowitz AR, Lesser GJ, Lee RH, Zehfus MH (1985) Hydrophobicity of amino acid residues in globular proteins. Science 229: 834–838

26. Meinshausen N, Bühlmann P (2010) Stability selection. J R Stat Soc Ser B Stat Methodol 72: 417–473
27. Kipf TN, Welling M (2017) Semi-supervised classification with graph convolutional networks. In: International conference on learning representations
28. Daberdaku S, Ferrari C (2019) Antibody interface prediction with 3D Zernike descriptors and SVM. Bioinformatics 35:1870–1876
29. Kunik V, Peters B, Ofran Y (2012) Structural consensus among antibodies defines the antigen binding site. PLoS Comput Biol 8: e1002388
30. Das S, Chakrabarti S (2021) Classification and prediction of protein–protein interaction interface using machine learning algorithm. Sci Rep 11:1–12
31. Quadrini M, Cavallin M, Daberdaku S, Ferrari C (2021) ProSPs: protein sites prediction based on sequence fragments. In: Nicosia G et al (eds) Machine learning, optimization, and data science. LOD 2021. Lecture notes in computer science. Springer, Cham, pp 568–580
32. Quadrini M, Merelli E, Piergallini R (2019) Label core for understanding RNA structure. In: Cazzaniga P, Besozzi D, Merelli I, Manzoni L (eds) Computational intelligence methods for bioinformatics and biostatistics. CIBB 2019. Lecture notes in computer science, vol 12313. Springer, Cham, pp 171–179
33. Quadrini M, Merelli E, Piergallini R (2019) Loop grammars to identify RNA structural patterns. Bioinformatics 3:302–309
34. Quadrini M, Tesei L, Merelli E (2020) ASPRAlign: a tool for the alignment of RNA secondary structures with arbitrary pseudoknots. Bioinformatics 36:3578–3579
35. Quadrini M, Tesei L, Merelli E (2019) An algebraic language for RNA pseudoknots comparison. BMC Bioinformatics 20:1–18
36. Quadrini M, Merelli E (2018) Loop-loop interaction metrics on RNA secondary structures with pseudoknots. Bioinformatics 3:29–37
37. Belenchia M, Rocchetti G, Maestri S, Cimadamore A, Montironi R, Santoni M, Merelli E (2021) Agent-based learning model for the obesity paradox in RCC. Front Bioeng Biotechnol 9:642760
38. Maestri S, Merelli E, Pettini M (2022) Agent-based models for detecting the driving forces of biomolecular interactions. Sci Rep 12:1–12
39. Piangerelli M, Maestri S, Merelli E (2020) Visualising 2-simplex formation in metabolic reactions. J Mol Graph Model 97:107576
40. Maestri S, Merelli E (2019) Algebraic characterisation of non-coding RNA. In: International meeting on computational intelligence methods for bioinformatics and biostatistics. Springer, Cham, pp 145–158
41. Maestri S, Merelli E (2019) Process calculi may reveal the equivalence lying at the heart of RNA and proteins. Sci Rep 9:1–9

Chapter 17

Computational Methods to Target Protein-Protein Interactions

Michał K. Jastrzębski

Abstract

The chapter emphasizes the importance of understanding protein-protein interactions in cellular mechanisms and highlights the role of computational modeling in predicting these interactions. It discusses sequence-based approaches such as evolutionary trace (ET), correlated mutation analysis (CMA), and subtractive correlated mutation (SCM) for identifying crucial amino acid residues, considering interface conservation or evolutionary changes. The chapter also explores methods like differential ET, hidden-site class model, and spatial cluster detection (SCD) for interface specificity and spatial clustering. Furthermore, it examines approaches combining structural and sequential methodologies and evaluates modeled predictions through initiatives like critical assessment of prediction of interactions (CAPRI). Additionally, the chapter provides an overview of various software programs used for molecular docking, detailing their search, sampling, refinement and scoring stages, along with innovative techniques and tools like normal mode analysis (NMA) and adaptive Poisson-Boltzmann solver (APBS) for electrostatic calculations. These computational and experimental approaches are crucial for unraveling protein-protein interactions and aid in developing potential therapeutics for various diseases.

Key words Protein-protein docking, Structure-based methods, Sequence-based methods

1 Introduction

Many cellular mechanisms are dependent on protein-protein interactions. These interactions encompass allosteric regulation, gene regulatory signal transduction, and signal transduction in signalling proteins, among others. Predicting and modeling inter-protein processes is crucial from a structural biology perspective. By understanding these processes, one can more accurately predict the biological functions that these structures perform in signaling cascades or in pathological mechanisms of function. It is also noteworthy that protein-protein modeling can elucidate binding affinity and specificity, the nature of the free energy funnels, and the effects of mutations. Furthermore, modeling at the atomic level of interfaces enables us to understand and potentially control changes in

Agnieszka A. Kaczor (ed.), *Protein-Protein Docking: Methods and Protocols*, Methods in Molecular Biology, vol. 2780,
https://doi.org/10.1007/978-1-0716-3985-6_17,

© The Author(s), under exclusive license to Springer Science+Business Media, LLC, part of Springer Nature 2024

binding kinetics, specificity, or selectivity of protein complex formation.

By comprehending the structure of a protein complex, one can leverage the existence of "hot spots" in the interfaces of the complex-specific regions critical for the energy of protein-protein binding. Consequently, the capability to model protein-protein structure complexes presents the potential to discover drugs that might counteract the negative effects of certain structures [1].

X-ray crystallography and other experimental methods can determine the structure of protein-protein complexes. However, they are not always feasible and come with limitations; they are expensive and time-consuming. Modeling protein-protein complexes offers a way to expand upon these methodologies.

2 Structure-Based Approaches

To predict the structures of macromolecules utilizing the available biochemical and biophysical data, researchers employ molecular modeling methods. These methods are broadly categorized into structure-based and sequence-based approaches.

Algorithms that facilitate molecular protein-protein docking analysis belong to a more extensive set of algorithms that leverage protein structure-based techniques. When the three-dimensional structures of the components are known, perhaps due to elucidation through crystallographic methods, it becomes possible to identify potential protein-protein interfaces. This identification process involves a computerized search of the interaction space between two protomers. However, the drawback of these methods is the limited number of known protein structures. As a counter, sequence-based approaches harness the fact that we have knowledge of far more amino acid sequences in proteins than their 3D structures. By pinpointing the residues present at the dimerization interface, the joining mechanism can be simulated.

Structure-based techniques encompass protein-protein docking, molecular dynamics (MD) simulations, electrostatic analysis using an adaptive Poisson-Boltzmann solver (APBS), normal mode analysis (NMA), and de novo protein design. On the other hand, sequence-based techniques concentrate on the analysis of protein sequences to identify evolutionary modifications in protein-protein interfaces. These methods can be further sub-divided based on whether they presuppose conservation of interfaces or acknowledge alterations in protein interfaces throughout evolutionary progression.

In this chapter, the emphasis will predominantly be on structure-based methods, particularly protein-protein docking. A comprehensive exploration of both structure- and sequence-based approaches is available in the article, "Oligomerization of G Protein-Coupled Receptors: Computational Methods" [2].

3 Protein-Protein Docking

The initial exploration of protein-protein docking dates to the 1970s, marked by Wodak and Janin's proposition to model the conformation of amino acids in protein side chains [3]. Contrary to popular perception, modern protein-protein docking, while not as prevalent as traditional small-molecule docking, is not significantly less efficient [4].

Protein-protein docking fundamentally relies on modeling interactions based on the shape complementarity of distinct protein unit regions. To streamline computationally intensive analyses, protein models and algorithms, particularly those focusing on docking regions with pivotal amino acids, are employed. Initial modeling was rigid, but as computational capabilities expanded and models became more refined, there has been a shift towards incorporating protein side-chain flexibility. This elasticity is paramount, as it influences conformational changes in structure, thereby modifying the docking region's accessibility.

The rapidity with which protein-protein docking yields results is a notable advantage over more time-intensive structural methods like MD simulation. Additionally, it is a relatively straightforward method rooted in fundamental physical principles, and it allows for the extraction of entire complex coordinates.

Predominant docking algorithms utilize atomic coordinates of two proteins, sourced from X-ray crystal structures of subunits, nuclear magnetic resonance-determined structures, Cryo-EM, or computational techniques such as homogeneous modeling. These proteins are generally treated as rigid entities, although newer algorithms consider some degree of chain flexibility. Interaction modeling produces a set of plausible solutions. Preliminary results undergo sampling to eliminate misleading solutions, followed by ranking processes, which may involve scoring or filtering [5].

Scoring functions play a pivotal role in protein-protein modeling. They are typically based on a weighted combination of diverse parameters, including chemical, physical, and structural shape factors. Of particular note are the elements of shape complementarity, electrostatics, atom-atom potential, and desolvation effects. For accurate complex structure scoring, it's imperative to harness as much empirical data as accessible. Yet, this does not always ensure dependable predictions. Presently, iterative solutions, like the Rosetta software, are being explored to address scoring challenges [6].

The CAPRI experiment serves as the benchmark for gauging the accuracy and reliability of utilized algorithms and models. In recent analyses, 42 protein-protein complex models were examined, the lion's share of which were modeled accurately, underscoring the immense potential of this modeling approach. A routine assumption in protein-protein docking is that protein complexation transpires in a biological solvent. This presents hurdles, especially when many potential protein complexes emerge in a membrane environment, as seen with G-protein coupled receptor (GPCR) complexes [5].

Flexible modeling gains significance with the increasing influence of flexible regions that constitute the protein-protein binding site. For instance, GPCR dimer formation critically involves transmembrane regions and intra- and extracellular loops, whose conformational shifts during complex formation aren't adequately mimicked in rigid models.

Contemporary protein-protein docking now often integrates advanced assumptions, algorithms, and software than traditional rigid-body docking. Advancements in techniques, the integration of bioinformatics tools, and growing computational prowess (highlighting the rising capabilities of graphics processing units (GPUs) over conventional central processing units (CPUs)) have enabled methodologies such as [7]:

- Soft docking techniques
- Protein interaction surface predictions
- Knowledge-based docking potentials
- Data-driven docking
- Side-chain flexibility modeling
- Re-scoring docking decoys
- Backbone flexibility modeling
- Interfacial water modeling
- Multimeric docking and docking servers

While fundamental protein-protein docking techniques can be accessed via free web servers, maximizing their potential often requires local software installation, such as the Rosetta server [8]. Linux is the preferred platform, given its open-source flexibility. MacOS is a secondary option, while Windows, although viable, may not support all features. In certain instances, specialized software becomes a prerequisite. Given the complexities in defining rotation and translation boundaries, comprehensive searches are typically warranted to elucidate potential complex characteristics.

Input files, which encompass ligands, protein residue ionization data, protein residue and atom coordinates, and pdb format heteroatom data, should ideally be manually refined to exclude superfluous information that might not be supported by specific programs [5].

4 All-Atom Molecular Dynamics Simulation

Molecular dynamics (MD) simulation is a computational method used to examine the dynamic properties of proteins and protein complexes.

One of the primary challenges in experimental studies of dynamics is the trade-off between achieving detailed structural information with high spatial resolution, as provided by X-ray crystallography, and capturing an appropriate time scale for the depiction of dynamic properties, such as that offered by fluorescence resonance energy transfer (FRET). All-atom MD simulation is a technique that provides both of these features, standing alongside protein-protein docking in its capability to offer spatial resolution and the necessary time scale [5].

MD simulations enable the exploration of dynamic properties, which is crucial when modeling macromolecules. These molecules may undergo spatial transformations over time, either through conformational shifts or during dimer system formations. Dynamics also play a pivotal role in discerning association and dissociation parameters.

However, a limitation of MD simulations lies in the time duration, which typically spans the nano- to microsecond range. While this range is adequate for studying minor conformational adjustments or protomer allosteric shifts (which transpire on the microsecond scale), it is insufficient for probing vast conformational transitions of protomers relative to each other due to the slower nature of these processes. Additionally, the need to craft a model of the complex being studied, especially for substantial protein transmembrane complexes, remains a challenge [2].

Recent research underscores the myriad conformational states of receptors [9], which influence their diverse responses to extracellular triggers and contribute to the receptor subtypes. While MD simulations might not capture millisecond processes, they can still offer insights into the molecular dynamics of G-protein-coupled receptor interactions influenced by factors such as solvents, ligands, membrane dynamics, or higher complex formation kinetics.

Noteworthy receptors currently under study include the heterodimer formed between the metabotropic glutamate receptor type 2 and the serotonergic receptor type 2A. MD analyses of these receptors revealed shifts in the binding pocket shape during protomer dimerization [10]. Furthermore, docking studies of lysergic acid diethylamide (LSD) and the neurotransmitter serotonin showcased that their dimerization induced an allosteric change, expanding the 5-HT_{2A}R binding pocket, which heightens the affinity of hallucinogenic compounds for the receptor, in contrast to the reduced glutamatergic activation of the mGluR2 receptor [11].

MD simulations also allow pharmacological analyses of how active compounds bind to molecular targets. For instance, Dror et al. [12] executed simulations for several beta-blockers, elucidating the initial association, drug entry mode, and final binding dynamics.

MD simulations face challenges, including the extension of analysis timeframes and the precise mapping of a protein's surrounding environment. By accurately depicting physiologically pertinent membrane parameters, it becomes feasible to increasingly define membrane effects crucial for detailed modeling of entities like GPCRs. An alternative strategy involves enhancing the MD technique through modifications. One such example is the coarse-grained MD (CGMD) approach, where atom arrangements are simplified by grouping them, facilitating longer simulation time scales compared to the traditional all-atom method. The newly-formed group should replicate the original atoms' properties as closely as possible. This atom-to-group reduction minimizes computational degrees of freedom and costs [13, 14].

5 Normal Mode Analysis (NMA)

Normal mode analysis (NMA) is a technique that models the carbon atoms of all amino acids in a protein as interconnected by springs, described by harmonic motion. It serves as an influential tool for predicting the motions of macromolecules. This method involves generating sets of parameters that help describe visualizations, including correlations between observed and modeled atomic displacements, the degree of motion collectivity, and the average fluctuation in distances.

Research indicates that the motion of many proteins can be approximated by considering perturbations primarily towards one or two low-frequency normal modes. Often, a single low-frequency normal mode (typically one of the three lowest frequencies) is adequate to guide the NMA. The assumption underlying the NMA approach is that the vibrational soft modes encapsulate the most extensive movements within the protein and are thus the most pertinent.

The applications of NMA span a broad spectrum in structural biology. It facilitates the study of protein conformational shifts, enables analysis of membrane models, and assists in determining protein structure by concurrently using X-ray crystallography and templates to phase diffraction data. NMA proves especially effective for larger proteins (those exceeding 1000 amino acids), and when the time scale of dynamic events surpasses the nano-microsecond range that MD simulations typically capture [15].

6 De Novo Protein Design

The objective of this method is to identify mutations that either stabilize or destabilize the protein dimer interface. It is employed to study the assembly of monomers into a protein macrostructure, aiming to comprehend the interactions among them and clarify receptor functionality [16].

The procedure initiates with the prediction of interfering mutations. These predictions consider the distance between carbon atoms in the flexible chain, incorporating specific sequence alterations within the protein. Additionally, an energy function is used, which accounts for the nature of the amino acid residues involved. The subsequent step aims to validate the accuracy of these predictions. In this phase, sequences are chosen from a prioritized list of sequence positions, which have the highest specificity for the flexible template. The selection process is guided by energy minimization.

7 Sequence-Based Approaches

By pinpointing the amino acid residues present at the protein-protein dimerization interface, the assembly process can be simulated. This technique represents a sequence-based approach. Two primary classes of methods emerge from this approach: one that presumes the interface remains conserved between subclasses, and another that posits the interface might undergo changes. A comprehensive review titled "Oligomerization of G Protein-Coupled Receptors: Computational Methods" delves deeply into sequence-based methods [2].

For methods premised on interface conservation, we can identify:

- Evolutionary trace (ET) method
- Correlated mutation analysis (CMA)
- Subtractive correlated mutation (SCM)

Conversely, for methods grounded in the belief that the protein-protein interface might alter during evolution, we can highlight:

- Differential evolutionary trace (diff. ET)
- Hidden-site class model of evolution
- Spatial cluster detection (SCD)

8 Evolutionary Trace (ET) Method

The evolutionary trace (ET) method is employed to predict and identify functionally crucial amino acid residues situated in a protein's active site. The procedure encompasses several steps: generating a set of potential sequences that align with a protein's functional interface; building a phylogenetic tree; categorizing sequences into groups; and pinpointing residues that are consistent within a specific sequence group. The foundational belief of this method is that amino acid residues essential for a protein's function experience fewer mutations compared to others [17].

However, this assumption introduces the primary limitation of the method: even minor mutations in critical residues can induce conformational or structural modifications in the receptor. Furthermore, such alterations can influence the emergence of numerous receptor subtypes. For instance, slight changes can impact processes like dimerization, leading to the development of various dopamine receptor subtypes.

9 Correlated Mutation Analysis (CMA)

CMA (correlated mutation analysis) is a sequence-based method that leverages the correlation between the positional trends of amino acid residues and a coordinated mutation, especially when the interface is preserved for structural or functional reasons. This method sheds light on inter-domain contacts since the two sites exhibiting a correlation are proximately situated in the protein's three-dimensional space.

Göbel et al. [18] employed this approach to derive correlations for known protein structures. However, when juxtaposing CMA outcomes with crystallographic data, the prediction accuracy scarcely surpassed 70%. It has been suggested that enhancing the results might be feasible either through the amalgamation of CMA and ET methods or by incorporating an auxiliary filtering algorithm.

10 Subtractive Correlated Mutation (SCM)

SCM is utilized on interconnected protein sequences originating from the same organism. The SCM method identifies which amino acid residues are at the interface between two proteins by excluding correlated residues in monomers that are recognized as components of other sequences.

An illustrative application of SCM is the multi-step filtering procedure used by Filizola and Weinstein [19] to eliminate false positives during their analysis of opioid receptor homodimerization. To pinpoint association through di- and oligomerization, only those residues on the receptor surface were considered. Residues that were spatially proximate but separated by more than six residues in sequence were excluded. Implementing such stringent criteria within the SCM approach enabled the accurate identification of the correct helices for the δ-opioid receptors and κ-opioid receptors.

11 Differential Evolutionary Trace (diff. ET)

The differential ET method assumes multiple iterations for the division of the phylogenetic tree. Through this, information on the differences between groups of sequences can be obtained, as interfaces are consistently preserved across subtypes. For each subtype, trace residues are distinct. To identify them, one subtracts the trace residues of all sequences from those of the target subtype. This results in residues specific to that subtype's role [20]. This method might enhance the precision of interface predictions by mapping the residues chosen for ET analysis.

11.1 Hidden-Site Class Model of Evolution

This model elucidates the evolutionary relationship between proteins by examining the rate of amino acid substitution over time. It addresses the variance in probabilities of yielding a functional effect at diverse positions within amino acid sequences of a protein when using a single substitution matrix. The hidden-site class model enhances this process by applying different substitution matrices to various parts of the protein sequence [21]. While this model effectively predicts functionally significant residues, its limitation lies in its inability to cluster these residues spatially.

11.2 Spatial Cluster Detection (SCD)

This method evaluates both the spatial location of detected residuals and the specificity of the interface subtype. The spatial positioning of the interface is believed to be conserved within the subtype. To make the model more straightforward, the three-dimensional structure is distilled into a thick tube. This tube is represented as a two-dimensional ring structure by projecting the residuals onto a plane that's perpendicular to the tube's axis [22].

11.3 Methods Combining Structural and Sequential Approaches

Combining both structure-based and sequence-based methods is of notable interest. Integrating these approaches allows for the exploitation of their benefits while mitigating their respective challenges. Typically, sequence-based methods pinpoint functionally pertinent amino acid residues at the protein-protein interface. These findings then inform structure-based methodologies. The adrenergic and CCR5 chemokine receptors have been explored using this combined approach [23, 24].

An exemplary application of this method was demonstrated by de Juan et al. [24]. They employed a combination of sequence-based algorithms: correlated mutation analysis (CMA), sequence space automatization methods (SSAM), and level entropy method (LEM) to identify lipid-exposed determinants and correlated residues on the receptor surface. This information was subsequently incorporated into a structure-based strategy using GRAMM software, which proposed 100 protein-protein models.

12 Comparing Modeled Predictions

Once the structural and energetic optimization of the protein-protein complex is finalized, the accuracy of the prediction needs assessment. It's vital to employ the precision of the utilized algorithm. A prevalent method for gauging prediction accuracy is the root mean square deviation (RMSD) between the predicted and the crystal structure of the complex. This is achieved by overlaying the coordinates of the test protein's three-dimensional structure with the computed positions of the backbone atoms.

To measure the reliability and accuracy of the algorithms in a consistent and reproducible manner, a test set of proteins was formulated, known as the Docking Benchmark 1.0, consisting of 59 test cases. This benchmark utilizes known structures of enzyme inhibitors, antigen-antibody systems, and other complex instances. An advancement of this test is the second generation, termed Benchmark 2.0, which encompasses 84 test cases. Here, it's crucial to identify situations where the rigid body presumption is insufficient, necessitating the consideration of flexible adjustments to the primary structure [6].

Scoring criteria that offer a numerical representation of the modeled structure's similarity utilize seven parameters. These include van der Waals attraction and repulsion interactions, short-range and long-range interactions (both attraction and repulsion), and the desolvation effect, governed by the rules of the energy-based weighted function (ETW).

The CAPRI (critical assessment of prediction of interactions) initiative and the "Modeling of Protein Interactions in Genomes" experiment evaluate docking algorithms and serve as a platform to compare results across groups. These are blind contests wherein, in contrast to Benchmark evaluations, structures remain undisclosed until the commencement of the test. Participating groups present their top ten predictions, ranked from the most to the least credible. The submitted structures are then juxtaposed with an authentic model using distinctly defined parameters [25].

CAPRI models are categorized as highly accurate, medium accuracy, acceptable, or incorrect. This categorization hinges on a trio of measures encompassing native and non-native contacts for the prediction. These measures are the fraction of native residue-residue contacts (fnat), the root mean square deviation of all carbon atoms post protein alignment, the ligand backbone root mean square displacement in the predicted versus the target structures computed after aligning the receptors of these structures (L-rms), and the root mean square displacement of the backbone interface residues in the predicted against the target complexes (I-rms) [26].

CAPRI predictions are formulated devoid of knowledge regarding the complex structure. However, dockers aren't ultimately tasked with affirming their predictions since the experimental structure is subsequently disclosed by the coordinators and researchers. Contrarily, in genuinely blind predictions without an experimental structure, experimental validation is imperative for models [27].

The CAPRI experiment offers an equitable platform for research groups, fostering the continuous refinement of ever-accurate docking algorithms. The effectiveness of this approach is underscored by the fact that, between 2001 and 2010, merely 6 out of 42 protein-protein complexes weren't successfully modeled in the CAPRI experiment.

The multitude of docking methods in the CAPRI experiment, rooted in the principles described above, are vast. Notable ones are consolidated in Table 1 [25].

13 Docking Software

Below is a summary of several programs employed for molecular docking, as outlined in Table 2 [1].

These programs encompass search, sampling, and refinement stages for each evaluated protein-protein docking tool. Detailed procedures or operational descriptions for specific molecular docking programs are elaborated upon [6, 26]. Furthermore, examples of programs for normal mode analysis (NMA) and electrostatic calculations are presented [5].

SymmDock The search stage initiates with geometric docking, focusing on matching local shape attributes and geometric blending. Subsequent filtering using biological structure data determines the binding residue. The model's flexibility permits varying extents of steric collisions and main chain flexibility using hinge bending regions. Refinement and scoring stages revolve around shape complementarities based on computed parameters.

Haddock The process starts with randomization of orientation and rigid body energy minimization for initial structure searches. Furthermore, semi-rigid simulated annealing is executed in torsion angle space, followed by refinement in Cartesian space using an explicit solvent. Additional filtering of the modeled structures

Table 1
Docking methods in the CAPRI experiments

No.	Predictor	Program	Algorithm
1	Scripps, US (Abagyan)	ICM-DISCO	Rigid body pseudo-Brownian MC with grid-based E function; refinement via grid-based biased probability MC minimization using ICM
2	Boston U. US (Camacho)	SmoothDock	Clustering, refinement and discrimination protocol; constrained minimization using CHARMM; Coulomb and solvation
3	Weizmann Inst. IL (Einstein)	MolFit	Weighted geometric docking with FTT followed by clustering and filtering. Scoring by weighted geometric complementarity
4	Cancer Research UK LRI (Fitzjohn/Bates)	Guided docking	Rigid body molecular mechanics using CHARMM22 force field. Flexible refinement step in some cases. Solvation used in final E ranking
5	Sheffield U. UK (Gardiner)	GAPDOCK	GA for sampling different relative rotations and clustering. Scoring based on shape correlation and buried area, as per CCP4 package; some clash checks
6	U. Washington US (Gray/Baker)	–	Monte Carlo and clustering, side-chain repacking, rigid body minimization. Fitted multiterm scoring function, mainly vdW orientation-dependent hydrogen bonding; implicit solvation
7	Aberdeen U. UK (Mustard)	CONCORD + HEX	Dock 500 essential dynamics structures from CONCORD using HEX
8	Columbia U. US (Norel)	PPD	Geometric hashing; multiple rescoring
9	Scripps US (Olson/Norledge)	Surfdock	Fourier correlation of spherical harmonics. Scoring by electrostatic, hydrophobic, van der Waals (buried surface area) interactions and shape complementarity. Clashes determined by overlap of the docked surfaces
10	Universidade Nova De Lisboa (Palma/Krippahl)	BiGGER (Chemera)	Binary grids; scoring with geometric, contacts counts, electrostatics and solvation, no clash checks
11	Aberdeen U. UK (Ritchie)	HEX	Spherical polar Fourier correlation of shape and electrostatics, followed by soft rigid body OPLS minimization
12	North Western U. US (Shoichet)	Northwestern DOCK	Search for hot-spot correspondences between receptor and ligand to calculate orientations, precalculation of ensemble of ligand conformations, receptor held rigid, energy evaluation using electrostatics and van der Waals terms
13			

(continued)

Table 1
(continued)

No.	Predictor	Program	Algorithm
	Imperial Coll. UK (Sternberg)	3D-DOCK MULTIDOCK	FFT; rescore: residue potentials; flexible refinement: mean-field side-chain multicopy, solution clustering
14	UCSD US (Ten Eyck)	DOT	FFT for shape complementarity and Poisson-Boltzmann electrostatics
15	Kitasato U. JP (Umeyama/Komatsu)	TSCF	Solvent cluster fitting; refinement by molecular mechanics and molecular dynamics using AMBER force field
16	SUNY/StonyBrook, US (Tovchigrechko, Vakser)	GRAMM	FFT for shape complementarity; softer potential for conformational changes; no clash check; symmetry of multimeric receptors used to enhance sampling; docking of ligand to overlapping fragments of receptor for speed.
17	U. Autonoma Madrid, SP (Valencia)	–	Neural network-based predictions of interactions sites, using information on related sequences
18	Boston U. US (Weng)	ZDOCK	FFT with scoring by pairwise shape complementarity, solvation and electrostatics; clustering
19	Tel Aviv U., IL NIH, US	PatchDock; BUDDA; PPD	Geometric docking: matching of local shape features and geometric hashing, fast geometric scoring and search, avoids exhaustive orientation search

leverages chemical shift perturbation (CSP), residual dipolar coupling (RDC), and other NMR data, such as diffusion anisotropy relaxation or mutagenesis. Docking flexibility is influenced by structure-based and side-chain assemblies, coupled with symmetry constraints.

Simulated annealing molecular dynamics (MD) and steepest descent optimization focus on elastic side chains and spines. Scoring is achieved by aggregating the intermolecular interaction energy, buried surface area, and desolvation energy.

ZDOCK This program's scanning algorithm employs an FFT-based grid search for optimal translational configurations. The structure search phase begins with an FFT study of the rigid body, followed by the ZDOCK angular search exploring the rotational space of the ligand's rigid body using uniformly distributed sets of Euler angles. The procedure is as follows: parsing the receptor and ligand files, discretizing the receptor and ligand after rotation, calculating the FFT to correlate the ligand and structure for tested Euler angles, and documenting the grid point position with

Table 2
A summary of software and corresponding algorithms used in molecular docking

No.	Software	Search algorithm	Scoring function
1	ATTRACT	Randomized search	LJ-type effective potentials and electrostatics
2	PatchDock/ SymmDock	Local shape match	Geometric shape complementarity
3	FTDOCK	FFT-based corelation	Shape complementarity and electrostatic interactions
4	GRAMM	FFT-based corelation	Shape and hydrophobic match
5	GRAMMX	FFT-based corelation	Soft Lennard-Jones potential, evolutionary conservation of predicted interface, statistical residue-residue preference, volume of the minimum, empirical binding free energy and atomic contact energy
6	MolFit	FFT-based corelation	Geometric complementarity, hydrophobic complementarity, and electrostatic interactions
7	DOT	FFT-based corelation	van der Waals and electrostatic energies
8	ZDOCK	FFT-based corelation	Shape complementarity, electrostatics, and knowledge-based pair potentials
9	PIPER	FFT-based corelation	Shape complementarity, electrostatic interactions, and knowledge-based pair potentials
10	SDOCK	FFT-based corelation	van der Waals attractive potential, geometric collision, electrostatic potential, and desolvation energy
11	ClusPro	FFT-based corelation	Shape complementarity, electrostatics and desolvation energy
12	HEX	SFT-based corelation	Surface complementarity and electrostatics
13	FRODOCK	SFT-based corelation	van der Waals, electrostatics, and desolvation
14	HADDOCK	Randomization of orientations and rigid body energy minimization followed by semi-rigid simulated annealing in torsion angle space, and finally refinement in Cartesian space with explicit solvent	Sum of intermolecular van der Waals, electrostatic, ambiguous interaction restraint energy, the buried surface area and the desolvation energy

the maximum result. Biological structure data is then filtered to identify blocking residues. ZDOCK scoring has evolved, with different versions employing various parameters like the grid-based shape complementarity (GSC), desolvation, electrostatics, and pairwise shape complementarity function (PSC).

RDOCK Energy minimization is accomplished using the molecular dynamics program *CHARMM* via a three-tiered approach. The initial step resolves collisions through energy minimization. Next, polar interactions are optimized by minimizing electrostatic interactions and internal energy while preserving neutral ion side chains. Finally, charged interactions are optimized through total energy minimization sans constraints.

Other programs employ innovative techniques. For instance, *RosettaDock* linearizes the short-range LJ potential for its scoring function analyzing van der Waals potential. The *SCWRL* program employs graph theory to expedite the search for optimal side-chain configurations, while *DOT* accumulates data related to van der Waals energy and Poisson-Boltzmann electrostatics for refined predictions.

GROMACS, a molecular simulation software, is utilized for normal mode analysis (NMA). Boasting superior performance on individual processors, it can analyze extensive systems. Advancements in multi-program/data approaches, algorithmic enhancement, and hand-coded routines have elevated NMA program development, juxtaposed with renowned software like *NAMD* or *Amber* [28]. Simpler NMA servers include *ElNemo* [29] and *NOMAD-Ref* [30]. ElNemo provides a user-friendly interface for ENM, offering comprehensive analyses with no size restrictions on proteins. Conversely, NOMAD-Ref allows extensive molecule normal mode calculations, presenting holistic, all-atom biomolecular structure representations.

APBS is a renowned program for electrostatic calculations, modeling biomolecular solvation via the Poisson-Boltzmann equation. Its extension, *APBSmem*, facilitates these computations in membrane-associated research and is widely accessible at no cost [31].

Acknowledgments

This research was funded under OPUS grant from National Science Center (NCN, Poland), grant number UMO-2021/43/B/NZ7/01732.

References

1. Kaczor AA, Bartuzi D, Stępniewski TM et al (2018) Protein–protein docking in drug design and discovery. Methods Mol Biol 1762:285–305
2. Selent J, Kaczor AA (2011) Oligomerization of G protein-coupled receptors: computational methods. Curr Med Chem 18(30):4588–4605
3. Wodak SJ, Janin J (1978) Computer analysis of protein-protein interaction. J Mol Biol 124(2): 323–342
4. Janin J (2010) Protein–protein docking tested in blind predictions: the CAPRI experiment. Mol BioSyst 6(12):2351
5. Kaczor AA, Selent J, Poso A (2013) Chapter 5: Structure-based molecular modeling approaches to GPCR oligomerization. In: Conn PM (ed) Methods in cell biology, vol 117. Academic Press, pp 91–104
6. Pierce B, Weng Z (2007) Structure prediction of protein complexes. In: Xu Y, Xu D, Liang J (eds) Computational methods for protein structure prediction and modeling, Biological and medical physics, biomedical engineering. Springer, New York
7. Ritchie D (2008) Recent progress and future directions in protein-protein docking. Curr Protein Pept Sci 9(1):1–15
8. Schueler-Furman O, Wang C, Bradley P et al (2005) Progress in modeling of protein structures and interactions. Science 310(5748): 638–642
9. Park PS-H (2012) Ensemble of G protein-coupled receptor active states. Curr Med Chem 19(8):1146–1154
10. Bruno A, Guadix AE, Costantino G (2009) Molecular dynamics simulation of the heterodimeric mGluR2/5HT_{2A} complex. An atomistic resolution study of a potential new target in psychiatric conditions. J Chem Inf Model 49(6):1602–1616
11. González-Maeso J, Ang RL, Yuen T et al (2008) Identification of a serotonin/glutamate receptor complex implicated in psychosis. Nature 452(7183):93–97
12. Dror RO, Pan AC, Arlow DH et al (2011) Pathway and mechanism of drug binding to G-protein-coupled receptors. Proc Natl Acad Sci 108(32):13118–13123
13. Periole X, Huber T, Marrink S-J et al (2007) G protein-coupled receptors self-assemble in dynamics simulations of model bilayers. J Am Chem Soc 129(33):10126–10132
14. Spijker P, Van Hoof B, Debertrand M et al (2010) Coarse grained molecular dynamics simulations of transmembrane protein-lipid systems. Int J Mol Sci 11(6):2393–2420
15. Davis IW, Murray LW, Richardson JS et al (2004) MOLPROBITY: structure validation and all-atom contact analysis for nucleic acids and their complexes. Nucleic Acids Res 32: W615–W619
16. Taylor MS, Fung HK, Rajgaria R et al (2008) Mutations affecting the oligomerization interface of G-protein-coupled receptors revealed by a novel de novo protein design framework. Biophys J 94(7):2470–2481
17. Lichtarge O, Bourne HR, Cohen FE (1996) An evolutionary trace method defines binding surfaces common to protein families. J Mol Biol 257(2):342–358
18. Göbel U, Sander C, Schneider R et al (1994) Correlated mutations and residue contacts in proteins. Proteins Struct Funct Genet 18(4): 309–317
19. Filizola M, Weinstein H (2002) Structural models for dimerization of G-protein coupled receptors: the opioid receptor homodimers. Biopolymers 66(5):317–325
20. Nemoto W, Toh H (2006) Membrane interactive alpha-helices in GPCRs as a novel drug target. Curr Protein Pept Sci 7(6):561–575
21. Reggio PH (2006) Computational methods in drug design: modeling G protein-coupled receptor monomers, dimers, and oligomers. AAPS J 8(2):E322–E336
22. Nemoto W, Toh H (2004) Prediction of interfaces for oligomerizations of G-protein coupled receptors. Proteins 58(3):644–660
23. Gouldson PR, Snell CR, Bywater RP et al (1998) Domain swapping in G-protein coupled receptor dimers. Protein Eng Des Sel 11(12):1181–1193
24. de Juan D, Mellado M, Rodriguez-Frade JM et al (2005) A framework for computational and experimental methods: identifying dimerization residues in CCR chemokine receptors. Bioinformatics 21(2):ii13–ii18
25. Méndez R, Leplae R, De Maria L et al (2003) Assessment of blind predictions of protein-protein interactions: current status of docking methods. Proteins 52(1):51–67
26. Kaczor AA, Selent J, Sanz F et al (2013) Modeling complexes of transmembrane proteins: systematic analysis of protein-protein docking tools. Mol Inform 32(8):717–733
27. Gray JJ (2006) High-resolution protein–protein docking. Curr Opin Struct Biol 16(2): 183–193

28. Hess B, Kutzner C, van der Spoel D et al (2008) GROMACS 4: algorithms for highly efficient, load-balanced, and scalable molecular simulation. J Chem Theory Comput 4(3): 435–447
29. Suhre K, Sanejouand Y-H (2004) ElNemo: a normal mode web server for protein movement analysis and the generation of templates for molecular replacement. Nucleic Acids Res 32(2):W610–W614
30. Lindahl E, Azuara C, Koehl P et al (2006) NOMAD-ref: visualization, deformation and refinement of macromolecular structures based on all-atom normal mode analysis. Nucleic Acids Res 34(2):W52–W56
31. Callenberg KM, Choudhary OP, de Forest GL et al (2010) APBSmem: a graphical interface for electrostatic calculations at the membrane. PLoS One 5(9):e12722

Chapter 18

In Silico Design, Synthesis, and Evaluation of PROTAC Against Hematopoietic Prostaglandin D Synthase

Hidetomo Yokoo, Norihito Shibata, and Yosuke Demizu

Abstract

Chemical protein knockdown technology using proteolysis-targeting chimeras (PROTACs) to hijack the endogenous ubiquitin-proteasome system is a powerful strategy to degrade disease-related proteins. This chapter describes in silico design of a hematopoietic prostaglandin D synthase (H-PGDS) degrader, PROTAC(H-PGDS), using a docking simulation of the ternary complex of H-PGDS/PROTAC/E3 ligase as well as the synthesis of the designed PROTAC(H-PGDS)s and evaluation of their H-PGDS degradation activity.

Key words In silico molecular design, Hematopoietic prostaglandin D synthase, PROTAC, Protein knockdown, Ubiquitin-proteasome system

1 Introduction

Proteolysis-targeting chimeras (PROTACs), which use the ubiquitin-proteasome system, have recently attracted attention as a chemical protein knockdown method [1, 2]. PROTACs are expected to be an innovative drug discovery strategy for diseases that are difficult to treat with existing drugs. Generally, a PROTAC is a bifunctional molecule consisting of two small-molecule ligands connected with a linker; one ligand binds to the protein of interest (POI) and the other binds to an E3 ubiquitin ligase. To discover highly active PROTACs, combinations of the POI ligand and the E3 ligase ligand as well as the appropriate linker need to be investigated. Furthermore, analysis of the ternary structure of the POI/PROTAC/E3 ligase is useful for the rational design and optimization of PROTACs. We have recently succeeded in developing **PROTAC (H-PGDS)-7** (**26**) against hematopoietic prostaglandin D synthase (H-PGDS) using a docking simulation of a ternary complex of H-PGDS/PROTAC/cereblon (CRBN, which is an E3 ligase) [3]. **PROTAC(H-PGDS)-7** (**26**) showed potent and

Agnieszka A. Kaczor (ed.), *Protein-Protein Docking: Methods and Protocols*, Methods in Molecular Biology, vol. 2780,
https://doi.org/10.1007/978-1-0716-3985-6_18,
© The Author(s), under exclusive license to Springer Science+Business Media, LLC, part of Springer Nature 2024

selective degradation activity of H-PGDS protein in KU812 cells. In this chapter, we describe the experimental methods to create a PROTAC with a focus on: (1) in silico design of PROTAC (H-PGDS)s, (2) synthesis of PROTAC(H-PGDS)s, and (3) H-PGDS protein degradation.

1.1 In Silico Design of PROTAC(H-PGDS)

Ligand docking simulation of TFC-007 [4], which is an H-PGDS inhibitor used in PROTAC(H-PGDS) design, is conducted using a co-crystal structure of the H-PGDS protein and the H-PGDS inhibitor F092 [5]. Docking simulation of the ternary complex of **PROTAC(H-PGDS)s**, H-PGDS and CRBN is performed using the docking software suite MOE PROTAC-Modeling Tools [6, 7] in which the calculation method is implemented. Molecular dynamics (MD) simulations are subsequently calculated using the stable structure obtained from the docking simulation.

1.2 Synthesis of PROTAC(H-PGDS)s

PROTAC(H-PGDS)s are composed of two biologically active scaffolds: one is a TFC-007 derivative binding to the H-PGDS, and the other is a pomalidomide binding to CRBN to induce ubiquitylation of the H-PGDS. The chimeric molecules **PROTAC(HPGDS)-1** (**21**), **-3** (**22**), **-4** (**23**), **-5** (**24**), **-6** (**25**), **-7** (**26**), and **-9** (**31**) with different lengths of linkers are synthesized.

1.3 H-PGDS Protein Degradation

The protein degradation activity of **PROTAC(H-PGDS)s** is evaluated by western blot analysis. The detailed protocols for western blot analysis are described below.

2 Materials

2.1 In Silico Design of PROTAC(H-PGDS)

1. Polyethylene glycol (PEG).
2. MOE (Molecular Operating Environment) (MOLSIS Inc., Tokyo, Japan).
3. MOE PROTAC-Modeling Tools (MOLSIS Inc.).
4. Co-crystal structure of the hematopoietic prostaglandin D synthase (H-PGDS) protein and F092 (PDB: 5YWX).
5. Co-crystal structure of the cereblon (CRBN) protein and pomalidomide (PDB: 4CI3).

2.2 Synthesis of PROTAC(H-PGDS)s

1. 1-(4-Nitrophenyl)-4-piperidinecarboxylic acid (**1**).
2. 1-(3-Dimethylaminopropyl)-3-ethylcarbodiimide (EDC).
3. 1-Hydroxybenzotriazole (HOBt).
4. *N*,*N*-Dimethylformamide (DMF).
5. 1-(*tert*-Butoxycarbonyl)piperazine.
6. Sodium hydrogen carbonate ($NaHCO_3$).

7. Sodium sulfate (Na_2SO_4).
8. Brine (ca. 25% aqueous NaCl).
9. Palladium on carbon (Pd/C).
10. Hydrogen gas (H_2).
11. Ethanol (EtOH).
12. 2-Phenoxypyrimidine-5-carboxylic acid.
13. Trifluoroacetic acid (TFA).
14. Dichloromethane (CH_2Cl_2).
15. Methanol (MeOH).
16. Potassium carbonate (K_2CO_3).
17. Ethyl acetate (EtOAc).
18. NH-silica gel 60 N, spherical, particle size 40 μm.
19. 2-(1*H*-Benzotriazole-1-yl)-1,1,3,3-tetramethyluronium hexafluorophosphate (HBTU).
20. *N*,*N*-Diisopropylethylamine (DIPEA).
21. Acetonitrile (MeCN).
22. Diethylamine (Et_2NH).
23. 4-Fluoro-thalidomide.
24. 4-Fluoronitrobenzene.
25. Tetrahydrofuran (THF).
26. Silica gel 60 N, spherical, particle size 40–100 μm.

2.3 H-PGDS Protein Degradation

1. 6-well culture plate.
2. Human chronic myelogenous leukemia KU812 cells [8] (JCRB0104, Japanese Collection of Research Bioresources).
3. Roswell Park Memorial Institute-1640 (RPMI-1640) complete medium: RPMI-1640 medium, 10% heat-inactivated fetal bovine serum, and 100 μg/mL kanamycin.
4. Phosphate-buffered saline (PBS): 2.68 mM KCl, 1.17 mM KH_2PO_4, 137 mM NaCl, and 8.10 mM Na_2HPO_4.
5. SDS lysis buffer: 0.1 M Tris–HCl at pH 7.5, 10% glycerol, and 1% SDS.
6. 4 × SDS-PAGE sample loading buffer: 0.4 M Tris–HCl at pH 7.5, 40% glycerol, 4% SDS, and 0.04% Bromophenol blue.
7. Running buffer: 25 mM Tris, 192 mM glycine, and 0.1% SDS.
8. Dithiothreitol (DTT).
9. Pierce™ BCA protein assay kit (Thermo Fisher Scientific, Waltham, MA, USA).
10. Polyacrylamide gel, 5–12.5% (DRC Co., Ltd., Tokyo, Japan).

11. Electrophoresis apparatus (ATTO, Tokyo, Japan).
12. Transfer buffer: 25 mM Tris, 192 mM glycine, and 20% MeOH.
13. PDVF membrane (0.45 μm).
14. Filter paper (Advantec, Chiba, Japan).
15. Transfer apparatus for western blotting including power supply (Bio-Rad, Hercules, CA, USA).
16. Blocking buffer: 5% skim milk, 50 mM Tris–HCl pH 7.5, 150 mM NaCl, and 0.05% Tween 20.
17. Anti-H-PGDS rabbit polyclonal antibody [9].
18. Anti-β-actin mouse monoclonal antibody (Sigma-Aldrich, St. Louis, MO, USA).
19. TTBS buffer: 50 mM Tris–HCl pH 7.5, 150 mM NaCl, and 0.05% Tween 20.
20. Anti-rabbit IgG, HRP-linked Whole Ab Donkey (Cytiva, Tokyo, Japan).
21. Anti-mouse IgG, HRP-linked Whole Ab Sheep (Cytiva).
22. CanGet Signal (Toyobo, Tokyo, Japan) (*see* **Note 1**).
23. Clarity Western ECL substrate (Bio-Rad).
24. LAS-3000 (Fujifilm, Tokyo, Japan).

3 Methods

3.1 *In Silico* Design of PROTAC(H-PGDS) (Figs. 1 and 2)

1. Conduct ligand docking simulation of TFC-007 using a co-crystal structure of the H-PGDS protein and F092 using the docking software suite MOE in which the calculation method is implemented (*see* **Note 2**).
2. Calculate the ternary complex of **PROTAC(H-PGDS)s**, H-PGDS and CRBN derived from co-crystal structures of H-PGDS-TFC-007 and CRBN-pomalidomide with various linkers using the docking software suite MOE PROTAC-Modeling Tools in which the calculation method is implemented.
3. Select docking poses of the ternary complex with criteria such as RMSD < 2.5 Å, RMSD_RF_PP < 670 Å^2, ligand_E < 3600 kcal/mol and ligand_E_PP < 670 Å^2.
4. Produce a two-dimensional interaction model with criteria such as H-bond < −0.5 kcal/mol, ionic < −0.5 kcal/mol, and distance >4.5 Å.
5. Add glutathione to the conformation from the docking simulation in the same pose as the crystal structure (PDB: 5YWX).

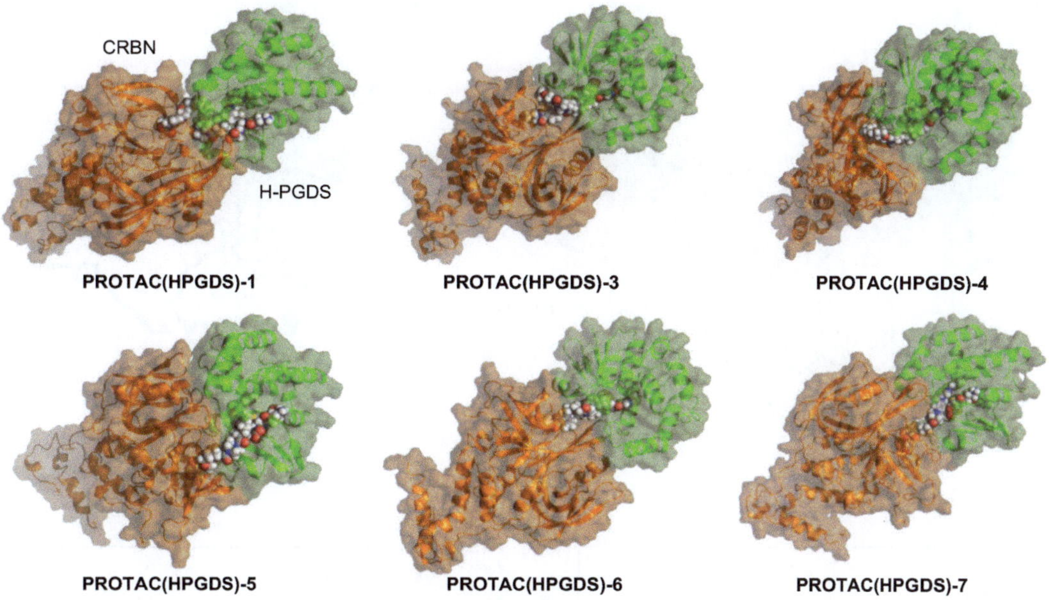

Fig. 1 Top pose of ternary complexes of **PROTAC(HPGDS)-1** (**21**), **-3** (**22**), **-4** (**23**), **-5** (**24**), **-6** (**25**), and **-7** (**26**). CRBN and H-PGDS is shown in orange and green, respectively. These complexes contained GSH in the ligand binding cavity of H-PGDS. (This figure was modified from Yokoo et al. [3])

6. Add water molecules and neutralize the system by the addition of NaCl at 0.1 mol/L.
7. Use a periodic boundary condition to carry out the simulation. In all cases, the cell size is larger than the protein by 10 Å. Control the temperature of the simulation by the NAMD [10] integrated in MOE.
8. Perform the initial energy minimization process of each simulation by standard protocols in MOE. Maintain the simulation speed at a 2 fs time step during all the simulations.
9. Save the system every 0.5 ps during the simulations. First, perform a 10 ps time evolution with a temperature 0 K (minimization step).
10. Next, perform a 100 ps time evolution with a temperature of 10–300 K and a tether of 0.5–100 Å (equilibration step).
11. Then, perform a 100 ps time evolution with a temperature of 300 K (production step).
12. After completion of the MD simulation for all systems, analyze the results in MOE. For each result, select the most stable structure in the production step and calculate the ligand interactions and potential energy.

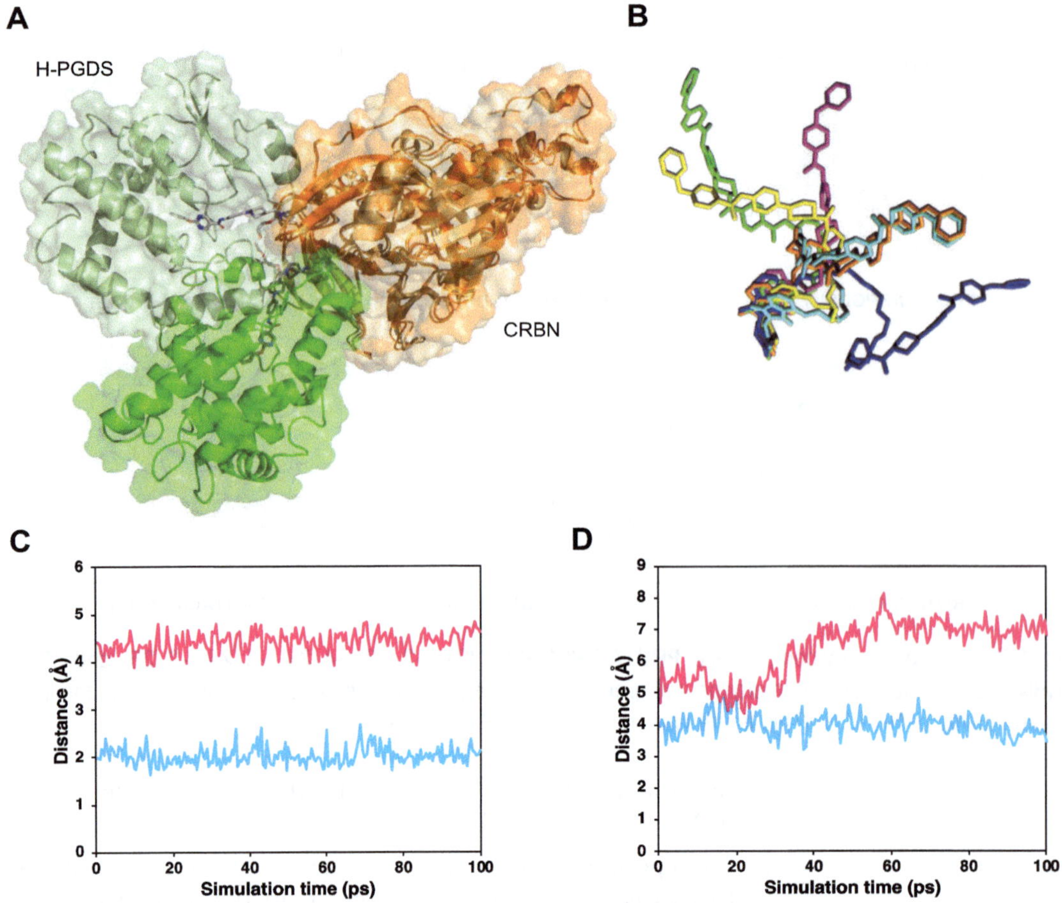

Fig. 2 (**a**) Overlay of two ternary complexes. A ternary complex of CRBN (dark orange), H-PGDS (dark green) and **PROTAC(HPGDS)-1** and a ternary complex of CRBN (light orange), H-PGDS (light green), and **PROTAC (HPGDS)-7** are aligned at each CRBN. (**b**) Ligands of ternary complexes such as **PROTAC(HPGDS)-1** (blue), **PROTAC(HPGDS)-3** (cyan), **PROTAC(HPGDS)-4** (green), **PROTAC(HPGDS)-5** (yellow), **PROTAC(HPGDS)-6** (orange), and **PROTAC(HPGDS)-7** (magenta). (**c**, **d**) The distance between the imide moiety of pomalidomide and the H380 of CRBN (**c**) or the pyrimidine moiety of TFC-007 and the W104 of H-PGDS (**d**) calculated from the simulations of the ternary complexes containing **PROTAC(HPGDS)-7** (cyan) and **PROTAC(HPGDS)-9** (magenta). (This figure was modified from Yokoo et al. [3])

3.2 Design and Synthesis of PROTAC (HPGDS)-1 (21), -3 (22), -4 (23), -5 (24), -6 (25), -7 (26), and -9 (31) (Fig. 3)

1. Preparation of compound **2**: Stir a mixture of EDC (7.2 mmol), HOBt (6.6 mmol), 1-(4-nitrophenyl)-4-piperidinecarboxylic acid (**1**; 6.0 mmol) and 1-(*tert*-butoxycarbonyl) piperazine (9.0 mmol) in DMF (8.0 mL) at room temperature for 6 h. Pour the reaction mixture into water, then collect the precipitate by filtration and concentrate in vacuo to obtain compound **2**, and use the next reaction without further purification.
2. Preparation of compound **3**: Stir the above residue **2** and 5% palladium on carbon (91.2 mg) in EtOH (60 mL) under a hydrogen atmosphere at room temperature for 12 h. Filter the

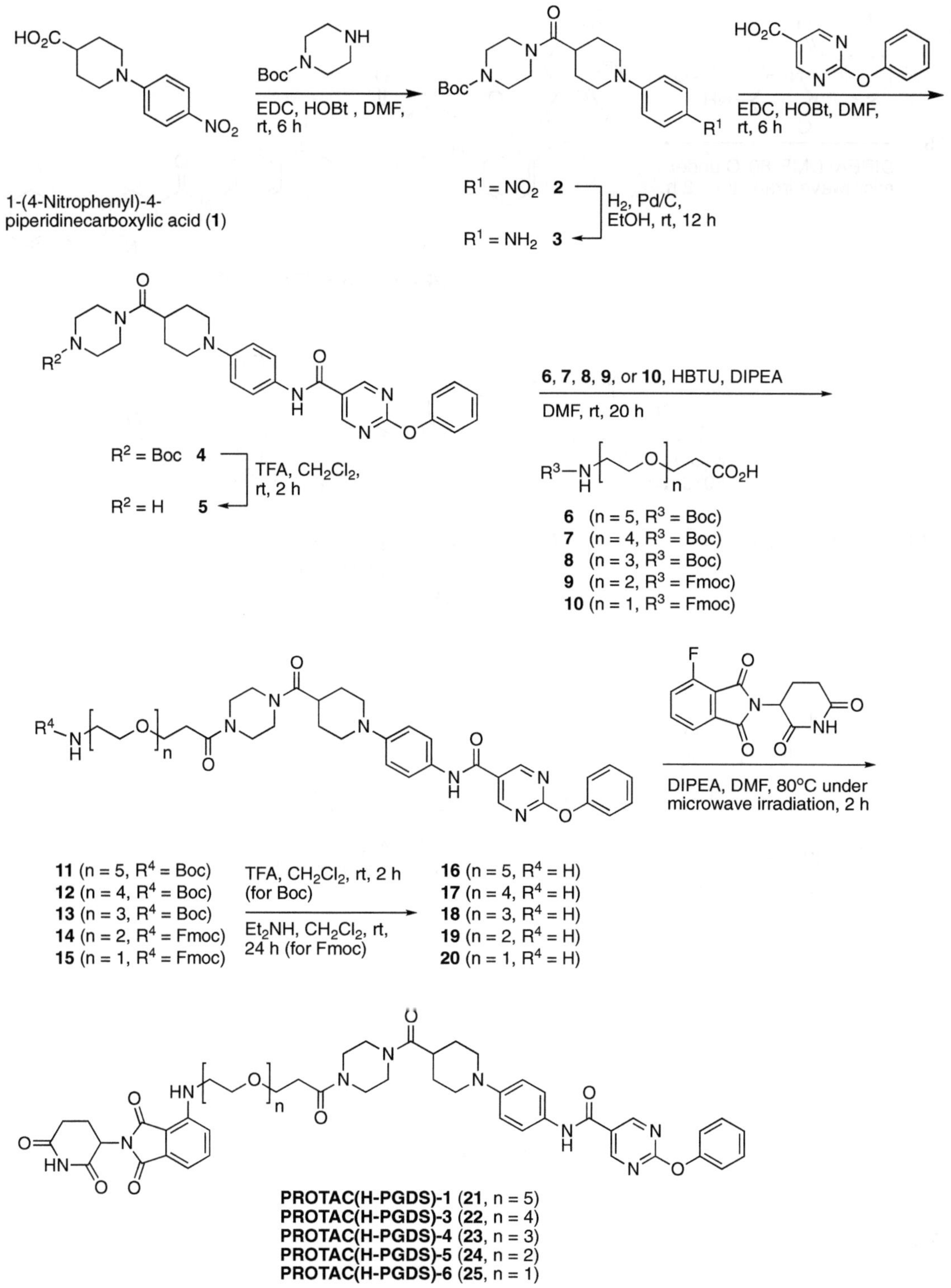

Fig. 3 Synthesis of **PROTACs(H-PGDS)**

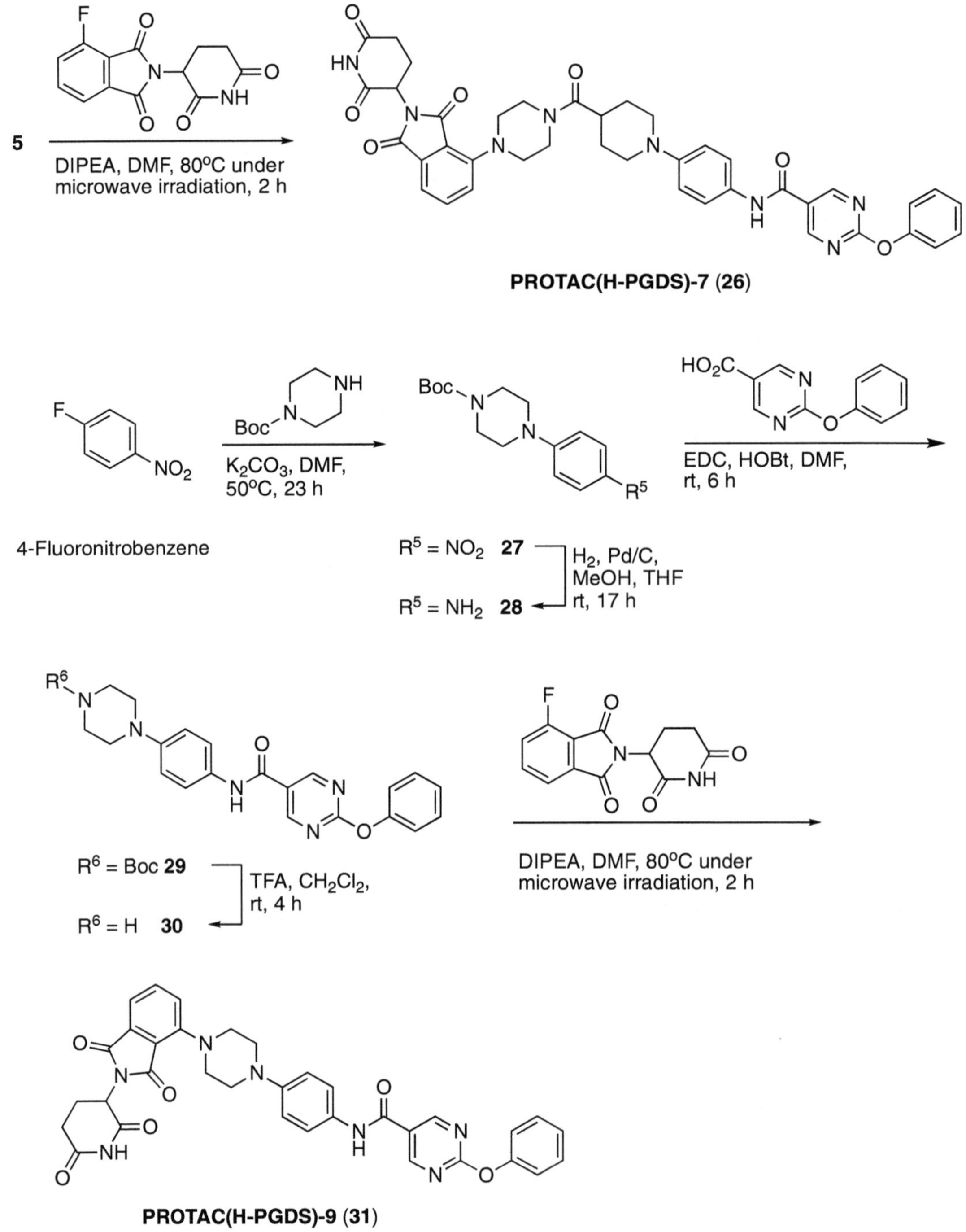

Fig. 3 (continued)

reaction mixture and concentrate to give compound **3** (92% from starting compound **1**) as a purple solid.

3. Preparation of compound **4**: Stir a mixture of **3** (5.4 mmol), 2-phenoxypyrimidine-5-carboxylic acid (5.4 mmol), EDC (6.5 mmol) and HOBt (6.0 mmol) in DMF (20 mL) at room temperature for 6 h. Pour the reaction mixture into water and collect the precipitate by filtration to give **4** (36%) as a gray solid.
4. Preparation of compound **5**: Stir a mixture of **4** (0.04 mmol) and TFA (0.4 mmol) in CH_2Cl_2 (1.0 mL) at room temperature for 2 h. Concentrate the reaction mixture and purify the residue by column chromatography (NH silica gel, CH_2Cl_2/MeOH = 10/1) to give **5** (>99%) as a colorless oil.
5. Preparation of compounds **11**, **12**, **13**, **14**, and **15**: Stir a mixture of **5** (0.04 mmol), **6**, **7**, **8**, **9**, or **10** (0.04 mmol), HBTU (0.07 mmol) and DIPEA (0.09 mmol) in DMF (2.0 mL) at room temperature for 20 h. Suspend the reaction mixture with CH_2Cl_2 and wash with water and saturated aqueous $NaHCO_3$. Dry the organic phase with Na_2SO_4 and then concentrate in vacuo. Purify the resulting residue by silica gel column chromatography (CH_2Cl_2/MeOH = 19/1 to 9/1) to give **11** (26%), **12** (47%), **13** (46%), **14** (34%), or **15** (92%) as a brownish oil.
6. Preparation of compounds **16**, **17**, and **18**: Stir a mixture of **11**, **12**, or **13** (0.04 mmol) and TFA (0.4 mmol) in CH_2Cl_2 (1.0 mL) at room temperature for 2 h. Concentrate the reaction mixture and use the residue for the next reaction without further purification.
7. Preparation of compounds **19** and **20**: Add Et_2NH/CH_2Cl_2 (1/3, v/v) to a solution of compound **14** or **15** in CH_2Cl_2. Stir the reaction mixture for 24 h at room temperature. Concentrate the reaction mixture in vacuo and then purify the residue by a short silica gel column (NH silica gel, CH_2Cl_2/MeOH = 1/0 to 19/1) to give **19** or **20** as a brownish oil.
8. Preparation of compounds **PROTAC(HPGDS)-1 (21)**, **-3 (22)**, **-4 (23)**, **-5 (24)**, and **-6 (25)**: Add 4-fluoro-thalidomide (0.06 mmol) and DIPEA (0.12 mmol) to a solution of compounds **16**, **17**, **18**, **19**, or **20** (0.04 mmol) in DMF (0.2 mL). Stir the reaction mixture for 40 min at 80 °C under microwave irradiation and for an additional 42 h at 50 °C. Cool the reaction mixture to room temperature, filter through a membrane filter and then purify the residue by preparative HPLC (gradient 30–60% MeCN-H_2O containing 0.1% TFA in 40 min) to give **21** (5% from **11**), **22** (3% from **12**), **23** (12% from **13**), **24** (4% from **14**), or **25** (24% from **15**) as a yellow solid.

9. Preparation of compound **PROTAC(HPGDS)-7 (26)**: Add 4-fluoro-thalidomide (0.06 mmol) and DIPEA (0.12 mmol) to a solution of compound **5** (0.04 mmol) in DMF (0.2 mL). Stir the reaction mixture at 80 °C for 2 h under microwave irradiation. Cool the reaction mixture to room temperature, filter through a membrane filter and then purify the residue by preparative HPLC (gradient 30–60% MeCN-H_2O containing 0.1% TFA in 40 min) to give **26** (35%) as a yellow solid.
10. Preparation of compound **27**: Add K_2CO_3 (2.4 mmol) to a solution of 4-fluoronitrobenzene (1.78 mmol) and 4-Boc-piperadine (300 mg, 1.61 mmol) in DMF (5 mL). Stir the mixture for 23 h at 50 °C. After cooling to room temperature, dilute the reaction mixture with EtOAc and then wash with water and brine. Dry the organic layer with Na_2SO_4 and concentrate in vacuo to give **27** (328 mg, 66%) as a yellow solid.
11. Preparation of compound **28**: Add 10% Pd on carbon (70 mg) to a solution of compound **27** (1.0 mmol) in MeOH (10 mL) and THF (2 mL). Stir the solution under a H_2 atmosphere for 17 h at room temperature. Filter the reaction mixture through a celite pad and concentrate the filtrate in vacuo. Purify the resulting residue by silica gel column chromatography (EtOAc/*n*-hexane = 2/3 to 3/2) to give compound **28** (87%) as a purple solid.
12. Preparation of compound **29**: Add 2-phenoxypyrimidine-5-carboxylic acid (0.9 mmol), EDC (0.9 mmol), and HOBt (0.9 mmol) to a solution of compound **28** (0.8 mmol) in DMF (4 mL). Stir the reaction mixture for 6 h at room temperature. Pour water into the reaction mixture and collect the resulting precipitate by filtration. Wash the precipitate with water and dry under vacuum to give compound **29** (71%) as a gray solid.
13. Preparation of compound **30**: Add TFA (0.5 mL) to a solution of compound **29** (0.1 mmol) in CH_2Cl_2 (1 mL). After stirring for 4 h at room temperature, concentrate the solvent and purify the resulting residue by silica gel column chromatography (NH silica gel, CH_2Cl_2/MeOH = 1/0 to 9/1) to give compound **30** (97%) as a light brown solid.
14. Preparation of compound **PROTAC(HPGDS)-9 (31)**: Add 4-fluoro-thalidomide (0.03 mmol) and DIPEA (0.3 mmol) to a solution of compound **30** (0.01 mmol) in DMF (0.2 mL). Stir the reaction mixture at 80 °C for 2 h under microwave irradiation. Cool the reaction mixture to room temperature, filter through a membrane filter and then purify the residue by preparative HPLC (gradient 30–60% MeCN-H_2O containing 0.1% TFA in 40 min) to give **31** (17%) as a yellow solid.

3.3 H-PGDS Protein Degradation (Fig. 4)

1. Culture KU812 cells in RPMI-1640 complete medium.
2. Seed cells at a density of 400,000 cells per well in a 6-well plate and treat the cells with the **PROTAC(H-PGDS)**s for 6 or 24 h.
3. Collect the cells in 1.5 mL tubes and centrifuge at 2,000 rpm for 5 min.
4. Aspirate the medium from the cell pellet and add 1 mL of cold PBS.
5. Centrifuge the cells at 2,000 rpm for 5 min, aspirate the PBS from the cell pellet, and add 100 μL of SDS lysis buffer.
6. Vortex the sample to dissolve the cell pellet and boil the sample at 100 °C for 5 min.
7. Place the cell lysate on ice for 1 min and subsequently spin down at 10,000 rpm at 4 °C for 5 min to remove undissolved debris.
8. Transfer the supernatant to a new 1.5 mL tube.
9. Determine protein concentrations by the BCA method [11] using the BCA protein assay kit (*see* **Note 3**). In brief, dilute the cell lysate or BSA standard, add working reagent for the BCA assay to each well in a 96-well plate, mix briefly, and incubate at 37 °C for 30 min. Measure the absorbance at 560 nm, using a microplate reader.
10. Dilute equal amounts of protein from each cell lysate to equal volumes with SDS lysis buffer. Then, add 4 × SDS-PAGE sample loading buffer and DTT (final concentration 50 mM) to the samples.
11. Vortex the samples to mix the contents and boil the samples at 100 °C for 5 min.
12. Mount a polyacrylamide gel in the electrophoresis apparatus and pour running buffer into the upper and lower chambers of the electrophoresis apparatus.
13. Load the samples on the polyacrylamide gel and separate the samples at 200 V until the dye (bromophenol blue) in the sample loading buffer reached the bottom of the polyacrylamide gel.
14. After SDS-PAGE, place a PVDF membrane (*see* **Note 4**) on the gel, to which the separated proteins in the gel are electrophoretically transferred at 4 °C for 3 h at 70 V in transfer buffer.
15. Gently shake the membrane with blocking buffer at room temperature for 1 h.
16. Gently shake the membrane with CanGet Signal buffer A containing anti-H-PGS rabbit polyclonal antibody (1/1,000 dilution) or the blocking buffer containing anti-β-actin mouse

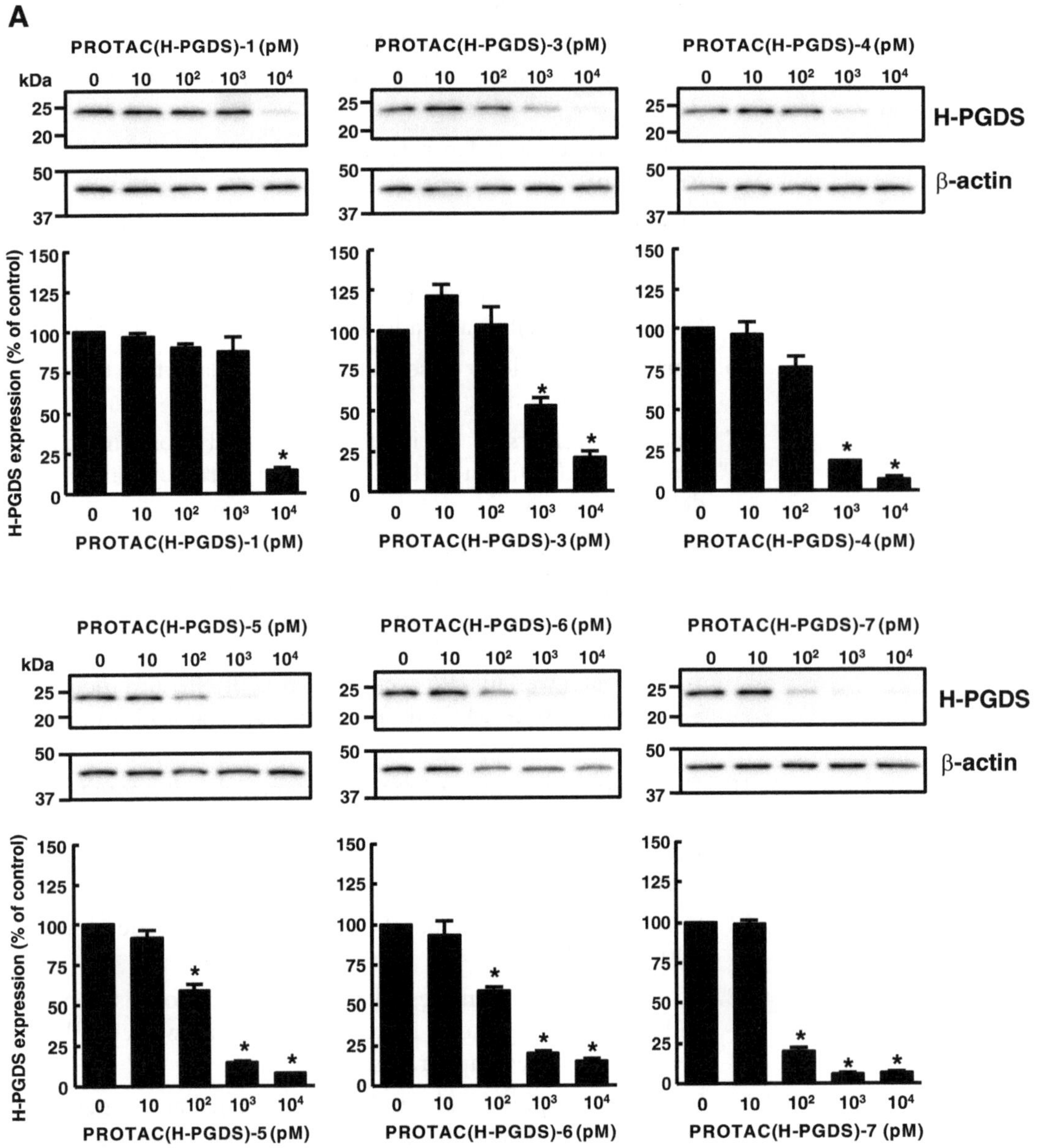

Fig. 4 Evaluation of the H-PGDS protein degradation activity of PROTAC (H-PGDS)s. KU812 cells are incubated with the indicated concentration of compounds for 6 h (**a**, **b**, and **d**), and 24 h (**c**). The H-PGDS/β-actin ratios are normalized by the vehicle control to 100. Data in the bar graph are the means ± SEM ($n = 3$). $*P < 0.05$ compared with the DMSO-treated control in a Dunnett's test. (This figure was modified from Yokoo et al. [3])

monoclonal antibody (1/2,000 dilution) at room temperature for 2 h.

17. Remove primary antibodies and wash the membrane four times with TTBS buffer for 5 min each.

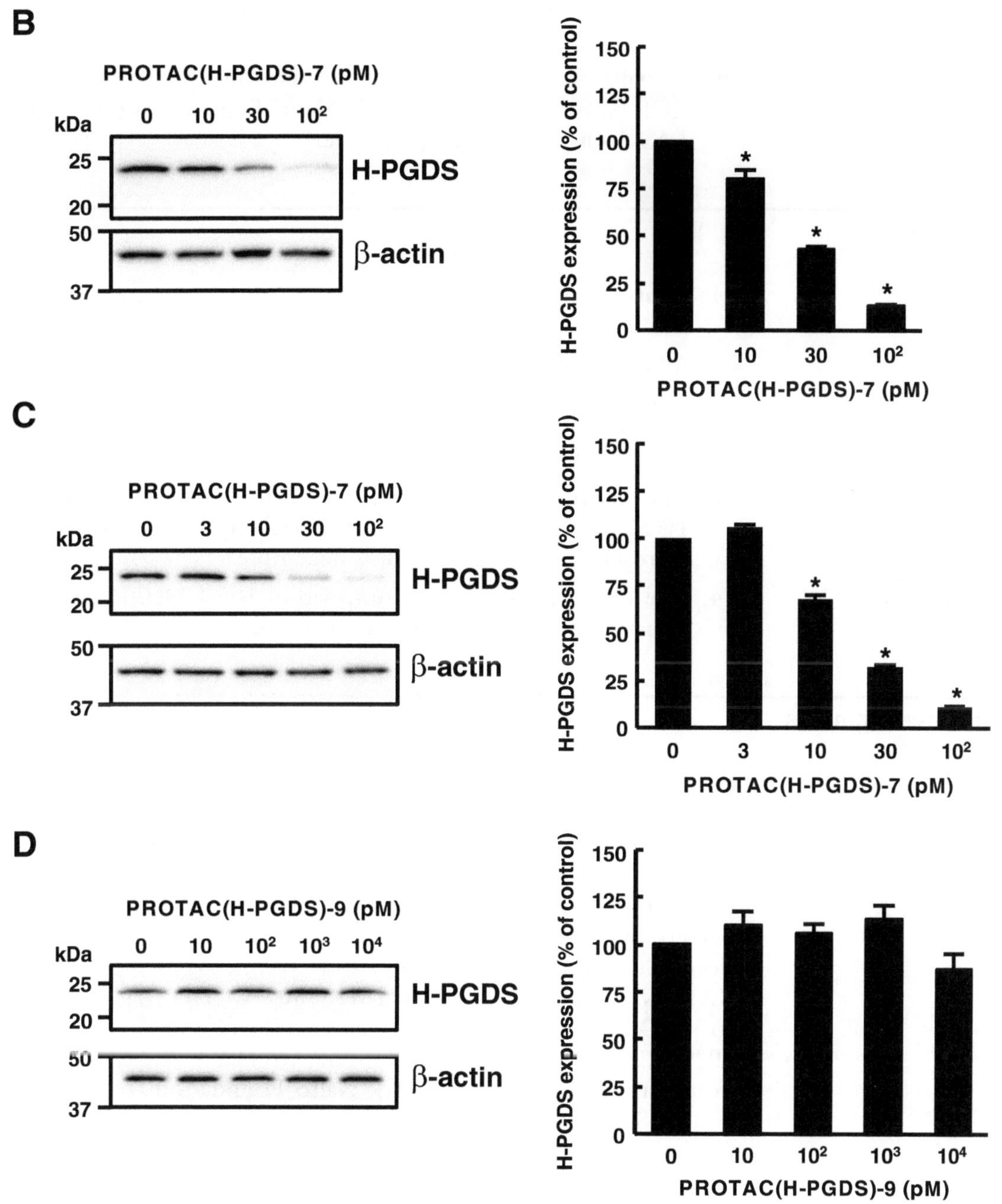

Fig. 4 (continued)

18. Gently shake the membrane with CanGet Signal buffer B containing anti-rabbit IgG linked HRP (1/5,000 dilution) or the blocking buffer containing anti-mouse IgG linked HRP (1/5,000 dilution) at room temperature for 1 h.

19. Remove secondary antibodies and wash the membrane four times with TTBS buffer for 5 min each.
20. Visualize the immunoreactive proteins using Clarity Western ECL substrate and quantify their light emission with a LAS-3000 lumino-image analyzer.

4 Notes

1. CanGet Signal can enhance the signal detected by the antibodies.
2. All the simulations and analysis are calculated using MOE (ver 2020.09, MOLSIS Inc.) with the Amber10: EHT force field. The docking software suite MOE PROTAC-Modeling Tools is used for the calculation of the ternary complexes [6, 7].
3. The reagent in the BCA protein assay kit reacts somewhat with 1% SDS, but the absorbance of the SDS lysis buffer generated by reacting with the reagent is negligible.
4. The PVDF membrane is immersed in 100% MeOH and then transferred into transfer buffer before use.

Acknowledgments

We would like to express our deepest appreciation to Dr. Kentaro Kamiya (MOLSIS Inc.) for his help with the PROTAC modeling and MD simulation. This study was supported in part by grants from the Japan Agency for Medical Research and Development (22mk0101197, 22fk0210110, and 22fk0310506 to Y.D.), Japan Society for the Promotion of Science and the Ministry of Education, Culture, Sports, Science and Technology (JSPS/MEXT KAKENHI Grant Number JP18H05502 to Y.D.; JP21K05320 to Y.D.; JP18K07311 to N.S.; JP21K06075 to N.S.; JP20K22711 to H.Y.; JP22K15257 to H.Y.), Takeda Science Foundation (to Y.D. and H.Y.), Naito Foundation (to Y.D. and H.Y.), Sumitomo Foundation (to Y.D.), Novartis Foundation (Japan) for the Promotion of Science (to Y.D.) and Chugai Foundation for Innovative Drug Discovery Science: C-FINDs (to H.Y.). We are grateful to the *Journal of Medicinal Chemistry*, American Chemical Society, for allowing the reproduction of Figures 2, S1, S4 and S11, which were originally published in the *Journal of Medicinal Chemistry* [3]. We thank Renee Mosi, PhD, from Edanz (https://jp.edanz.com/ac) for editing a draft of this manuscript.

References

1. Lai AC, Crews CM (2017) Induced protein degradation: an emerging drug discovery paradigm. Nat Rev Drug Discov 16:101–114
2. Burslem GM, Crews CM (2020) Proteolysis-targeting chimeras as therapeutics and tools for biological discovery. Cell 181:102–114
3. Yokoo H, Shibata N, Endo A et al (2021) Discovery of a highly potent and selective degrader targeting hematopoietic prostaglandin D synthase via in silico design. J Med Chem 64:15868–15882
4. Nabe T, Kuriyama Y, Mizutani N et al (2011) Inhibition of hematopoietic prostaglandin D synthase improves allergic nasal blockage in guinea pigs. Prostaglandins Other Lipid Mediat 95:27–34
5. Takaya D, Inaka K, Omura A et al (2018) Characterization of crystal water molecules in a high-affinity inhibitor and hematopoietic prostaglandin D synthase complex by interaction energy studies. Bioorg Med Chem 26: 4726–4734
6. Drummond ML, Williams CI (2019) In silico modeling of PROTAC-mediated ternary complexes: validation and application. J Chem Inf Model 59:1634–1644
7. Drummond ML, Henry A, Li H, Williams CI (2020) Improved accuracy for modeling PROTAC-mediated ternary complex formation and targeted protein degradation via new in silico methodologies. J Chem Inf Model 60: 5234–5254
8. Kishi K (1985) A new leukemia cell line with Philadelphia chromosome characterized as basophil precursors. Leuk Res 9:381–390
9. Mohri I, Eguchi N, Suzuki K et al (2003) Hematopoietic prostaglandin D synthase is expressed in microglia in the developing postnatal mouse brain. Glia 42:263–274
10. Phillips JC, Braun R, Wang W et al (2005) Scalable molecular dynamics with NAMD. J Comput Chem 26:1781–1802
11. Smith PK, Krohn RI, Hermanson GT et al (1985) Measurement of protein using bicinchoninic acid. Anal Biochem 150:76–85

Correction to: Protein-Protein Docking: Methods and Protocols

Agnieszka A. Kaczor

Correction to:
Chapters 8 and 9 in: Agnieszka A. Kaczor (ed.), *Protein-Protein Docking: Methods and Protocols*, Methods in Molecular Biology, vol. 2780,
https://doi.org/10.1007/978-1-0716-3985-6

Chapters 8 and 9 of this book were inadvertently published without the Acknowledgement sections.

The acknowledgement section has now been updated in both the chapters, which is as follows: **This work was supported by Polish Ministry of Science and Higher Education (financial support for statutory activity of National Medicines Institute).**

The updated versions of these chapters can be found at
https://doi.org/10.1007/978-1-0716-3985-6_8
https://doi.org/10.1007/978-1-0716-3985-6_9

Agnieszka A. Kaczor (ed.), *Protein-Protein Docking: Methods and Protocols*, Methods in Molecular Biology, vol. 2780,
https://doi.org/10.1007/978-1-0716-3985-6_19,
© The Author(s), under exclusive license to Springer Science+Business Media, LLC, part of Springer Nature 2025

Index

Agnieszka A. Kaczor (ed.), *Protein-Protein Docking: Methods and Protocols*, Methods in Molecular Biology, vol. 2780, https://doi.org/10.1007/978-1-0716-3985-6,
© The Editor(s) (if applicable) and The Author(s), under exclusive license to Springer Science+Business Media, LLC, part of Springer Nature 2024

L

M

P

R

S

T

U

MIX
Papier aus verantwortungsvollen Quellen
Paper from responsible sources
FSC® C105338

If you have any concerns about our products,
you can contact us on
ProductSafety@springernature.com

In case Publisher is established outside the EU,
the EU authorized representative is:
Springer Nature Customer Service Center GmbH
Europaplatz 3, 69115 Heidelberg, Germany

Printed by Libri Plureos GmbH
in Hamburg, Germany